T0223932

CASUAL CALCULUS

3 Volumes

A Friendly Student Companion

3 Volumes

CASUAL CALCULUS

A Friendly Student Companion

Kenneth H Luther

Valparaiso University, USA

World Scientific

NEW JERSEY · LONDON · SINGAPORE · BEIJING · SHANGHAI · HONG KONG · TAIPEI · CHENNAI · TOKYO

Published by

World Scientific Publishing Co. Pte. Ltd.

5 Toh Tuck Link, Singapore 596224

USA office: 27 Warren Street, Suite 401-402, Hackensack, NJ 07601

UK office: 57 Shelton Street, Covent Garden, London WC2H 9HE

British Library Cataloguing-in-Publication Data
A catalogue record for this book is available from the British Library.

CASUAL CALCULUS: A FRIENDLY STUDENT COMPANION
(In 3 Volumes)

ISBN 978-981-124-263-2 (set_hardcover)
ISBN 978-981-124-264-9 (set_paperback)
ISBN 978-981-124-265-6 (set_ebook for institutions)
ISBN 978-981-124-266-3 (set_ebook for individuals)

ISBN 978-981-122-392-1 (vol. 1_hardcover)
ISBN 978-981-122-488-1 (vol. 1_paperback)
ISBN 978-981-122-393-8 (vol. 1_ebook for institutions)
ISBN 978-981-122-394-5 (vol. 1_ebook for individuals)

ISBN 978-981-124-197-0 (vol. 2_hardcover)
ISBN 978-981-124-198-7 (vol. 2_paperback)
ISBN 978-981-124-199-4 (vol. 2_ebook for institutions)
ISBN 978-981-124-211-3 (vol. 2_ebook for individuals)

ISBN 978-981-122-395-2 (vol. 3_hardcover)
ISBN 978-981-122-489-8 (vol. 3_paperback)
ISBN 978-981-122-396-9 (vol. 3_ebook for institutions)
ISBN 978-981-122-397-6 (vol. 3_ebook for individuals)

For any available supplementary material, please visit
https://www.worldscientific.com/worldscibooks/10.1142/11927#t=suppl

Printed in Singapore

Dedicated to my on-line summer students for the last decade, whose feedback encouraged me to continue developing this content even though we all had access to "regular" Calculus texts.

Preface

Welcome back! I hope that you are visiting Casual Calculus Volume 2 because you found Volume 1 helpful.

The Preface of Volume 1 gave the set-up for the work as a whole, across three volumes, so I won't repeat it here. I will just reiterate some of the organizational items in case some readers are jumping directly into this Volume 2.

The large structure is:

- Volume 1 contains Chapters 1–6, which correspond to a standard first semester of Calculus, ending with the Fundamental Theorem of Calculus.
- Volume 2 contains Chapters 7–12, which go with a standard second semester of single-variable Calculus.
- Volume 3 contains Chapters 13–18, which match what is often Calculus 3 (Multivariable Calculus).

In this Volume 2, you may notice a slightly different traffic pattern in the presentation of infinite series as compared to regular Calculus texts. We use local linear approximation to motivate polynomial approximations, and use that to step into the idea of general Taylor Polynomials, before finally opening the floodgates to Taylor Series. If this is your first visit to the second half of single-variable Calculus, none of that will have any meaning to you yet.

The section-by-section set up of this book is as follows: Each topic of content begins with a narrative section that leads you through the main ideas and presents examples along the way. After each Example is a "You Try It" problem that's very similar to the example. What I hope you do, as you're reading, is to stop after an Example you think you understood, and immediately try your hand at the associated YTI problem. The solutions to all YTI problems are at the end of the very section they're shown in — so, if you think you've succeeded at the YTI problem, go check the solution to be sure. Or, if you get stuck on the YTI problem, then go look at its solution to get a hint. Once you've completed a section, you'll see the YTI problems collected, along with a set of Practice Problems and Challenge Problems. The Practice Problems should be similar to the YTI Problems, but you're getting them all at once, and so you don't necessarily know which specific technique to use or which Example to follow — you have to think about it! The solutions to the Practice Problems are at the back end of the book. So while they're all available, they are more physically separated from the section they come from; the idea is that you might be inclined to rely on them a bit less, although they are still there when you need them. Then finally, the Challenge Problems are a bit tougher than the others, and you can use those to see if you're successfully synthesizing the ideas you've seen in their section. Solutions to Challenge Problems come after those to the Practice Problems.

At many locations in the text, I will pose some "Food For Thought" (FFT) based on an open question left unanswered. These little puzzlers are bracketed by the symbol 🍽 (it's supposed to be a fork, plate, and knife).

To keep you focused on problem solving, some derivations or other more theoretical discussions are held off until the end of a section. I have always been a fan of heavy metal, and I see jumping into a derivation or proof as the mathematical version of jumping into a mosh pit: you're mostly there to sing along, but every once in a while, you have to wade in and get a bit bruised. So, each of these more mathematically violent discussions are set off with a subheading of, "Into the Pit!!"

As we get into it, I will tell you in all honesty that this second half of single-variable Calculus (usually offered as Calculus II) is my least favorite of the three semesters, and the majority of my students have felt the same way. It turns out that Calculus III is more of a continuation of Calculus I, while Calculus II is sort of a long appendix to Calc I which contains assorted applications of single variable Calculus. However, it all does pay off in other areas — particularly if you intend to study differential equations, or return to a deeper study of Calculus (via Advanced Calculus or Real Analysis). If nothing else, you get to draw pretty pictures while we're exploring polar coordinates in the final chapter.

Contents

Chapter 7

The Integration Dojo

7.1   Integration by Substitution (Review)

Introduction

If you've been going through the prior chapters of this text, then you will have encountered the technique of integration by substitution twice — in Secs. 4.6 (indefinite integrals) and 6.5 (definite integrals). If you have jumped directly to this current section or one of its neighbors and have not encountered integration by substitution yet, then you should go read up about it in Secs. 4.6 and 6.5. However, what we often call "u-substitution" could very well be the most ubiquitous integration technique there is. So, given that this Chapter is all about integration techniques, it seems like a good idea to start if off with a recap of integration by substitution. Here comes a rapid fire set of examples, plus a list of problems for practice. (The only problems found in this section are Practice Problems. They come with complete solutions as always.)

Review of Integration by Substitution

The basic idea of substitution is to take an integrand that was likely produced by the chain rule — so that you are looking at

$$\int f'(g(x)) \cdot g'(x) \, dx$$

and "hide" the inner function $g(x)$ and its derivative $q'(x)$ so that the integral looks simpler:

$$\int f(u) \, du$$

To start off this review, here's an easy one.

$\boxed{\textbf{EX 1}}$ Find $\int (2x - 3)^5 \, dx$ and evaluate $\int_1^2 (2x - 3)^5 \, dx$

This case *barely* needs a substitution, and once you get used to forms like this, you may not need to go through a formal substitution. The antiderivative $(2x - 3)^5$ looks like a power function with some clutter. If we rename $2x - 3$ as u, this cleans up the integral a bit, although incompletely:

$$\int (2x - 3)^5 \, dx \implies \int u^5 \, dx$$

We have *substituted* u for $2x - 3$, but the substitution is not finished. What we're really doing is implementing a *coordinate transformation*, and trading the coordinate (variable) x for u via the rule $2x - 3 = u$. But we can't toss out x in favor of u without also tossing out dx in favor of du. This change must also obey the rule we've designated, $2x - 3 = u$. If $2x - 3 = u$, then $d/dx(2x - 3) = \frac{d}{dx}(u)$, or $2 = du/dx$. Since we're all friends here and we've agreed before[1] to treat the derivative du/dx as a fraction, we can rewrite $2 = du/dx$ as $dx = du/2$. Now we have the entire substitution. We can present the pieces as:

$$2x - 3 = u \quad ; \quad dx = \frac{1}{2} du$$

The complete makeover and solution of the indefinite integral looks like this:

$$\int (2x - 3)^5 \, dx = \int (u)^5 \cdot \frac{1}{2} \, du = \frac{1}{2} \int u^5 \, du = \frac{1}{2} \left(\frac{1}{6} u^6 + C \right) = \frac{1}{12} u^6 + C$$

Then we reverse the substitution and return u to $2x - 3$ so that:

$$\int (2x - 3)^5 \, dx = \frac{1}{12}(2x - 3)^6 + C$$

(If you need a refresher on what's going on with the $+C$, then go check out Sec. 4.6.) For the definite integral, we must also replace the limits of integration during the substitution process; the limits 1 and 2 are values of x; we must find the corresponding values of u to complete the substitution. But with $u = 2x - 3$, then,

$$x = 1 \to u = -1$$
$$x = 2 \to u = 1$$

---

[1] in the discussion of the Chain Rule

All the other parts of the substitution follow just as for the indefinite integral:

$$\int_1^2 (2x - 3)^5 \, dx = \frac{1}{2} \int_{-1}^1 (u)^5 \, du = \frac{1}{12} u^6 \Big|_{-1}^1 = \frac{1}{2} \left((1)^6 - (-1)^6 \right) = 0$$

Do you remember what it means when a definite integral comes out to a value of 0? Perhaps Fig. 7.1 will help. ∎

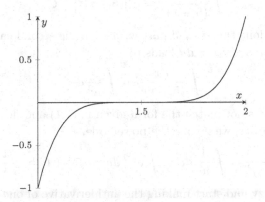

Fig. 7.1   The function $f(x) = (2x - 3)^5$ on $[1, 2]$.

Here is an example about how the natural logarithm function can pop up out of nowhere. But it is also a cautionary tale. In Sec. 5.4, we talked about how students can tend to over-use L-Hopital's Rule; "When you have a hammer, every problem looks like a nail." The antiderivative formula

$$\int \frac{1}{u} \, du = \ln |u| + C \tag{7.1}$$

is another tool which often becomes a hammer that students use to turn too many problems into nails. A common error is for students to extend "the antiderivative of one over u is the natural log of u" to "the antiderivative of one over anything is always the natural log of that thing". It is important to remember that (7.1) only holds when the integrand really reduces to $1/u$ upon substitution.

EX 2 Find $\displaystyle\int \frac{dx}{5+x}$ and $\displaystyle\int \frac{dx}{(5+x)^2}$.

For the first antiderivative, we use $5 + x = u$, and so also $dx = du$, to convert the integral as follows:

$$\int \frac{dx}{5+x} = \int \frac{du}{u} = \ln|u| + C = \ln|5+x| + C$$

Here is where it's easy to get into trouble; if you get all excited about this first result and leap directly to

$$\int \frac{dx}{(5+x)^2} = \ln|(5+x)^2| + C$$

then you have done this second part wrong. In the second case, the substitution $5 + x = u$ and $dx = du$ leads to

$$\int \frac{dx}{(5+x)^2} = \int \frac{1}{u^2}\, du$$

This integral does not match the form given in (7.1) and does not return a natural log. Rather, we invoke the power rule.

$$\int \frac{dx}{(5+x)^2} = \int \frac{1}{u^2}\, du = \int u^{-2}\, du = u^{-1} + C = \frac{1}{5+x} + C$$

So don't go crazy and start making the antiderivative of one over anything into the natural log of that thing. Equation (7.1) is not a hammer that turns every integral of one-over-something into a nail. ∎

This next problem is just a normal example with no extra sermonizing needed.

EX 3 Find $\displaystyle\int x\sqrt{4+2x^2}\, dx$ and evaluate $\displaystyle\int_0^{\sqrt{6}} x\sqrt{4+2x^2}\, dx$

Our makeover $4 + 2x^2 = u$, from which $dx = du/4$. Then,

$$\int x\sqrt{4+2x^2}\, dx = \frac{1}{4}\int \sqrt{u}\, du = \frac{1}{4}\left(\frac{2}{3}u^{3/2} + C\right) = \frac{1}{6}(4+2x^2)^{3/2} + C$$

For the definite integral, let's also change the limits of integration. With $4 + 2x^2 = u$, we get:

$$x = 0 \rightarrow u = 4$$
$$x = \sqrt{6} \rightarrow u = 16$$

so that

$$\int_0^{\sqrt{6}} x\sqrt{4+2x^2}\, dx = \frac{1}{4}\int_4^{16} \sqrt{u}\, du = \frac{1}{4}\left(\frac{2}{3}u^{3/2}\right)\Bigg|_4^{16}$$

$$= \frac{1}{6}\left(16^{3/2} - 4^{3/2}\right) = \frac{1}{6}\left(4^3 - 2^3\right)$$

$$= \frac{1}{6}(56) = \frac{28}{3} \quad \blacksquare$$

EX 4 Find $\int \sin x \cos x\, dx$ with $\sin x = u$ and then with $\cos x = u$.

With $\sin x = u$ and so also $\cos x\, dx = du$, we get

$$\int \sin x \cos x\, dx = \int u\, du = \frac{1}{2}u^2 + C = \frac{1}{2}\sin^2 x + C$$

With $\cos x = u$ and so also $\sin x\, dx = -du$, we get

$$\int \sin x \cos x\, dx = \int u\,(-du) = -\frac{1}{2}u^2 + C = -\frac{1}{2}\cos^2 x + C$$

𝄞𝄞 FFT: The same integral appears to have led to two different results. How can this be true? 𝄞𝄞 $\quad\blacksquare$

EX 5 Without actually solving, state which of these definite integrals will give a result of 0, then solve it to confirm.

$$\int_0^2 q e^{q^2}\, dq \quad , \quad \int_{-1}^1 t e^{t^2}\, dt \quad , \quad \int_{-1}^0 x e^{x^2}\, dx$$

First, note that whether we're using q, t, or x as the variable of integration is completely irrelevant. Don't be distracted by that; the story is all in the endpoints. A definite integral gives a final value of 0 if there are equal sized but oppositely sized areas contributing to the total area "under" the curve; that is, which graph and endpoints together present a region of integration which has exactly half below the horizontal axis and half above? The only one which could possibly do that is the middle one. With $t^2 = u$ and so also $t\, dt = du/2$, and with endpoints changed as follows:

$$t = -1 \rightarrow u = 1$$
$$t = 1 \rightarrow u = 1$$

we get

$$\frac{1}{2}\int_1^1 e^u\, du = \frac{1}{2}(e^1 - e^1) = 0$$

Actually, as soon as we saw the new endpoints, we could have claimed the answer is zero. 🔘 FFT: Why is it always true that as long as $f(x)$ is defined at $x = a$, we have $\displaystyle\int_a^a f(x)\,dx = 0$? 🔘   ■

Finally, please bear with a last bit of discussion. First, regarding problem types:

You have been asked to *find* indefinite integrals — that is, to come up with an entire class of functions — made distinct from each other by an arbitrary constant — which would all produce the given expression as a derivative; you have been asked to *evaluate* definite integrals, which require a final numerical value most often taken to mean the area "under" the curve of the given function along an interval delineated by the endpoints of integration. Be sure to understand the difference between *finding* antiderivatives and *evaluating* definite integrals.

Second, regarding the method for definite integrals:

Another (unfortunately) common way I've seen people get through definite integrals is to avoid changing the limits of integration during the substitution by reverting to the original variable at the end and applying the original limits. Let's re-do the definite integral in EX 1 using this method, and see how it can possible go wrong. Using, as before, $2x - 3 = u$ and so $dx = du/2$, here's a first attempt:

$$\int_1^2 (2x-3)^5\,dx = \frac{1}{2}\int_1^2 u^5\,du = \frac{1}{12}u^6\Big|_1^2 = \frac{1}{12}(2^6 - 1^6) = \frac{63}{12}$$

This first attempt is incorrect. Do you see why? The limits were never changed, and the original limits were values of x, not values of u. By plugging in 1 and 2 into u^6, we've mixed apples and oranges.

So that's no good, but it's unfortunately an all-too-common error when using substitution for definite integrals. Don't do that!

Here's a second attempt, which does resolve to the correct final value:

$$\int_1^2 (2x-3)^5\,dx = \frac{1}{2}\int_1^2 u^5\,du = \frac{1}{12}u^6\Big|_1^2$$

$$= \frac{1}{12}(2x-3)^6\Big|_1^2 = \frac{1}{12}\left((1)^6 - (-1)^6\right) = 0$$

But while the result is good, there are two steps in the middle that are still invalid. The integral in the second phrase of the chain has values of the variable x (1,2) placed on an integral displaying a variable of u; like in the first (incorrect) attempt, this is mixing apples and oranges. The very next phrase, where $u^6/12$ is apparently waiting to have endpoints of $1, 2$ plugged in, is even worse: it claims: "Hey, we're about to plug in 1 and 2, which are values of x, into spots currently held by u." Now, this fixed in the next step by returning u to the original $2x - 3$, but still ... yuck. I suppose you could make this a bit better by writing the sequence of operations as:

$$\int_1^2 (2x - 3)^5 \, dx = \frac{1}{12} \int_{(1)}^{(2)} u^5 \, du = \frac{1}{12} u^6 \Big|_{(1)}^{(2)}$$

$$= \frac{1}{12} (2x - 3)^6 \Big|_1^2 = \frac{1}{12} \left((1)^6 - (-1)^6 \right) = 0$$

Here, at least the parentheses around the endpoints in the middle help indicate that they are not values of u, but that's still pretty bad. So sure, in this second attempt, we end up with the right number at the end, but doing definite integrals this way is open to too many errors — not only are there notational problems, but it's just too easy to forget to return u to $2x - 3$ and end up back in our first attempt shown above.

Overall, when doing definite integrals with substitution, you have two procedural choices:

(1) Assign a substitution, carry out the substitution *without changing the endpoints*, complete the antiderivative, then return to the original variable before applying the endpoints via the Fundamental Theorem of Calculus, OR

(2) Assign a substitution, carry out the substitution and *change the endpoints too*, then complete the antiderivative and apply the Fundamental Theorem of Calculus with the new endpoints

If you are encountering this content for a second time and got used to opting for the former method up until now, I encourage you to reconsider and opt for the second method. You may think the first method a shortcut to save work, but it's not. Yes, it saves you work on replacing the endpoints, but it requires more work at the end of the problem where you're forced to return to the original variable. Plus, you're committing grave sins against mathematical notation along the way. So remember, only you can prevent misuse of substitution!

Integration By Substitution (Review) — Problem List

This review section has only Practice Problems.

Integration By Substitution (Review) — Practice Problems

Try these as you get the hang of the You Try It problems. Solutions to these problems are available in Sec. A.1.1.

(1) Find $\int \sqrt{2-x}\,dx$ and evaluate $\int_0^2 \sqrt{2-x}\,dx$.

(2) Find $\int \dfrac{x}{1+x^4}\,dx$ and evaluate $\int_0^{\sqrt[4]{3}} \dfrac{x}{1+x^4}\,dx$.

(3) Find $\int \cos\theta \sin^3\theta\,d\theta$ and evaluate $\int_{-\pi/2}^{\pi/2} \cos\theta \sin^3\theta\,d\theta$.

(4) Evaluate $\int_0^{\pi/3} \dfrac{\sin t}{\cos^2 t}\,dt$.

(5) Find $\int \dfrac{3x}{(x^2+1)^4}\,dx$.

(6) Evaluate $\int_0^{10} \cot\theta \tan\theta\,d\theta$.

(7) Find $\int \dfrac{\sec^2\theta}{1+\tan\theta}\,d\theta$ and $\int \dfrac{\sec^2\theta}{(1+\tan\theta)^2}\,d\theta$.

(8) Evaluate $\int_0^{\pi} e^{3\cos x} \sin x\,dx$.

(9) Evaluate $\int_0^1 xe^{-x^2}\,dx$.

(10) Evaluate $\int_0^{\pi/2} \cos x \sin(\sin x)\,dx$.

(11) Find $\int \dfrac{1}{\sqrt{x}(1+\sqrt{x})}\,dx$.

(12) Consider $\int_0^1 \dfrac{x+a}{x+b}\,dx$, where a can be any real number, but b is restricted. State the necessary restriction on b and evaluate the integral.

(13) Find the area under the function $y = e^x/(1+e^{2x})$ from $x = 0$ to $x = 1$. Your answer must be in exact form.

7.2 Integration by Parts

Introduction

You have seen two fundamental differentiation techniques which, when you play a Reverse card on them, become antidifferentiation techniques. There is a power rule for antiderivatives that reverses to a power rule for derivatives (Sec. 3.5). Integration by substitution (Secs. 4.6, 6.5, 7.1) rolls back the chain rule. We will now see *integration by parts*, which is associated with the product rule.

Also at this point, let's agree to talk a bit more loosely, so that we can use the word "integrating" to mean "finding the antiderivative of", even though — technically — integrating is associated with Riemann Sums.

Reversing the Product Rule

You should remember that the product rule for the derivative of two functions $f(x)$ and $g(x)$ says (see Sec. 4.3 for review):

$$\frac{d}{dx}\left(f(x)g(x)\right) = f(x) \cdot g'(x) + f'(x) \cdot g(x)$$

Let's see what happens when we reverse the product rule, i.e. when we integrate both sides:

$$\int \left(\frac{d}{dx}(f(x)g(x))\right) dx = \int \left(f(x) \cdot g'(x)\right) dx + \int \left(f'(x) \cdot g(x)\right) dx$$

On the left hand side, we just see the antiderivative of a derivative — and those two processes undo each other (there's a legality here that's being overlooked, though!), so we get

$$f(x)g(x) = \int \left(f(x) \cdot g'(x)\right) dx + \int \left(f'(x) \cdot g(x)\right) dx$$

A rearrangement then gives

$$\int f(x)g'(x)\, dx = f(x)g(x) - \int f'(x)g(x)\, dx$$

Did we just undo the product rule? Is it time to celebrate? Well, no. What we did do, though, is still useful: we now have a *conversion formula* that allows us to trade one integral for another. Ideally, we'll use this to trade a harder integral for a simpler one.

Useful Fact 7.1. *As long as both antiderivatives exist, we can write*

$$\int f(x)g'(x)\,dx = f(x)g(x) - \int f'(x)g(x)\,dx \qquad (7.2)$$

This rule is called **integration by parts**.

The name "integration by *parts*" is suitable because you get to decide which part of your integral matches which part of the formula. That is, if you are integrating an expression that is written as a product, then you get to assign which component of the product is $f(x)$ and which is $g'(x)$. If you make a good choice, the integral can be traded for a simpler one. The integration by parts formula allows us to trade $\int f(x)g'(x)\,dx$ for $\int f'(x)g(x)\,dx$. In other words, we can *shift the derivative term*. If you have an integral in which you believe, "Gosh, this would be easier if $f(x)$ was $f'(x)$ instead!", then integration by parts allows you to make that switch. For example, wouldn't you rather integrate something with an x instead of with an x^2? With integration by parts, you can sometimes make that change — but there is a price to this choice. While you trade $f(x)$ for $f'(x)$, you also trade $g'(x)$ for $g(x)$. Sometimes that trade is worth making.

When you break an integral up using integration by parts, there are four *parts* to be considered: $f(x)$, $g(x)$, $f'(x)$, and $g'(x)$. You select two ($f(x)$ and $g'(x)$), compute the other two ($f'(x)$ and $g(x)$), then apply Eq. (7.2).

EX 1 Find $\int x \sin x\,dx$.

It would be really great if this antiderivative was simply

$$\int x \sin x\,dx = -\frac{1}{2}x^2 \cos x + C$$

Sadly, it's not. Since our integrand (remember, that's what we call the expression being integrated) is a product, then integration by parts may be a good option. Let's match up our integral to the left side of Eq. (7.2)

$$\int f(x)g'(x)\,dx \leftrightarrow \int x \sin x\,dx$$

We have to assign x and $\sin x$ to be $f(x)$ and $g'(x)$. Which gets to be which? The strategy is to make the better selection between two choices:

- We can trade x for its derivative 1, at the cost of trading $\sin x$ for its antiderivative $-\cos x$, OR
- We can trade $\sin x$ for its derivative $\cos x$, at the cost of trading x for its antiderivative $x^2/2$

The best choice is pretty clear. Trading a sine for its derivative or antiderivative really does not change the complexity of the problem. But changing x to either 1 or $x^2/2$ makes a big difference. Would you rather trade the given integrand for $1 \cdot (-\cos x)$ or for $(x^2/2)\sin x$? The latter is worse than the current problem, so it's definitely the bad choice.

Let's make our assignments as follows:
$$f(x) = x \qquad \text{and} \qquad g'(x) = \sin x$$
From these choices, we then compute
$$f'(x) = 1 \qquad \text{and} \qquad g(x) = -\cos x$$
(Notice that we had to do a tiny antiderivative to go from $g'(x)$ to $g(x)$!) Now, we'll convert the integral using Eq. (7.2):
$$\int f(x)g'(x)\,dx = f(x)g(x) - \int f'(x)g(x)\,dx$$
$$\int x \sin x\,dx = (x)(-\cos x) - \int (1)(-\cos x)\,dx$$
$$= -x\cos x + \int \cos x\,dx$$

This has not *solved* the integral for us, but it has converted the integral we could not do into one that's really easy. Since we know the integral on the right side, we can finish:
$$\int x \sin x\,dx = -x\cos x + \int \cos x\,dx = -x\cos x + \sin x + C$$
That's it! ∎

Like with integration by substitution, success at integration by parts requires a good imagination and perseverance. The big open questions are

(1) How do you know if your choice of parts is the right one?
(2) What happens if you still can't solve the new integral even if it's easier?
(3) Can you use integration by parts if there's only one function inside the integral?

(4) What about definite integrals?

Each question will be answered with some examples.

$\boxed{\textbf{EX 2}}$ Find $\displaystyle\int x\cos x\,dx$.

This is almost identical to EX 1, but it's here to illustrate how quickly you can discover you've made a poor choice during integration by parts. This is one of those times when the pep talk from the Preface comes in handy. You will fail to solve problems at first, while you're leaning this technique. You may have to try two or more times on any given problem. So what? If you pick the wrong assignment for integration by parts, the math police do not come to get you. You are not penalized with a fine. At worst, you lost a bit of time, but you gained some important experience that will help you make better choices in later problems. Don't tell yourself, "I can't do this problem." Tell yourself, "I can't do this problem ... yet."

Okay, enough of the coaching. Let's start this example with the *wrong* choice. Suppose we make the assignments

$$f(x) = \cos x \qquad \text{and} \qquad g'(x) = x$$

From these choices, we then compute

$$f'(x) = -\sin x \qquad \text{and} \qquad g(x) = \frac{1}{2}x^2$$

And now let's load up the integration by parts formula:

$$\int f(x)g'(x)\,dx = f(x)g(x) - \int f'(x)g(x)\,dx$$

$$\int x\cos x\,dx = (\cos x)\left(\frac{1}{2}x^2\right) - \int (-\sin x)\left(\frac{1}{2}x^2\right)\,dx$$

$$= \frac{1}{2}x^2\cos x + \frac{1}{2}\int x^2\sin x\,dx$$

Whoops! We just converted an integrand of $x\sin x$ into $x^2\cos x$. Our integral got even harder! So that was definitely the wrong choice of parts, and we found that out pretty quickly. As the saying goes, "no harm, no foul". We move on and try the other choice,

$$f(x) = x \qquad \text{and} \qquad g'(x) = \cos x$$

from which we compute

$$f'(x) = 1 \qquad \text{and} \qquad g(x) = \sin x$$

and then using (7.2),

$$\int x\cos x\,dx = x\sin x - \int (1)\sin\,dx = x\sin x + \cos x + C$$

We made the wrong choice at first, we realized it quickly, then we tried the other choice. No sweat. ∎

$\boxed{\textbf{EX 3}}$ Find $\int x^2 \ln x\,dx$.

Suppose we choose

$$f(x) = x^2 \qquad \text{and} \qquad g'(x) = \ln x$$

From these choices, we'd need to compute $f'(x)$ and $g(x)$. Well, $f'(x) = 2x$, that's easy. But to find $g(x)$, we need to know the antiderivative of $\ln x$. At this point, we don't know what it is! (We will soon, though.) So this choice of parts won't work. Again, the wrong choice was quickly diagnosed. Let's try the other choice:

$$f(x) = \ln x \qquad \text{and} \qquad g'(x) = x^2$$

From these assignments, we then compute

$$f'(x) = \frac{1}{x} \qquad \text{and} \qquad g(x) = \frac{1}{3}x^3$$

And now let's load up (7.2), $\int f(x)g'(x)\,dx = f(x)g(x) - \int f'(x)g(x)\,dx$:

$$\int x^2 \ln x\,dx = (\ln x)\left(\frac{1}{3}x^3\right) - \int \left(\frac{1}{x}\right)\left(\frac{1}{3}x^3\right)\,dx$$

$$= \frac{1}{3}x^3 \ln x - \frac{1}{3}\int x^2\,dx = \frac{1}{3}x^3 \ln x - \frac{1}{3}\left(\frac{1}{3}x^3\right) + C$$

$$= \frac{1}{3}x^3\left(\ln x - \frac{1}{3}\right) + C$$

Note that in this process we ramped up an x^2 to an x^3, but the point of integration by parts is not necessarily to make each *individual* piece get easier, but rather that the collective result of *all* changes makes the whole integral easier. ∎

How about one more brief example, showing proper choices of parts, as well as a bit of streamlining of the solution.

EX 4 Find $\int xe^{x/2}\,dx$.

We choose $f(x)$ and $g'(x)$ from the original integral and then compute the resulting $f'(x)$ and $g(x)$. Let's try:

$$\text{choose} \rightarrow \quad f(x) = e^{x/2} \qquad g'(x) = x$$
$$\text{compute} \rightarrow \quad f'(x) = \tfrac{1}{2}e^{x/2} \qquad g(x) = \tfrac{1}{2}x^2$$

Then, using (7.2),

$$\int xe^{x/2}\,dx = (e^{x/2})\left(\frac{1}{2}x^2\right) - \int \left(\frac{1}{2}e^{x/2}\right)\left(\frac{1}{2}x^2\right)\,dx$$

$$= \frac{1}{2}x^2 e^{x/2} - \frac{1}{4}\int x^2 e^{x/2}\,dx$$

Sigh. Once again, we've made the wrong initial choice and have created a new integral that's harder than the first! So we'll backtrack and make the other choice:

$$\text{choose} \rightarrow \quad f(x) = x \qquad g'(x) = e^{x/2}$$
$$\text{compute} \rightarrow \quad f'(x) = 1 \qquad g(x) = 2e^{x/2}$$

And using (7.2), $\int f(x)g'(x)\,dx = f(x)g(x) - \int f'(x)g(x)\,dx$:

$$\int xe^{x/2}\,dx = (x)(2e^{x/2}) - \int (1)(2e^{x/2})\,dx = 2xe^{x/2} - 2\int e^{x/2}\,dx$$

$$= 2xe^{x/2} - 2(2e^{x/2}) + C = 2e^{x/2}(x-2) + C \quad \blacksquare$$

You Try It

(1) Find $\int x\cos 5x\,dx$.

The second open question is: What if you apply integration by parts, generate a new integral that's easier than the original, but you still can't solve it? As long as you're making progress, don't give up! Sometimes, you have to use integration by parts more than once, and that can easily be seen with a brief example:

EX 5 Find $\int x^2 e^{-x}\,dx$.

We choose

$$f(x) = x^2 \qquad \text{and} \qquad g'(x) = e^{-x}$$

then compute

$$f'(x) = 2x \qquad \text{and} \qquad g(x) = -e^{-x}$$

and apply integration by parts:

$$\int x^2 e^{-x}\, dx = x^2(-e^{-x}) - \int (2x)(-e^{-x})\, dx = -x^2 e^{-x} + 2\int xe^{-x}\, dx \tag{7.3}$$

We have successfully converted the original integral into one that's a bit easier. But the new integral on the right side also needs integration by parts to be solved! That's no big deal, we just treat the new integral as a brand new problem, get the result, then feed that result back into (7.3).

With the choice of $f(x) = x$ and $g'(x) = e^{-x}$, so that we also have $f'(x) = 1$ and $g(x) = -e^{-x}$, we get:

$$\int xe^{-x}\, dx = -xe^{-x} - \int (1)(-e^{-x}\, dx = -xe^{-x} + \int e^{-x}\, dx = -xe^{-x} - e^{-x} + C$$

Handing this back to (7.3),

$$\int x^2 e^{-x}\, dx = -x^2 e^{-x} + 2\left(-xe^{-x} - e^{-x} + C\right) = -x^2 e^{-x} - 2xe^{-x} - 2e^{-x} + C$$

▣ FFT: Based on how this example went, can you predict how many times you'd need to use integration by parts to solve $\int x^3 \cos x\, dx$? ▣

■

You Try It

(2) Find $\int x^2 \cos 2x\, dx$.

(3) Find $\int e^{-\theta} \cos 2\theta\, d\theta$.

The third open question is: does the expression you're integrating have to have two easily identifiable parts? The answer is NO, since any expression can always be considered as multiplied by 1, and that multiple of 1 can also be used as one of the parts! You have been waiting patiently for many chapters to learn the antiderivative of $\ln x$, and this "trick" finally allows us to find it.

EX 6 Find $\int \ln x \, dx$.

This does not have two immediately visible parts. But if we rewrite it as:

$$\int 1 \cdot \ln x \, dx$$

then we can choose

$$f(x) = \ln x \qquad \text{and} \qquad g'(x) = 1$$

and compute

$$f'(x) = \frac{1}{x} \qquad \text{and} \qquad g(x) = x$$

Integration by parts then gives:

$$\int \ln x \, dx = (\ln x)(x) - \int \frac{1}{x} \cdot x \, dx = x \ln x - \int (1) \, dx = x \ln x - x + C$$

and now we know $\int \ln x \, dx = x \ln x - x + C$. ■

You Try It

  (4) Find $\int \ln(2x + 1) \, dx$

The final open question is: What about definite integrals? Fortunately, nothing much changes, as long as you remember to apply the endpoints of integration to all terms:

Useful Fact 7.2. *When applied to a definite integral, integration by parts is written like this:*

$$\int_a^b f(x)g'(x) \, dx = f(x)g(x) \Big|_a^b - \int_a^b f'(x)g(x) \, dx$$

EX 7 Evaluate $\int_1^2 \frac{\ln x}{\sqrt{x}} \, dx$.

We can choose

$$f(x) = \ln x \qquad \text{and} \qquad g'(x) = \frac{1}{\sqrt{x}}$$

and compute

$$f'(x) = \frac{1}{x} \qquad \text{and} \qquad g(x) = 2\sqrt{x}$$

Integration by parts then gives:

$$\int_1^2 \frac{\ln x}{\sqrt{x}}\, dx = (\ln x)(2\sqrt{x})\Big|_1^2 - \int_1^2 \frac{1}{x} \cdot (2\sqrt{x})\, dx$$

$$= 2\sqrt{x}\ln x\Big|_1^2 - 2\int_1^2 \frac{1}{\sqrt{x}}\, dx = 2\sqrt{x}\ln x\Big|_1^2 - 2(2\sqrt{x})\Big|_1^2$$

$$= 2\sqrt{x}(\ln x - 2)\Big|_1^2 = \left(2\sqrt{2}(\ln 2 - 2)\right) - \left(2\sqrt{1}(\ln 1 - 2)\right)$$

$$= \left(2\sqrt{2}(\ln 2 - 2)\right) - 2(0 - 2) = 2\sqrt{2}(\ln 2 - 2) + 4 \quad \blacksquare$$

Different Notation

There are often two styles of notation given for integration by parts. The one I prefer is used in these notes:

$$\int f(x)g'(x)dx = f(x)g(x) - \int g(x)f'(x)dx$$

Sometimes, though, a shorter, tidier version is given:

$$\int u\, dv = uv - \int v\, du$$

The links are fairly obvious:

$$f(x) \to u \quad f'(x) \to du \quad g(x) \to v \quad g'(x) \to dv$$

You can use whichever you end up liking better. I prefer the one that explicitly shows you all the variable dependencies of the functions.

Finally, now that you've seen many examples and (hopefully) solved some problems, I hope there is one question nagging you. In integration by parts, we choose from an integral a part $g'(x)$ and compute the corresponding $g(x)$ by finding its antiderivative. Did you notice we never put $+C$ into $g(x)$? We wait and $+C$ to the final result at the very end of the problems. But you should be skeptical: if $g(x)$ is the antiderivative of $g'(x)$, shouldn't we put a $+C$ into $g(x)$ when we first compute it? See if you can figure out why it's not needed.

Integration by Parts — Problem List

Integration by Parts — You Try It

These appeared above; solutions begin on the next page.

(1) Find $\int x \cos 5x \, dx$.

(3) Find $\int e^{-\theta} \cos 2\theta \, d\theta$.

(2) Find $\int x^2 \cos 2x \, dx$.

(4) Find $\int \ln(2x + 1) \, dx$.

Integration by Parts — Practice Problems

Try these as you get the hang of the You Try It problems. Solutions to these problems are available in Sec. A.1.2.

(1) Find $\int x^5 \ln x \, dx$.

(4) Find $\int \sin^{-1} x \, dx$.

(2) Find $\int t \sin 2t \, dt$.

(5) Find $\int x^2 \cosh(x) \, dx$.

(3) Find $\int x^2 \sin \pi x \, dx$.

(6) Evaluate $\int_0^1 \dfrac{x}{e^{2x}} \, dx$.

Integration by Parts — Challenge Problems

Try these problems to test your skills with the ideas in this section. Solutions to these problems are available in Sec. B.1.1.

(1) Find $\int e^{2x} \sin 3x \, dx$.

(2) Evaluate $\int_0^{1/2} \cos^{-1} x \, dx$.

(3) A reduction formula is a formula that reduces the complexity of an integral by reducing the power on one of its terms. Use integration by parts to derive the reduction formula

$$\int \cos^n x \, dx = \frac{1}{n} \cos^{n-1} x \sin x + \frac{n-1}{n} \int \cos^{n-2} x \, dx$$

Integration by Parts — You Try It — Solved

---

(1) Find $\int x \cos 5x \, dx$.

☐ The best choice is to assign / compute like so:

$$\text{choose} \;\rightarrow\quad f(x) = x \qquad g'(x) = \cos 5x$$
$$\text{compute} \;\rightarrow\quad f'(x) = 1 \qquad g(x) = \tfrac{1}{5} \sin 5x$$

Then, using (7.2),

$$\int x \cos 5x \, dx = \frac{1}{5} x \sin 5x - \frac{1}{5} \int \sin 5x \, dx$$

$$= \frac{1}{5} x \sin 5x - \frac{1}{5} \left(-\frac{1}{5} \cos 5x \right) + C$$

$$= \frac{1}{5} x \sin 5x + \frac{1}{25} \cos 5x + C \quad \blacksquare$$

---

(2) Find $\int x^2 \cos 2x \, dx$.

☐ The best choice is:

$$\text{choose} \;\rightarrow\quad f(x) = x^2 \qquad g'(x) = \cos 2x$$
$$\text{compute} \;\rightarrow\quad f'(x) = 2x \qquad g(x) = \tfrac{1}{2} \sin 2x$$

so that with (7.2),

$$\int x^2 \cos 2x \, dx = \frac{1}{2} x^2 \sin 2x - \int x \sin 2x \, dx$$

Well, integration by parts once at least made it easier, but we're not done; the new integral needs to be evaluated. For the new one,

$$\text{choose} \;\rightarrow\quad f(x) = x \qquad g'(x) = \sin 2x$$
$$\text{compute} \;\rightarrow\quad f'(x) = 1 \qquad g(x) = -\tfrac{1}{2} \cos 2x$$

so that

$$\int x \sin 2x \, dx = -\frac{1}{2} x \cos 2x + \frac{1}{2} \int \cos 2x \, dx$$

$$= -\frac{1}{2} x \cos 2x + \frac{1}{4} \sin 2x + C$$

Putting it all together,

$$\int x^2 \cos 2x \, dx = \frac{1}{2} x^2 \sin 2x - \left(-\frac{1}{2} x \cos 2x + \frac{1}{4} \sin 2x + C \right)$$

$$= \frac{1}{2} x^2 \sin 2x + \frac{1}{2} x \cos 2x - \frac{1}{4} \sin 2x + C \quad \blacksquare$$

---

(3) Find $\int e^{-\theta} \cos 2\theta \, d\theta$.

☐ In this one, an integral is found without the integral actually being solved! It's all smoke and mirrors with integration by parts, but hey, it works. Watch how it happens.

$$\text{choose} \rightarrow \quad f(\theta) = \cos 2\theta \qquad g'(\theta) = e^{-\theta}$$
$$\text{compute} \rightarrow \quad f'(\theta) = -2\sin 2\theta \qquad g(\theta) = -e^{-\theta}$$

so that with (7.2),

$$\int e^{-\theta} \cos 2\theta \, d\theta = -e^{-\theta} \cos 2\theta - 2 \int e^{-\theta} \sin 2\theta \, d\theta$$

Using integration by parts again for the new integral,

$$\text{choose} \rightarrow \quad f(\theta) = \sin 2\theta \qquad g'(\theta) = e^{-\theta}$$
$$\text{compute} \rightarrow \quad f'(\theta) = 2\cos 2\theta \qquad g(\theta) = -e^{-\theta}$$

so that (ignoring the $+C$ for the moment),

$$\int e^{-\theta} \sin 2\theta \, d\theta = -e^{-\theta} \sin 2\theta + 2 \int e^{-\theta} \cos 2\theta d\theta$$

and going back to the original integral,

$$\int e^{-\theta} \cos 2\theta \, d\theta = -e^{-\theta} \cos 2\theta - 2 \left(-e^{-\theta} \sin 2\theta + 2 \int e^{-\theta} \cos 2\theta \, d\theta \right)$$

$$\int e^{-\theta} \cos 2\theta \, d\theta = -e^{-\theta} \cos 2\theta + 2e^{-\theta} \sin 2\theta - 4 \int e^{-\theta} \cos 2\theta \, d\theta$$

$$5 \int e^{-\theta} \cos 2\theta \, d\theta = -e^{-\theta} \cos 2\theta + 2e^{-\theta} \sin 2\theta + C$$

$$\int e^{-\theta} \cos 2\theta \, d\theta = -\frac{1}{5} e^{-\theta} \cos 2\theta + \frac{2}{5} e^{-\theta} \sin 2\theta + C$$

Voila!   ■

---

(4) Find $\int \ln(2x + 1) \, dx$.

☐ Here, there are not two parts to choose from, so we go with

$$\text{choose} \rightarrow \quad f(x) = \ln(2x + 1) \qquad g'(x) = 1$$
$$\text{compute} \rightarrow \quad f'(x) = \frac{2}{2x+1} \qquad g(x) = x$$

Then, using (7.2),

$$\int \ln(2x + 1) \, dx = x \ln(2x + 1) - \int \frac{2x}{2x + 1} \, dx$$

That new integral looks a bit strange, but consider the substitution $u = 2x + 1$, then the numerator is $u - 1$ and $dx = du/2$, so we have

$$\int \frac{2x}{2x+1} \, dx = \frac{1}{2} \int \frac{u-1}{u} \, du = \frac{1}{2} \int \left(1 - \frac{1}{u}\right) du$$

$$= \frac{1}{2} \left(u - \ln u\right) + C = \frac{1}{2} \left(2x + 1 - \ln(2x+1)\right) + C$$

$$= x - \frac{1}{2} \ln(2x+1) + C$$

Finally, then

$$\int \ln(2x+1) \, dx = x \ln(2x+1) - \left(x - \frac{1}{2} \ln(2x+1)\right) + C$$

$$= \frac{1}{2}(2x+1) \ln(2x+1) - x + C \quad \blacksquare$$

7.3   The Calc 2 Boss Fight: Integration by Partial Fractions

Introduction

In our pairing of antidifferentiation techniques to differentiation techniques, we have the power rule as a counterpart to the power rule (of course), integration by substitution as a counterpart to the chain rule, and integration by parts as a counterpart to the product rule. What's missing? The quotient rule! Unfortunately (or fortunately, depending on your point of view), there is no commonly used rule for antiderivatives that's a direct counterpart to the quotient rule. The closest we come is *integration by partial fractions* — this technique does not reverse the quotient rule, but does help us find the antiderivatives of some functions presented as quotients. More specifically, integration by partial fractions can help find the antiderivatives of some rational functions, which are functions in which both the numerator and denominator are polynomials.

The secret to integration by partial fractions is to recognize that rational functions can have equivalent forms: one nice and compact, and one more dispersed — but easier to integrate. Our strategy is to "unbundle" the compact version of the function into the more dispersed version, and this is often possible when the denominator of the rational function has nice factors. In our work here, the algebraic manipulations will be more irritating than the calculus part of it[2]; in fact, the vast majority of the problems will end up requiring variations on these three integral forms, which should be familiar:

$$\int \frac{1}{ax + b}\, dx = \frac{1}{a}\ln|ax + b| + C \tag{7.4}$$

$$\int \frac{x}{x^2 + d^2}\, dx = \frac{1}{2}\ln(x^2 + d^2) + C \tag{7.5}$$

$$\int \frac{1}{x^2 + s^2}\, dx = \frac{1}{s}\tan^{-1}\left(\frac{x}{s}\right) + C \tag{7.6}$$

Note that the third one has an alternate (less tidy) form,

$$\int \frac{1}{x^2 + S}\, dx = \frac{1}{\sqrt{S}}\tan^{-1}\left(\frac{x}{\sqrt{S}}\right) + C$$

---

[2]It's always the damn algebra that gets you.

You should know by now how all of these types of integrals work out. Be sure you are adept with them. In particular, be sure you can identify the difference between (7.5) and (7.6).

We will solve upcoming problems so that we get the calculus part out of the way first, then concentrate on the tedious algebra.

Mathematical Dynamite

Would you believe that

$$\frac{4x+5}{x^2+x-2} \quad \text{and} \quad \frac{1}{x+2} + \frac{3}{x-1} \tag{7.7}$$

are the same function? It's true! Back when you learned algebra, you were probably asked to do things like, "Write the following as a single term, using a common denominator:"

$$\frac{1}{x+2} + \frac{3}{x-1}$$

And then you would proceed as follows:

$$\frac{1}{x+2} + \frac{3}{x-1} = \frac{1}{x+2} \cdot \frac{x-1}{x-1} + \frac{3}{x-1} \cdot \frac{x+2}{x+2}$$

$$= \frac{x-1}{(x+2)(x-1)} + \frac{3(x+2)}{(x+2)(x-1)} = \frac{x-1+3x+6}{(x+2)(x-1)}$$

$$= \frac{4x+5}{(x+2)(x-1)} = \frac{4x+5}{x^2+x-2}$$

Now we've confirmed that the two expressions in (7.7) do indeed represent the same function. With that in mind, which of the following integration problems would you rather solve?

$$\int \frac{4x+5}{x^2+x-2}\,dx \quad \text{or} \quad \int \left(\frac{1}{x+2} + \frac{3}{x-1}\right) dx$$

Clearly, the one on the right is preferable, because we know right away that

$$\int \left(\frac{1}{x+2} + \frac{3}{x-1}\right) dx = \ln|x+2| + 3\ln|x-1| + C$$

But this also means that

$$\int \frac{4x+5}{x^2+x-2}\,dx = \ln|x+2| + 3\ln|x-1| + C$$

because the integrands in both integrals are the same function, just in different form (compact versus dispersed).

This is the idea behind integration by partial fractions: given a nasty looking rational function, can we stick mathematical dynamite into the function to "blow it up" into a bunch of smaller pieces? If so, we can do the integration of the dispersed form instead. Here are some other examples of rational functions blown up into smaller pieces:

$$\frac{1}{x^2 - 1} \overset{kaboom!}{=} \frac{1}{2(x-1)} - \frac{1}{2(x+1)}$$

$$\frac{10}{(x-1)(x^2+9)} \overset{kaboom!}{=} \frac{1}{x-1} - \frac{x+1}{x^2+9}$$

$$\frac{1}{(x+5)^2(x-1)} \overset{kaboom!}{=} \frac{-1/6}{(x+5)^2} - \frac{1/36}{x+5} + \frac{1/36}{x-1}$$

Since we can "easily" integrate each separate term on the right sides of those expressions, we are also able to find the integral of the left sides.

The Common Denominator is Common Denominators

The strategy in this method is to factor the denominator of the given rational expression, and recognize that the expression can be dispersed into smaller pieces using those factors as denominators. To get the idea, let's just work with some numbers first. Of course you know that we can add $1/2$ and $2/5$ by finding a common denominator:

$$\frac{1}{2} + \frac{2}{5} = \frac{5}{10} + \frac{4}{10} = \frac{9}{10}$$

But what about the reverse? Could we start with $9/10$ and break it apart in reverse order? Sure! Since $10 = 2 \cdot 5$, we know that the denominator of 10 could be a common denominator of fractions with individual denominators of 2 and 5. So, we could pose that

$$\frac{9}{10} = \frac{?}{2} + \frac{?}{5}$$

Notice that we didn't worry about a third term with a denominator of 10 because if there was such a term, that term itself could be split into terms with denominators of 2 and 5. So what we're really looking for are *prime factors* of the denominator.

Here are some other templates for "blowing up" a single fraction into multiple individual terms (note we assume the original value is fully simplified, and any common factors between the numerator and denominator are already gone):

$$\frac{3}{14} = \frac{3}{2 \cdot 7} \overset{kaboom!}{=} \frac{?}{2} + \frac{?}{7}$$

$$\frac{11}{15} = \frac{11}{3 \cdot 5} \overset{kaboom!}{=} \frac{?}{3} + \frac{?}{5}$$

$$\frac{7}{12} = \frac{7}{2^2 \cdot 3} \overset{kaboom!}{=} \frac{?}{2} + \frac{?}{2^2} + \frac{?}{3}$$

$$\frac{13}{16} = \frac{13}{2^4} \overset{kaboom!}{=} \frac{?}{2} + \frac{?}{2^2} + \frac{?}{2^3}$$

Note in the last two cases that we factored the denominator down so that it included only powers of prime numbers; any other intermediate number (like 8 in the final case) are ignored because those terms (such as ?/8) could themselves be broken down.

If we had a number greater than one, we would separate the 1, and only "blow up" the fractional part that's less than one:

$$\frac{17}{12} = 1 + \frac{5}{12} = 1 + \frac{5}{2^2 \cdot 3} \overset{kaboom!}{=} 1 + \frac{?}{2} + \frac{?}{2^2} + \frac{?}{3}$$

If this makes sense with numbers, you're ready to move on to functions! When we "blow up" a rational function, we are performing **partial fraction decomposition**. Here's an example of a starting template for partial fraction decomposition of $1/(x^2 - 1)$:

$$\frac{1}{x^2 - 1} = \frac{1}{(x-1)(x+1)} \overset{kaboom!}{=} \frac{?}{x-1} + \frac{?}{x+1}$$

Our job is then to discover the mystery numerators. Of course, we should do things properly and get rid of those question marks that currently represent mystery numerators. Since the mystery numerators are *unknown*, let's assign "unknowns". Here, we need two of them. A proper partial fraction decomposition of $1/(x^2 - 1)$ would look like this:

$$\frac{1}{x^2 - 1} = \frac{A}{x-1} + \frac{B}{x+1} \qquad (7.8)$$

And if you agree with that partial fraction decomposition template, then you should also agree that

$$\int \frac{1}{x^2 - 1}\, dx = \int \frac{A}{x-1} + \frac{B}{x+1}\, dx = A\ln|x-1| + B\ln|x+1| + C$$

All we need to know is the values of A and B that put it all together (remember, C is the usual arbitrary constant). See, the Calculus part of it is easy!

So let's get busy. A rational function that's ready for partial fraction decomposition is (a) fully simplified, and (b) has its denominator factored down to irreducible contributing pieces. Let's assume the degree of the numerator is less than the degree of the denominator.[3] Given a rational function, the general strategy to form a *partial fraction decomposition template* is this:

- Factor the original denominator down to irreducible linear and quadratic pieces, or powers of linear and quadratic pieces. (This means that, for example, if you decide one of the factors is $x^2 + 2x + 1$, you're not done yet, because that is equal to $(x+1)^2$.)
- Build a template for partial fraction decomposition that allows one term for each piece of the reduced denominator; in each contributing term of the template, mystery numerators are assigned as follows:

  - if the denominator is a linear function or a power of a linear function, allow the numerator to be a constant, like A or B, e.g.

  $$\frac{7x}{(x-3)^2} = \frac{A}{x-3} + \frac{B}{(x-3)^2}$$

  - if the denominator is an irreducible quadratic function (or a power of such), allow the numerator to be a linear function, such as $Cx + D$, e.g.

  $$\frac{x^2}{(x+2)(x^2+3)} = \frac{A}{x+2} + \frac{Bx+C}{x^2+3}$$

Once we have a proper partial fraction decomposition template for a rational function, we have a pretty good idea what the indefinite integral of the function will look like; we just don't know (yet) the values of the mystery

---

[3]See remark at the end of this section.

numerators. Here are some examples, complete up to the point where we need to discover values of mystery numbers.

$\boxed{\textbf{EX 1A}}$ Start finding $\displaystyle\int \frac{x-9}{(x+5)(x-2)}\,dx$.

A partial fraction decomposition template for the integrand is,

$$\frac{x-9}{(x+5)(x-2)} \overset{kaboom!}{=} \frac{A}{x+5} + \frac{B}{x-2}$$

and so

$$\int \frac{x-9}{(x+5)(x-2)}\,dx = \int \frac{A}{x+5} + \frac{B}{x-2}\,dx$$
$$= A\ln|x+5| + B\ln|x-2| + C$$

where we still need to figure out the values of A and B that tie it all together (C is the constant of integration). ■

$\boxed{\textbf{EX 2A}}$ Start finding $\displaystyle\int \frac{1}{(x+5)^2(x-1)}\,dx$.

First, let's disassemble the integrand. Remember to account for ALL possible denominators which could contribute to given denominator:

$$\frac{1}{(x+5)^2(x-1)} \overset{kaboom!}{=} \frac{A}{x+5} + \frac{B}{(x+5)^2} + \frac{C}{x-1}$$

Then,

$$\int \frac{1}{(x+5)^2(x-1)}\,dx = \int \frac{A}{x+5} + \frac{B}{(x+5)^2} + \frac{C}{x-1}\,dx$$
$$= A\ln|x+5| - \frac{B}{x+5} + C\ln|x-1| + D$$

(where A, B, C are to be determined, and D is the generic arbitrary constant of integration). ■

$\boxed{\textbf{EX 3A}}$ Start finding $\displaystyle\int \frac{10}{(x-1)(x^2+9)}\,dx$.

First, let's disassemble the integrand. Since that second part of the denominator is an irreducible quadratic, the numerator of its mystery term should be a linear function, and we have:

$$\frac{10}{(x-1)(x^2+9)} = \frac{A}{x-1} + \frac{Bx+C}{x^2+9}$$

So using the basic integral types shown in (7.4), (7.5), and (7.6),

$$\int \frac{10}{(x-1)(x^2+9)}\, dx = \int \frac{A}{x-1} + \frac{Bx+C}{x^2+9}\, dx$$

$$= \int \frac{A}{x-1}\, dx + \int \frac{Bx}{x^2+9}\, dx + \int \frac{C}{x^2+9}\, dx$$

$$= A\ln|x-1| + \frac{B}{2}\ln(x^2+9) + \frac{C}{3}\tan^{-1}\frac{x}{3} + D$$

(where A, B, C are to be determined, and D is the generic arbitrary constant of integration). ∎

This process, in which we break apart a rational function into its least common denominator components so that we can integrate it piece by piece, is called **integration by partial fractions**.

In all three of the above examples, the calculus part of things is over with! Now on to the tedious algebra, where we must find all those mystery values from numerators in our templates. Let's see how this works with the very first case we had, in Eq. (7.8). We must discover the values of A and B for which,

$$\frac{1}{x^2-1} = \frac{A}{x-1} + \frac{B}{x+1}$$

There are two methods we can use. Both methods begin with a multiplication of the full equation at hand by the entire denominator. For the template in (7.8), we multiply both sides by the full denominator:

$$\frac{1}{x^2-1}\cdot(x^2-1) = \left(\frac{A}{x-1} + \frac{B}{x+1}\right)\cdot(x^2-1)$$

which leads, with some canceling, to:

$$1 = A(x+1) + B(x-1) \tag{7.9}$$

Since we have two unknowns to find, you might expect we need to somehow generate two equations out of this one expression. We have two options.

Method 1: Match coefficients. Let's gather the terms on both sides of (7.9) according to powers of x (we'll add more to the left side for clarity):

$$1 = A(x+1) + B(x-1)$$

$$0x + 1 = (A+B)x + (A-B)$$

The coefficients of x on both sides must match, and so must the constant terms. This gives us two equations in two unknowns:

$$A + B = 0$$
$$A - B = 1$$

And we can solve these using any method that seems convenient. In this case, we get $A = -B$ from the top equation, and can pass that along to the second to get $(-B) - B = 1$, or $B = -1/2$. And then, $A = -B = 1/2$.

Method 2: Clever selection of values of x. Eq. (7.9) must hold for *all* possible values of x. We can use this to our advantage. If you look at $1 = A(x + 1) + B(x - 1)$ and understand that plugging in any value of x generates one equation in A and B that must hold, then there are two values of x just *begging* to be plugged in. Specifically, if we plug $x = 1$ in to (7.9), we get $1 = 2A + 0$, or $A = 1/2$. If we try $x = -1$ we get $1 = 0 - 2B$ which gives us that $B = -1/2$. Technically we have now received two equations in two unknowns, but the equations themselves display the solution values. These are the same values as we found with Method 1.

Picking up in our break-down of the original function as shown in Eq. (7.8), we get

$$\frac{1}{x^2 - 1} = \frac{1/2}{x - 1} + \frac{(-1/2)}{x + 1}$$

so that,

$$\int \frac{1}{x^2 - 1} \, dx = \int \frac{1/2}{x - 1} + \frac{(-1/2)}{x + 1} \, dx = \frac{1}{2} \ln |x - 1| - \frac{1}{2} \ln |x + 1| + C$$

By properties of logarithms, we can simplify this answer further:

$$\int \frac{1}{x^2 - 1} \, dx = \frac{1}{2} \left(\ln |x - 1| - \ln |x + 1| \right) + C$$

$$= \frac{1}{2} \ln \left| \frac{x - 1}{x + 1} \right| + C$$

Common error alert! Please, please, please be sure you remember (and understand *why*),

$$\int \frac{1}{x^2 - 1} \, dx \quad \text{is NOT equal to} \quad \ln |x^2 - 1| + C$$

Of all the integration mistakes that cause people to lose points, this could very well be the worst one in all of Calculus.

🍴 FFT: Here is one last remark before we finish up EX 1A, 2A, and 3A from above. When finding values of mystery numerators in partial fraction decompositions, you are free to use either method shown above. But the more unknowns there are, the worse Method 1 gets. For example, if you have four unknown values in your template, Method 1 generates a full system of four equations in four unknowns. Do you even remember how to solve such a thing? Method 2 can lead to several equations, but they're individually simpler because you are trying to eliminate terms by your chosen values of x. On the other hand, Method 2 comes with an inherent "technical foul". Do you know what it is? Consider Eq. (7.9) and the values of x we selected to produce solvable expressions for those unknowns. Think about those values as compared to the domain of the function being integrated. Why is this not a deal-breaker? 🍴

EX 1B Finish finding $\displaystyle\int \frac{x-9}{(x+5)(x-2)}\, dx$.

In EX 1A, we posed the template

$$\frac{x-9}{(x+5)(x-2)} = \frac{A}{x+5} + \frac{B}{x-2}$$

Multiplying by the denominator $(x+5)(x-2)$ gives

$$x - 9 = A(x-2) + B(x+5)$$

Remember, this equation must hold for ANY value of x. Specifically, if we select $x = 2$, this equation simplifies to $-7 = 7B$ so that $B = -1$. If we select $x = -5$, this simplifies to $-14 = -7A$ so that $A = 2$. We can now place these two values into the template result from EX 1A:

$$\int \frac{x-9}{(x+5)(x-2)}\, dx = \int \frac{A}{x+5} + \frac{B}{x-2}\, dx$$
$$= A\ln|x+5| + B\ln|x-2| + C$$

to give

$$\int \frac{x-9}{(x+5)(x-2)}\, dx = 2\ln|x+5| - \ln|x-2| + C$$

If you like using properties of logs, this result can be tidied up:

$$\int \frac{x-9}{(x+5)(x-2)}\, dx == \ln \frac{|x+5|^2}{|x-2|} + C \quad \blacksquare$$

You Try It

(1) Find $\displaystyle\int \frac{x}{x^2 - 5x + 6}\, dx$.

EX 2B Finish finding $\displaystyle\int \frac{1}{(x+5)^2(x-1)}\,dx$.

In EX 2A, we posed the template

$$\frac{1}{(x+5)^2(x-1)} = \frac{A}{x+5} + \frac{B}{(x+5)^2} + \frac{C}{x-1}$$

Multiplying by the denominator $(x+5)^2(x-1)$ gives

$$1 = A(x+5)(x-1) + B(x-1) + C(x+5)^2$$

Remember, this equation must hold for ANY value of x. Specifically, if we select $x = 1$, we can eliminate the terms with A and B:

$$1 = C(1+5)^2 \rightarrow C = \frac{1}{36}$$

Another "obvious" clever value is $x = -5$:

$$1 = B(-5-1) \rightarrow B = -\frac{1}{6}$$

There are no other immediately obvious values of x to use. But we are free to select *any* value of x. What's the easiest number out there? Zero! With $x = 0$ and the values of B and C that are already known,

$$1 = A(0+5)(0-1) - \frac{1}{6}(0-1) + \frac{1}{36}(0+5)^2$$

$$1 = -5A + \frac{1}{6} + \frac{25}{36}$$

$$5A = -\frac{36}{36} + \frac{6}{36} + \frac{25}{26}$$

$$A = \frac{1}{5}\left(-\frac{5}{36}\right) = -\frac{1}{36}$$

Now with values of A, B, C known, we can complete EX 2A:

$$\int \frac{1}{(x+5)^2(x-1)}\,dx = A\ln|x+5| - \frac{B}{x+5} + C\ln|x-1| + D$$

$$= -\frac{1}{36}\ln|x+5| + \frac{1}{6(x+5)} + \frac{1}{36}\ln|x-1| + D$$

and again, if you like properties of logs, this compresses a bit:

$$\int \frac{1}{(x+5)^2(x-1)}\,dx = \frac{1}{6(x+5)} + \frac{1}{36}\ln\frac{|x-1|}{|x+5|} + D \quad \blacksquare$$

You Try It

(2) Find $\displaystyle\int \frac{3x^2 - x + 1}{x^3 - x^2}\,dx$.

EX 3B Finish finding $\int \dfrac{10}{(x-1)(x^2+9)}\,dx$.

In EX 3A, we posed the template
$$\frac{10}{(x-1)(x^2+9)} = \frac{A}{x-1} + \frac{Bx+C}{x^2+9}$$
Multiplying by the denominator $(x-1)(x^2+9)$ gives
$$10 = A(x^2+9) + (Bx+C)(x-1) = A(x^2+9) + Bx(x-1) + C(x-1)$$
We have two immediately "clever" values of x to use:
$$x = 1 \longrightarrow A = 1$$
$$x = 0 \longrightarrow C = -1$$
We still need another value of x, so how about $x = -1$? With the values of A and C we now have, this gives $10 = 10 + 2B + 2$, so that $B = -1$. Now with values of A, B, C known, we can complete EX 3A:
$$\int \frac{10}{(x-1)(x^2+9)}\,dx = A\ln|x-1| + \frac{B}{2}\ln(x^2+9) + \frac{C}{3}\tan^{-1}\frac{x}{3} + D$$
$$= \ln|x-1| - \frac{1}{2}\ln(x^2+9) - \frac{1}{3}\tan^{-1}\frac{x}{3} + D \quad \blacksquare$$

Following the last two examples, it is important to emphasize the difference between denominators that are powers of linear functions, like $(x+a)^2$, and irreducible quadratics, like $x^2 + a^2$. These two ARE NOT the same thing, and the "mystery numerators" you need to attach to them are different! Numerators for powers of linear functions only need to be constants, like A. Numerators for irreducible quadratics need to be *linear functions* like $Ax + B$.

Also, feel free to develop your own strategy for the order in which you solve these problems. In EX 1, 2, and 3, I've set up a partial fraction decomposition template and produced the result of the integral *before* having the specific values of the unknown constants to put in place. I prefer to integrate without the clutter of possibly ugly numerical values in the way. But you may prefer to find the unknowns first, and then tackle the integral. Do what works best for you.

You Try It

(3) Find $\int \dfrac{x+1}{x(x^2+1)}\,dx$.

(4) Find $\int \dfrac{1}{x\sqrt{x+1}}\,dx$. (Hint: Use the substitution $u = \sqrt{x+1}$.)

Definite Integrals

All of the examples so far have been indefinite integrals; for definite integrals, nothing changes — except for the application of the endpoints of integration via the Fundamental Theorem of Calculus, once the general antiderivative is determined. Let's look ahead to Sec. 7.6 by including a slight twist on the use of the Fundamental Theorem.

EX 4 Only one of the two definite integrals allows use of the Fundamental Theorem of Calculus. Identify which, and solve it.

$$\int_0^1 \frac{6x}{(x+1)(x^2-4)}\,dx \quad \text{or} \quad \int_0^3 \frac{6x}{(x+1)(x^2-4)}\,dx$$

Remember that the Fundamental Theorem of Calculus requires that the function being integrated (the integrand) is continuous within the interval of integration. The integrand in this function has discontinuities at (in numerical order), $x = -2$, $x = -1$, and $x = 2$. Sadly, the interval of integration in the second integral, from $x = 0$ to $x = 3$, crosses over one of these discontinuities. Therefore, the integrand is not continous on that interval, and we cannot use the Fundamental Theorem of Calculus. And so, the first integral is the proper one. (Integrals which do not directly allow use of the Fundamental Theorem are called "Improper Integrals", and we'll look at them in Sec. 7.6).

Let's get to work on the first integral. Unlike EX 1, 2, and 3 above, I'll get the values of the necessary partial fraction constants before doing the integral, just for more variety. Here is the set-up, the cross multiplication, and the selection of "clever" values for x:

$$\frac{6x}{(x+1)(x^2-4)} = \frac{A}{x+1} + \frac{B}{x-2} + \frac{C}{x+2}$$
$$6x = A(x-2)(x+2) + B(x+1)(x+2) + C(x+1)(x-2)$$
$$x = 2 \to 12 = 12B \qquad\qquad \to \quad B = 1$$
$$x = -2 \to -12 = C(-1)(-4) \qquad \to \quad C = -3$$
$$x = -1 \to -6 = A(-3)(1) \qquad\qquad \to \quad A = 2$$

So,

$$\int_0^1 \frac{6x}{(x+1)(x^2-4)}\,dx = \int_0^1 \left(\frac{2}{x+1} + \frac{1}{x-2} - \frac{3}{x+2}\right)\,dx$$

$$= \left.\left(2\ln|x+1| + \ln|x-2| - 3\ln|x+2|\right)\right|_0^1$$

$$= (2\ln 2 + \ln 1 - 3\ln 3) - (2\ln 1 + \ln 2 - 3\ln 2)$$

$$= (2\ln 2 - 3\ln 3) - (-2\ln 2)$$

$$= 4\ln 2 - 3\ln 3 = \ln\frac{16}{27} \quad \blacksquare$$

You Try It

(5) Only one of the two definite integrals allows use of the Fundamental Theorem of Calculus. Identify which, and solve it.

$$\int_1^5 \frac{x+1}{x(x^2+1)}\,dx \qquad \text{or} \qquad \int_{-1}^1 \frac{x+1}{x(x^2+1)}\,dx$$

To close out, let's resolve two lingering issues. First, here's a respose to the "Food For Thought" from above. The values of x we select to resolve constants in the partial fraction decomposition are often going to be the very values which are not in the domain of the original integrand (go back and take a look at EX 1B, 2B, 3B). We are relying on the idea that our partial fraction decomposition needs to hold for *all* possible values of x", where what we should understand is that the decomposition and resulting equations must hold for *almost* all values of x (excluding the discontinuities of the original integrand). But because the functions we deal with here are relatively benign, the values of the unknowns we find from any selected values of x are going to be the same, so if we're selecting *suspicious* values, that's OK. We're all friends here, and we will allow this technicality to be ignored.

Second, we restricted ourselves in this section to rational functions in which the degree of the numerator is less than the degree of the denominator. If that's not the case, then polynomial division can be brought in. If $\deg(P) \geq \deg(Q)$, then the rational function $P(x)/Q(x)$ can be rewritten with polynomial division as

$$\frac{P(x)}{Q(x)} = r(x) + \frac{p(x)}{q(x)}$$

where $r(x)$ is a standard polynomial of degree one or more, and deg$p$ < deg$q$. The polyomial $r(x)$ can be integrated as usual, and the remaining rational function $p(x)/q(x)$ can be integrated using the methods shown in this section. We won't include these cases here, because polynomial division is an algebraic detour that would distract us from the main goal.

Quick Reference

Here's a summary of how these problems work. Given an antiderivative (or definite integral) of a rational function,

- Factor the denominator; recognize that your function can be considered the result of a least-common-denominator combination of several terms. Identify ALL possible denominators that could combine to give the ofiginal.
- Set up an equation that shows the decomposition of your function into smaller terms, one term for each possible original denominator. Your job is (eventually) to discover the numerators of all the terms that make the combination work.
- If a term has a linear function or power of a linear function as its denominator, assign a constant, such as A, as its mystery numerator.
- If a term has an irreducible quadratic as its denominator, assign linear function, such as $Ax + B$, as its mystery numerator.
- If you prefer it, solve the integral using the partial fraction decomposition before having specific values to plug in. Or, find the constants before doing the integration ... your choice!
- To find the constants: multiply both sides of your partial fraction decomposition by the original denominator to clear out all fractions and leave one big equation with multiple unknowns A, B, ...
- Find the values of A, B, etc. that make it all work, either by
  - Picking clever values of x to plug into the equation and eliminate constants
  - Matching coefficients of the variable on the left and right sides of this new equation
- If you're working with a definite integral, apply the endpoints via the FTOC.

Integration by Partial Fractions — Problem List

Integration by Partial Fractions — You Try It

These appeared above; solutions begin on the next page.

(1) Find $\displaystyle\int \frac{x}{x^2 - 5x + 6}\, dx$.

(2) Find $\displaystyle\int \frac{3x^2 - x + 1}{x^3 - x^2}\, dx$.

(3) Find $\displaystyle\int \frac{x + 1}{x(x^2 + 1)}\, dx$.

(4) Find $\displaystyle\int \frac{1}{x\sqrt{x + 1}}\, dx$. (Hint: Use the substitution $u = \sqrt{x + 1}$.)

(5) Only one of the two definite integrals allows use of the Fundamental Theorem of Calculus. Identify which, and solve it.

$$\int_1^5 \frac{x + 1}{x(x^2 + 1)}\, dx \qquad \text{or} \qquad \int_{-1}^1 \frac{x + 1}{x(x^2 + 1)}\, dx$$

Integration by Partial Fractions — Practice Problems

Try these as you get the hang of the You Try It problems. Solutions to these problems are available in Sec. A.1.3.

(1) Find $\displaystyle\int \frac{1}{x^2 + 3x - 4}\, dx$.

(2) Find $\displaystyle\int \frac{x^2}{(x - 3)(x + 2)^2}\, dx$.

(3) Find $\displaystyle\int \frac{x^2 - x + 6}{x^3 + 3x}\, dx$.

(4) Evaluate $\displaystyle\int_0^1 \frac{x - 1}{x^2 + 3x + 2}\, dx$.

Integration by Partial Fractions — Challenge Problems

Try these problems to test your skills with the ideas in this section. Solutions to these problems are available in Sec. B.1.2.

(1) Find $\displaystyle\int \frac{1}{x^2(x - 1)^2}\, dx$.

(2) Find $\displaystyle\int \frac{x^2 - 2x - 1}{(x - 1)^2(x^2 + 1)}\, dx$.

(3) Find $\displaystyle\int \frac{e^{2x}}{e^{2x} + 3e^x + 2}\, dx$. (Hint: Use the substitution $u = e^x$.)

Integration by Partial Fractions — You Try It — Solved

---

(1) Find $\displaystyle\int \frac{x}{x^2 - 5x + 6}\, dx$.

□ First, let's blow up the integrand. The components of the denominator are linear functions, so their numerators can be constants:

$$\frac{x}{x^2 - 5x + 6} = \frac{x}{(x-3)(x-2)} \overset{kaboom!}{=} \frac{A}{x-3} + \frac{B}{x-2} \qquad (7.10)$$

So if nothing else, we know that

$$\int \frac{x}{x^2 - 5x + 6}\, dx = \int \frac{A}{x-3} + \frac{B}{x-2}\, dx = A\ln|x-3| + B\ln|x-2| + C$$

Now we need to find A and B. Multiplying (7.10) by the denominator $(x-3)(x-2)$ gives

$$x = A(x-2) + B(x-3)$$

Remember, this equation must hold for ANY (shhhh, don't tell!) value of x. Specifically, if we select $x = 2$, this simplifies to $2 = -B$ so that $B = -2$. If we select $x = 3$, this simplifies to $A = 3$. So,

$$\int \frac{x}{(x-3)(x-2)}\, dx = 3\ln|x-3| - 2\ln|x-2| + C \quad \blacksquare$$

---

(2) Find $\displaystyle\int \frac{3x^2 - x + 1}{x^3 - x^2}\, dx$.

□ The components of the denominator are (powers of) linear functions, so their partial fraction numerators can be constants:

$$\frac{3x^2 - x + 1}{x^3 - x^2} = \frac{3x^2 - x + 1}{x^2(x-1)} = \frac{A}{x} + \frac{B}{x^2} + \frac{C}{x-1} \qquad (7.11)$$

and so,

$$\int \frac{3x^2 - x + 1}{x^3 - x^2}\, dx = \int \frac{A}{x} + \frac{B}{x^2} + \frac{C}{x-1}\, dx$$
$$= A\ln|x| - \frac{B}{x} + C\ln|x-1| + D$$

To find A, B, and C we can multiply (7.11) by the denominator $x^2(x-1)$ to give

$$3x^2 - x + 1 = Ax(x-1) + B(x-1) + Cx^2$$

The following values of x lead to values of constants as shown:

$$x = 0 \rightarrow B = -1$$
$$x = 1 \rightarrow C = 3$$
$$x = 11 \rightarrow = 2A + (-1) + (3)(4) \rightarrow A = 0$$

Putting it together,

$$\int \frac{3x^2 - x + 1}{x^3 - x^2} \, dx = \frac{1}{x} + 3\ln|x - 1| + C \quad \blacksquare$$

---

(3) Find $\int \dfrac{x + 1}{x(x^2 + 1)} \, dx$.

☐ First, let's blow up the integrand. The denominator contains both a linear function x and an irreducible quadratic $x^2 + 1$; their partial fraction template is:

$$\frac{x + 1}{x(x^2 + 1)} \overset{kaboom!}{=} \frac{A}{x} + \frac{Bx + C}{x^2 + 1} = \frac{A}{x} + \frac{Bx}{x^2 + 1} + \frac{C}{x^2 + 1} \quad (7.12)$$

so that

$$\int \frac{x + 1}{x(x^2 + 1)} \, dx = \int \frac{A}{x} + \frac{Bx}{x^2 + 1} + \frac{C}{x^2 + 1} \, dx$$

$$= A\ln|x| + \frac{B}{2}\ln(x^2 + 1) + C\tan^{-1}(x) + D$$

To find A, B, and C we can multiply (7.12) by the denominator $x(x^2+1)$ to give

$$x + 1 = A(x^2 + 1) + (Bx + C)(x)$$

Let's find the constants A, B, and C by matching coefficients this time. Regrouping the right hand side, we get

$$x + 1 = Ax^2 + A + Bx^2 + Cx$$
$$0x^2 + x + 1 = (A + B)x^2 + (C)x + A$$

so that matching coefficients of x^2, x and the constants on each side, we see that:

$$A + B = 0 \quad \text{from matching the coefficients of } x^2$$
$$C = 1 \quad \text{from matching the coefficients of } x$$
$$A = 1 \quad \text{from matching the constants}$$

The result of these equations is $A = 1$, $B = -1$, $C = 1$. Altogether,

$$\int \frac{x + 1}{x(x^2 + 1)} \, dx = \ln|x| - \frac{1}{2}\ln(x^2 + 1) + \tan^{-1}(x) + D \quad \blacksquare$$

(4) Find $\displaystyle\int \frac{1}{x\sqrt{x+1}}\,dx$.

□ This problem requires a substitution first. Select $u = \sqrt{x+1}$, then $u^2 = x+1$, so that $x = u^2 - 1$ and $dx = 2u\,du$. This gives

$$\int \frac{1}{x\sqrt{x+1}}\,dx = \int \frac{2u}{(u^2-1)u}\,du = \int \frac{2}{(u+1)(u-1)}\,du$$

This new integral requires partial fractions (big surprise, huh?). Fortunately, it's not too bad. For variety, let's find all the constants before we do the integration. The partial fraction decomposition is:

$$\frac{2}{(u+1)(u-1)} = \frac{A}{u+1} + \frac{B}{u-1}$$

Multiplying both sides by the full denominator gives

$$2 = A(u-1) + B(u+1)$$

Selecting $u = 1$ gives $B = 1$, and selecting $u = -1$ gives $A = -1$. So,

$$\int \frac{2}{(u+1)(u-1)}\,du = \left(\int \frac{-1}{u+1} + \frac{1}{u-1} \right) du$$

$$= -\ln|u+1| + \ln|u-1| + C = \ln \frac{|u-1|}{|u+1|} + C$$

Reversing the substitution,

$$\int \frac{1}{x\sqrt{x+1}}\,dx = \ln \frac{|\sqrt{x+1}-1|}{|\sqrt{x+1}+1|} + C \quad\blacksquare$$

---

(5) Only one of the two definite integrals allows use of the Fundamental Theorem of Calculus. Identify which, and solve it.

$$\int_1^5 \frac{x+1}{x(x^2+2)}\,dx \qquad \text{or} \qquad \int_{-1}^1 \frac{x+1}{x(x^2+2)}\,dx$$

□ The interval $[-1, 1]$ contains the location of a discontinuity of the integrand ($x = 0$), therefore we cannot use the Fundamental Theorem of Calculus for the second integral. For the first, we have an irreducible quadratic as part of the denominator, so its partial fraction numerator may be a linear function. With

$$\frac{x+1}{x(x^2+2)} = \frac{A}{x} + \frac{Bx+C}{x^2+2} = \frac{A}{x} + \frac{Bx}{x^2+2} + \frac{C}{x^2+2} \qquad (7.13)$$

we have a general antiderivative

$$F(x) = \int \frac{A}{x} + \frac{Bx}{x^2 + 2} + \frac{C}{x^2 + 2} \, dx$$

$$= A \ln |x| + \frac{B}{2} \ln(x^2 + 2) + \frac{C}{\sqrt{2}} \tan^{-1}\left(\frac{x}{\sqrt{2}}\right) + D$$

To find A, B, and C we can multiply the left portions of (7.13) by the denominator $x(x^2 + 2)$ to give

$$x + 1 = A(x^2 + 2) + (Bx + C)(x)$$

$$x = 0 \rightarrow A = \frac{1}{2}$$

$$x = 1 \rightarrow 2 = \frac{3}{2} + B + C$$

$$x = -1 \rightarrow 0 = \frac{3}{2} + B - C$$

If we add the two equations generated from $x = 1$ and $x = -1$, we find $2 = 3 + 2B$, or $B = -1/2$. Then borrowing the third equation, $0 = 3/2 - 1/2 - C$, or $C = 1$. With $A = 1/2$, $B = -1/2$, and $C = 1$, we're ready to go. Choosing the antiderivative with arbitrary constant $D = 0$, we now know the specifics of the antiderivative,

$$F(x) = \frac{1}{2} \ln |x| - \frac{1}{4} \ln(x^2 + 2) + \frac{1}{\sqrt{2}} \tan^{-1}\left(\frac{x}{\sqrt{2}}\right) \qquad (7.14)$$

The Fundamental Theorem of Calculus tells us that

$$\int_1^5 \frac{x + 1}{x(x^2 + 2)} \, dx = F(5) - F(1)$$

where $F(x)$ is the general antiderivative. We have (and these are messy!)

$$F(5) = \frac{1}{2} \ln 5 - \frac{1}{4} \ln 27 + \frac{1}{\sqrt{2}} \tan^{-1}\left(\frac{5}{\sqrt{2}}\right)$$

$$= \ln \sqrt{5} - \frac{3}{4} \ln 3 + \frac{1}{\sqrt{2}} \tan^{-1}\left(\frac{5}{\sqrt{2}}\right)$$

$$F(1) = \frac{1}{2} \ln 1 - \frac{1}{4} \ln 3 + \frac{1}{\sqrt{2}} \tan^{-1}\left(\frac{1}{\sqrt{2}}\right)$$

$$= -\frac{1}{4} \ln 3 + \frac{1}{\sqrt{2}} \tan^{-1}\left(\frac{1}{\sqrt{2}}\right)$$

and so,

$$\int_1^5 \frac{x+1}{x(x^2+2)}\, dx = F(5) - F(1)$$

$$= \ln\sqrt{5} - \ln 3 + \frac{1}{\sqrt{2}}\left[\tan^{-1}\left(\frac{5}{\sqrt{2}}\right) - \tan^{-1}\left(\frac{1}{\sqrt{2}}\right)\right]$$

$$= \ln\frac{\sqrt{5}}{3} + \frac{1}{\sqrt{2}}\left[\tan^{-1}\left(\frac{5}{\sqrt{2}}\right) - \tan^{-1}\left(\frac{1}{\sqrt{2}}\right)\right]$$

That was pretty terrible. ∎

7.4 Trigonometric Integrals

Introduction

Trigonometric integrals are, well, integrals involving trigonometric functions — but with the implication that the integrands are more complicated than just, say, $\sin x$. The strategy will be to use suitable trigonometric identities to put the integrand into a form for which calls for substitution. You will need to recall the following possibilities for substitution with trig functions:

$$u = \sin x \to du = \cos x \, dx$$
$$u = \cos x \to du = -\sin x \, dx$$
$$u = \tan x \to du = \sec^2 x \, dx$$
$$u = \sec x \to du = \sec x \tan x \, dx$$

Also remember your favorite trig identities,

$$\sin^2 x + \cos^2 x = 1 \qquad (7.15)$$
$$\sec^2 x - \tan^2 x = 1 \qquad (7.16)$$

The integrals that we encounter will be of three types:

(1) Integrals with $\sin x$, $\cos x$.
(2) Integrals with $\sec x$, $\tan x$.
(3) Integrals that can be converted to one of the the first two types by identities, such as $\csc x = 1/\sin x$.

Integrals With Sines and Cosines

Remember that sines can be converted to cosines (or vice versa) when they come grouped in even powers, because we can use the identity $\sin^2 x + \cos^2 x = 1$. So, if you have an integral with sines and cosines in it, then:

Useful Fact 7.3. *If the power of the sine or cosine term is odd, borrow one of them to join with dx in a substitution, and convert the remaining (now even powered) group of them using identity (7.15).*

EX 1 Find $\int \sin^5 x \cos^4 x \, dx$.

The sine term in the integrand has an odd power, so let's borrow one a $\sin x$, move it over with dx, and present the new version of the integral:

$$\int \sin^4 x \cos^4 x \cdot \sin x \, dx$$

Next, we convert the rest of the sines into cosines:

$$\int (1 - \cos^2 x)^2 \cos^4 x \sin x \, dx$$

Now the integral is all set up for

$$\cos x = u \qquad \text{with} \qquad \sin x \, dx = -du$$

giving

$$\int (1 - \cos^2 x)^2 \cos^4 x \sin x \, dx = -\int (1 - u^2)^2 u^4 \, du = -\int (u^4 - 2u^6 + u^8) \, du$$

$$= -\frac{1}{5} u^5 + \frac{2}{7} u^7 - \frac{1}{9} u^9 + C$$

$$= -\frac{1}{5} \cos^5 x + \frac{2}{7} \cos^7 x - \frac{1}{9} \cos^9 x + C \quad \blacksquare$$

You Try It

(1) Find $\int \sin^3 x \cos^2 x \, dx$.

Useful Fact 7.4. *If the powers of the sines and cosines are even, use one of the identities*

$$\sin^2 x = \frac{1}{2}(1 - \cos 2x) \qquad \text{or} \qquad \cos^2 x = \frac{1}{2}(1 + \cos 2x)$$

(these are called half-angle identities).

EX 2 Evaluate $\int_0^\pi \sin^2 x \, dx$.

There are no cosines to use in a substitution, and we have an even power of sine. Using a half-angle identity from Useful Fact 7.4,

$$2 \sin^2 x = 1 - \cos 2x \qquad \text{or} \qquad \sin^2 x = \frac{1}{2}(1 - \cos 2x)$$

we can get

$$\int_0^\pi \sin^2 x\,dx = \int_0^\pi \frac{1}{2}\left(1 - \cos 2x\right)\,dx = \frac{1}{2}\int_0^\pi \left(1 - \cos 2x\right)\,dx$$

and then the brief substitution

$$2x = u \qquad ; \qquad dx = \frac{1}{2}\,du$$

gives (don't forget to change the limits, too!)

$$\frac{1}{2}\int_0^\pi \left(1 - \cos 2x\right)\,dx = \frac{1}{2}\int_0^{2\pi} \left(1 - \cos u\right)\cdot\frac{1}{2}\,du$$

$$= \frac{1}{4}\int_0^{2\pi} \left(1 - \cos u\right)\,du = \frac{1}{4}(u - \sin u)\Big|_0^{2\pi}$$

$$= \frac{1}{4}(2\pi - \sin 2\pi) - \frac{1}{4}(0 - \sin 0) = \frac{\pi}{2} \quad \blacksquare$$

You Try It

(2) Find $\displaystyle\int_0^{\pi/2} \cos^2\theta\,d\theta$.

Integrals With Tangents and Secants

Remember that tangents can be converted to secants (or vice versa) when they come grouped in even powers, because we can use the identity (7.16). So, if you have an integral with tangents and secants in it, then:

Useful Fact 7.5. *If the power of secant is even, move TWO of them over with the dx for use in a substitution, and convert the remaining (even powered) group of secants to tangents using identity (7.16).*

If the power of tangent is odd, save a sec x tan x *for use in a substitution, and convert the remaining (even powered) group of tangents to secants using identity (7.16).*

$\boxed{\textbf{EX 3}}$ Find $\displaystyle\int \tan^6 x \sec^4 x\,dx$.

Here we see that the secant term has an even power, so let's borrow two of those secants and rewrite this as

$$\int \tan^6 x \sec^2 x \cdot \sec^2 x\,dx$$

Next, we convert the rest of the secants into tangents:

$$\int \tan^6 x(1 + \tan^2 x) \sec^2 x \, dx$$

Now the integral is all set up for

$$\tan x = u \qquad \text{with} \qquad \sec^2 x \, dx = du$$

giving

$$\int \tan^6 x(1 + \tan^2 x)\sec^2 x \, dx = \int u^6(1 + u^2) \, du = \int (u^6 + u^8)du$$

$$= \frac{1}{7}u^7 + \frac{1}{9}u^9 + C$$

$$= \frac{1}{7}\tan^7 x + \frac{1}{9}\tan^9 x + C \quad \blacksquare$$

$\boxed{\text{EX 4}}$ Find $\int \tan^3 x \sec^4 x \, dx$.

We have an odd power of tangent, so we regroup and convert as suggested above, by borrowing both a tangent and a secant for dx:

$$\int \tan^3 x \sec^4 x \, dx = \int \tan^2 x \sec^3 x \cdot (\sec x \tan x) \, dx$$

$$= \int (\sec^2 x - 1) \sec^3 x \cdot (\sec x \tan x) \, dx \ldots$$

Now we introduce

$$\sec x = u \quad ; \qquad \sec x \tan x \, dx = du$$

to carry on the integral as

$$\ldots = \int (u^2 - 1)u^3 \, du = \int (u^5 - u^3) \, du = \frac{1}{6}u^6 - \frac{1}{4}u^4 + C$$

$$= \frac{1}{6}\sec^6 x - \frac{1}{4}\sec^4 x + C \quad \blacksquare$$

You Try It

(3) $\int \sec^2 x \tan^3 x \, dx$. Use *both* of the strategies listed in Useful Fact 7.5 (it matches both cases), and make sure you can explain the apparent difference in outcome.

Other Cases

The examples and strategies above don't cover all the possible cases. But some of the rest are solved by being clever. One way to be clever is to use the "trigonometric panic button": when in doubt, convert everything to sine and cosine. Also, be alert to the need to combine these techniques with other techniques.

You Try It

(4) Find $\int \cos^4 x \tan^3 x \, dx$.

(5) Find $\int \tan^2 x \, dx$.

Another way to be "clever" is simply to know what's worked for someone else before. Math has many "oral histories" in which one person at some point did something really clever, and that scheme gets passed down. It saves us all from reinventing the wheel.

EX 5 Find $\int \sec x \, dx$ by multiplying up and down by $\sec x + \tan x$ and seeing where that takes you.

Yes, this is another "make it look worse before it looks better" scenario. By taking the hint, we get:

$$\int \sec x \, dx = \int \sec x \cdot \frac{\sec x + \tan x}{\sec x + \tan x} \, dx = \int \frac{\sec x \tan x + \sec^2 x}{\sec x + \tan x} \, dx$$

The sudden reordering of the terms in the numerator helps us to see how this happens to be a perfect set up for $u = \sec x + \tan x$ and so $du = \sec x \tan x + \sec^2 x \, dx$, which gets us

$$\int \frac{du}{u} = \ln|u| + C$$

Reversing the substitution,

$$\int \sec x \, dx = \ln|\sec x + \tan x| + C \qquad (7.17)$$

This equation is numbered because it's a good one to flag for reference later on. ∎

You Try It

(6) Find $\int \sec^3 x \, dx$ using integration by parts with $f(x) = \sec x$ and $g'(x) = \sec^2 x$.

Quick Reference

For $\int \sin^m x \cos^n x \, dx$:

- If the power of the sine or cosine term is odd, borrow one of them to join with dx in a substitution, and convert the remaining (now even powered) group of them using identity $\sin^2 x + \cos^2 x = 1$.
- If the powers of the sines and cosines are even, use a half-angle identity, $\sin^2 x = \frac{1}{2}(1 - \cos 2x)$ or $\cos^2 x = \frac{1}{2}(1 + \cos 2x)$.

For $\int \tan^m x \sec^n x \, dx$:

- If the power of secant is even, move TWO of them over with the dx for use in a substitution, and convert the remaining (even powered) group of secants to tangents using identity (7.16).

- If the power of tangent is odd, save a $\sec x \tan x$ for use in a substitution, and convert the remaining (even powered) group of tangents to secants using identity (7.16).

Otherwise, be clever with an identity.

Trigonometric Integrals — Problem List

Trigonometric Integrals — You Try It

These appeared above; solutions begin on the next page.

(1) Find $\int \sin^3 x \cos^2 x \, dx$.

(2) Evaluate $\int_0^{\pi/2} \cos^2 \theta \, d\theta$.

(3) Find $\int \sec^2 x \tan^3 x \, dx$. Use *both* of the strategies listed in Useful Fact 7.5 (it matches both cases), and make sure you can explain the apparent difference in outcome.

(4) Find $\int \cos^4 x \tan^3 x \, dx$.

(5) Find $\int \tan^2 x \, dx$.

(6) Find $\int \sec^3 x \, dx$ using integration by parts with $f(x) = \sec x$ and $g'(x) = \sec^2 x$.

Trigonometric Integrals — Practice Problems

Try these as you get the hang of the You Try It problems. Solutions to these problems are available in Sec. A.1.4.

(1) Find $\int \sin^6 x \cos^3 x \, dx$.

(2) Find $\int_0^{\pi/2} \sin^2 2\theta \, d\theta$.

(3) Find $\int \sec^6 t \, dt$.

(4) Evaluate $\int_0^{\pi/4} \sec^4 \theta \tan^4 \theta \, d\theta$.

(5) Use a strategy similar to that of EX 5 to find $\int \csc x \, dx$.

(6) Recall that hyperbolic trig functions come with identities quite similar to (but not always exactly the same as) regular trig functions. Find and use such an identity to solve $\int \sinh^3(x) \cosh^2(x) \, dx$.

Trigonometric Integrals — Challenge Problems

Try these problems to test your skills with the ideas in this section. Solutions to these problems are available in Sec. B.1.3.

(1) Evaluate $\displaystyle\int_0^{\pi/2} \cos^5 x \, dx$.

(2) Find $\displaystyle\int x \cos^2 x \, dx$.

(3) Find $\displaystyle\int \tan^3(2x) \sec^5(2x) \, dx$.

Trigonometric Integrals — You Try It — Solved

---

(1) Find $\int \sin^3 x \cos^2 x \, dx$.

☐ There is an odd power of $\sin x$, so we we borrow (set aaside) one of the sines to go with dx, and convert the remaining sines to cosines:

$$\int \sin^3 x \cos^2 x \, dx = \int \sin^2 x \cos^2 x \cdot \sin x \, dx$$

$$= \int (1 - \cos^2 x) \cos^2 x \cdot \sin x \, dx \dots$$

Then introducing

$$\cos x = u \quad ; \quad \sin x \, dx = -du$$

we can carry on:

$$\dots = -\int (1 - u^2)(u^2) \, du = \int (u^4 - u^2) \, du$$

$$= \frac{1}{5} u^5 - \frac{1}{3} u^3 + C = \frac{1}{5} \cos^5 x - \frac{1}{3} \cos^3 x + C \quad ■$$

---

(2) Evaluate $\int_0^{\pi/2} \cos^2 \theta \, d\theta$.

☐ This calls for one of the half-angle formulas in Useful Fact 7.4,

$$\int_0^{\pi/2} \cos^2 \theta \, d\theta = \int_0^{\pi/2} \frac{1}{2}(1 + \cos 2\theta) \, d\theta$$

$$= \left(\frac{1}{2} \theta + \frac{1}{4} \sin 2\theta \right) \Big|_0^{\pi/2} = \frac{\pi}{4} \quad ■$$

---

(3) Find $\int \sec^2 x \tan^3 x \, dx$.

☐ We have an odd power of tangent, and follow the appropriate strategy — borrow an entire "$\sec x \tan x$" to go with dx, and convert the rest of the tangents to secants:

$$\int \sec^2 x \tan^3 x \, dx = \int \sec x \tan^2 x \cdot (\sec x \tan x) \, dx$$

$$= \int \sec x (\sec^2 x - 1) \cdot (\sec x \tan x) \, dx$$

We introduce

$$\sec x = u \quad ; \quad \sec x \tan x \, dx = du$$

and carry on:

$$\cdots = \int u(u^2 - 1)\,du = \int (u^3 - u)\,du = \frac{1}{4}u^4 - \frac{1}{2}u^2 + C$$
$$= \frac{1}{4}\sec^4 x - \frac{1}{2}\sec^2 x + C$$

Now we hit the reset button, and see that the original integrand also has an even power of secant, so we could have applied the other strategy, by which we borrow a $\sec^2 x$ to go with dx:

$$\int \sec^2 x \tan^3 x\,dx = \int \tan^3 x \cdot \sec^2 x\,dx \ldots$$

Now we are ready for:

$$\tan x = u \quad ; \quad \sec^2 x\,dx = du$$

and can continue

$$\cdots = \int u^3\,du = \frac{1}{4}u^4 + C = \frac{1}{4}\tan^4 x + C$$

The two results look very different, but they both represent the same antiderivative. How different are they? ∎

---

(4) Find $\displaystyle\int \cos^4 x \tan^3 x\,dx$.

□ In the integrand $\cos^4 x \tan^3 x$, we have a mix of cosines and tangents, which doesn't match any of our cases. So we press the "trigonometric panic button" and convert everything to sines and cosines:

$$\int \cos^4 x \tan^3 x\,dx = \int \cos^4 x \cdot \frac{\sin^3 x}{\cos^3 x}\,dx = \int \cos x \sin^3 x\,dx \ldots$$

which is now ready for a substitution

$$\sin x = u \quad ; \quad \cos x\,dx = du$$

giving

$$\cdots = \int u^3\,du = \frac{1}{4}u^4 + C = \frac{1}{4}\sin^4 x + C \quad ∎$$

---

(5) Find $\displaystyle\int \tan^2 x \, dx$.

☐ For $\displaystyle\int \tan^2 x \, dx$, we can't apply any of the above strategies for dealing with tangents, since there are no secants to use in a substitution. So, what about converting everything into sines and cosines:

$$\int \tan^2 x \, dx = \int \frac{\sin^2 x}{\cos^2 x} \, dx = \int \frac{1 - \cos^2 x}{\cos^2 x} \, dx$$

$$= \int (\sec^2 x - 1) \, dx = \tan x - x + C \quad \blacksquare$$

---

(6) Find $\displaystyle\int \sec^3 x \, dx$ using integration by parts with $f(x) = \sec x$ and $g'(x) = \sec^2 x$.

☐ Let's make our assignments as instructed:

$$f(x) = \sec x \qquad \text{and} \qquad g'(x) = \sec^2 x$$

From these choices, we then compute

$$f'(x) = \sec x \tan x \qquad \text{and} \qquad g(x) = \tan x$$

so that

$$\int \sec \cdot \sec^2 x \, dx = \sec x \tan x - \int (\sec x \tan x) \tan x \, dx$$

$$= \sec x \tan x - \int \sec x \tan^2 x \, dx$$

By using the identity (7.16), we can expand the integral on the right side; this then puts us in a position where we have two instances of the integral we seek — so we can solve for the integral without actually solving the integral! In the process, we use the integral formula from (7.17):

$$\int \sec \cdot \sec^2 x \, dx = \sec x \tan x - \int \sec x (\sec^2 x - 1) \, dx$$

$$\int \sec \cdot \sec^2 x \, dx = \sec x \tan x - \int \sec x \sec^2 x \, dx + \int \sec x \, dx$$

$$\int \sec^3 x \, dx = \sec x \tan x - \int \sec x \sec^2 x \, dx + \ln|\sec x + \tan x|$$

$$2 \int \sec^3 x \, dx = \sec x \tan x + \ln|\sec x + \tan x|$$

Solving for the integral we want and accounting for the arbitrary constant,

$$\int \sec^3 x \, dx = \frac{1}{2} \left(\sec x \tan x + \ln |\sec x + \tan x| \right) + C \quad \blacksquare \qquad (7.18)$$

7.5   Surprise Boss Fight: Trigonometric Substitution

Introduction

In a few previous sections, we've encountered the mathematical rule of thumb, "sometimes you have to make something worse before you can make it look better." Trigonometric substitution is an integral technique which takes that to the extreme. To solve certain antiderivatives that have no trigonometric functions in them, our strategy is going to be to introduce trigonometric functions *on purpose* so that we can take advantage of trigonometric identities. That seems like a very silly plan, but it actually works!

A Crazy Strategy

Consider the expression $\sqrt{4 - x^2}$ and tinker with it by substituting $2\sin\theta$ for x. At first glance, you'd think this is a silly thing to do. But watch what happens:

$$\sqrt{4 - x^2} = \sqrt{4 - (2\sin\theta)^2} = \sqrt{4 - 4\sin^2\theta} = \sqrt{4(1 - \sin^2\theta)}$$
$$= 2\sqrt{1 - \sin^2\theta} = 2\sqrt{\cos^2\theta} = 2\cos\theta$$

In the long run, that substitution actually made the expression a lot simpler! So suppose we wanted to do the integral

$$\int \sqrt{4 - x^2}\, dx$$

We can't evaluate this integral using techniques we know so far. But we just saw above a substitution that we can use to simplify (or at least change) the function being integrated. If we're going to make the substitution $x = 2\sin\theta$ in this integral, we also need to convert dx; but if $x = 2\sin\theta$, then $dx = 2\cos\theta\, d\theta$. The full substitution then becomes

$$x = 2\sin\theta \quad ; \quad dx = 2\cos\theta d\theta$$

we can convert the integral (using the conversion of the function itself from above):

$$\int \sqrt{4 - x^2}\, dx = \int \sqrt{4 - (2\sin\theta)^2}(2\cos\theta\, d\theta)$$
$$= \int (2\cos\theta)(2\cos\theta\, d\theta) = 4\int \cos^2\theta\, d\theta$$

and while this integral isn't the easiest one in the world, at least we can do it using a half-angle identity:

$$4 \int \cos^2 \theta \, d\theta = 4 \int \left(\frac{1}{2}(1 + \cos 2\theta) \right) d\theta \qquad (7.19)$$

$$= 2 \left(\theta + \frac{1}{2} \sin 2\theta \right) + C = 2\theta + \sin 2\theta + C \qquad (7.20)$$

This is unfinished so far, because — just like in previous substitution problems — we have to return to the original variable. This is the extra "interesting" part of trigonometric substitution. Reverting from θ back to x is not just a matter of direct replacement. Our substitution was $x = 2\sin\theta$, but now we need to convert the *different* terms θ and $\sin 2\theta$ back into x's. Good grief!

The substitution $x = 2\sin\theta$ still gives us the relationship between θ and x, we just have to do a little work to figure out the details that will convert *theta* itself (and $\sin 2\theta$) back to x. First, since $x = 2\sin\theta$, then $\theta = \sin^{-1}(x/2)$. That's not pretty, but it is what it is.

But then there's $\sin 2\theta$. First, we can employ a known trig identity, $\sin 2\theta = 2\sin\theta\cos\theta$. From the original substitution, we know that $\sin\theta = x/2$, but what about $\cos\theta$? Believe it or not, this is actually pretty easy. Let's load the data from "$\sin\theta = x/2$" into a right triangle, as shown in Fig. 7.2(A). We have an angle (θ), the opposite side length x, and the hypotenuse length 2. A quick run through the Pythagorean theorem then gives us the length of the adjacent side as $\sqrt{4 - x^2}$, as now shown in Fig. 7.2(B). With the entire triangle labelled, we now know all six trig functions of θ; in particular,

$$\cos\theta = \frac{\sqrt{4 - x^2}}{2}$$

Continuing where we left off in Eq. (7.19), it's time to reverse the substitution using the information we just developed:

$$2\theta + \sin 2\theta + C = 2\theta + 2\sin\theta\cos\theta + C$$

$$= 2\sin^{-1}\frac{x}{2} + 2\left(\frac{x}{2}\right)\left(\frac{\sqrt{4 - x^2}}{2}\right) + C$$

$$= 2\sin^{-1}\frac{x}{2} + \frac{x}{2} \cdot \sqrt{4 - x^2} + C$$

(A)   (B)

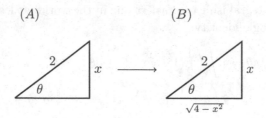

Fig. 7.2   Encoding $\sin\theta = x/2$ into a right triangle.

Putting it all together, from the original problem statement to Eq. (7.19) to this final result,

$$\int \sqrt{4 - x^2}\, dx = \ldots = 4 \int \cos^2\theta\, d\theta = \ldots$$

$$= 2\theta + \sin 2\theta + C = 2\sin^{-1}\frac{x}{2} + \frac{x}{2}\cdot\sqrt{4-x^2} + C$$

Finally! We're all done now.

Because a lot of things happened, let's summarize the process we just went through. Solutions for integrals like this do get lengthy, however it helps to see them as a sequence of smaller problems rather than one large problem:

- We used the substitution $x = 2\sin\theta$ (and so also $dx = 2\cos\theta\, d\theta$) to convert the integral:

$$\int \sqrt{4 - x^2}\, dx = \ldots = 4\int \cos^2\theta\, d\theta$$

- We solved the new integral as a trigonometric integral (Sec. 7.4):

$$4\int \cos^2\theta\, d\theta = \ldots = 2\theta + \sin 2\theta + C = 2\theta + 2\sin\theta\cos\theta + C$$

- We used our superior knowledge of trigonometry (triangles) to turn all the θ's back into x's:

$$2\theta + 2\sin\theta\cos\theta + C = 2\sin^{-1}\frac{x}{2} + \frac{x}{2}\cdot\sqrt{4-x^2} + C$$

giving the final result

$$\int \sqrt{4 - x^2}\, dx = 2\sin^{-1}\frac{x}{2} + \frac{x}{2}\cdot\sqrt{4-x^2} + C$$

Fortunately, integrals that require substitution with trigonometric functions come in only a few forms, and there are specific strategies for each form. Here are examples broken down by type.

Integrals Containing $\sqrt{a^2 - x^2}$

Useful Fact 7.6. *When the integral contains the term $\sqrt{a^2 - x^2}$, use the substitution*

$$x = a\sin\theta \qquad ; \qquad dx = a\cos\theta\, d\theta$$

The detailed example completed above is an integral of this type. The next example will pose the general case. At this point, having seen integration by parts, partial fractions, trigonometric integrals, and now trigonometric substitution, we have the tools to create many of the results you may have seen if you've ever browsed a table of integrals — such as in the back of a proper calculus textbook.

$\boxed{\text{EX 1}}$ Find $\displaystyle\int \sqrt{a^2 - x^2}\, dx$.

As this is the very case indicated in Useful Fact 7.6, we need to make the substitution

$$x = a\sin\theta \qquad ; \qquad dx = a\cos\theta\, d\theta$$

This changes the function much like above,

$$\sqrt{a^2 - x^2} = \sqrt{a^2 - a^2\sin^2\theta} = \sqrt{a^2\cos^2\theta} = a\cos\theta$$

The resulting integral (don't forget to replace dx too!) is:

$$\int \sqrt{a^2 - x^2}\, dx = \int a\cos\theta(a\cos\theta\, d\theta) = a^2 \int \cos^2\theta\, d\theta \ldots$$

This is now a trigonometric integral, and our strategies from Sec. 7.4 suggest a half-angle formula will help:

$$\ldots = a^2 \int \left(\frac{1}{2}(1 + \cos 2\theta)\right) d\theta = \frac{a^2}{2}\int (1 + \cos 2\theta)\, d\theta$$

$$= \frac{a^2}{2}\left(\theta + \frac{1}{2}\sin 2\theta\right) + C = \frac{a^2}{2}\theta + \frac{a^2}{4}\sin 2\theta + C$$

$$= \frac{a^2}{2}\theta + \frac{a^2}{2}\sin\theta\cos\theta + C$$

(we used the identity $\sin 2\theta = 2\sin\theta\cos\theta$ in the final step). Now it's time to reverse the substitution and send θ back to x. From the original substitution $x = a\sin\theta$, we have $\theta = \sin^{-1}(x/a)$. Encoding this information into a right triangle (use Fig. 7.2, with a in place of 2), we then have $\cos\theta = \sqrt{a^2 - x^2}/a$. Altogether,

$$\int \sqrt{a^2 - x^2}\, dx = \frac{a^2}{2}\sin^{-1}\frac{x}{a} + \frac{a^2}{2}\cdot\frac{x}{a}\cdot\frac{\sqrt{a^2 - x^2}}{a} + C$$

$$= \frac{a^2}{2}\sin^{-1}\frac{x}{a} + \frac{1}{2}x\sqrt{a^2 - x^2} + C$$

The final result is, rearranged,

$$\int \sqrt{a^2 - x^2}\, dx = \frac{1}{2}\left\{x\sqrt{a^2 - x^2} + a^2\sin^{-1}\frac{x}{a}\right\} + C \qquad (7.21)$$

🔘 FFT: If you ask for the solution to this integral from the on-line re-source Wolfram Alpha, the answer is reported as

$$\int \sqrt{a^2 - x^2}\, dx = \frac{1}{2}\left\{x\sqrt{a^2 - x^2} + a^2\tan^{-1}\frac{x}{\sqrt{a^2 - x^2}}\right\} + C$$

which does not look exactly the same as our result in (7.21). How can this be explained? 🔘 ∎

You Try It

(1) Find $\displaystyle\int \frac{1}{x^2\sqrt{25 - x^2}}\, dx$.

Integrals Containing $\sqrt{x^2 - a^2}$

Useful Fact 7.7. *When the integral contains the term $\sqrt{x^2 - a^2}$, use the substitution*

$$x = a\sec\theta \qquad ; \qquad dx = a\sec\theta\tan\theta\, d\theta$$

$\boxed{\textbf{EX 2}}$ Find $\displaystyle\int \frac{\sqrt{x^2 - 1}}{x}\, dx$.

The root term here matches the form $\sqrt{x^2 - a^2}$ with $a = 1$, so assign the substitution

$$x = \sec\theta \qquad ; \qquad dx = \sec\theta\tan\theta\, d\theta$$

and then the integral becomes

$$\int \frac{\sqrt{x^2 - 1}}{x}\, dx = \int \frac{\sqrt{\sec^2\theta - 1}}{\sec\theta}\cdot(\sec\theta\tan\theta\, d\theta)$$

$$= \int \frac{\tan\theta}{\sec\theta}(\sec\theta\tan\theta\, d\theta) = \int \tan^2\theta\, d\theta$$

This trigonometric integral was solved in You Try It 5 of Sec. 7.4:

$$\int \tan^2 \theta \, d\theta = \int \frac{\sin^2 \theta}{\cos^2 \theta} \, d\theta = \int \frac{1 - \cos^2 \theta}{\cos^2 \theta} \, d\theta$$

$$= \int (\sec^2 \theta - 1) \, d\theta = \tan \theta - \theta + C$$

Now we need to return θ's to x's. Our substitution was $x = \sec \theta$, which also means $\cos \theta = 1/x$. Immediately, then,

$$\theta = \cos^{-1} \frac{1}{x}$$

To get $\tan \theta$, we load data from $\cos \theta = \dfrac{1}{x}$ into a right triangle around the angle θ: the side adjacent to θ has length 1 and the hypotenuse has length x — which means the final (opposite) side has length $\sqrt{x^2 - 1}$. With a complete triangle, we then have

$$\tan \theta = \frac{\sqrt{x^2 - 1}}{1} = \sqrt{x^2 - 1}$$

(A) (B)

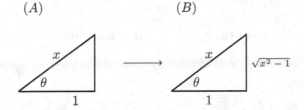

Fig. 7.3   Encoding $\cos \theta = 1/x$ into a right triangle.

This is all shown in Fig. 7.3. Therefore,

$$\tan \theta - \theta + C = \sqrt{x^2 - 1} - \cos^{-1} \frac{1}{x} + C$$

and altogether,

$$\int \frac{\sqrt{x^2 - 1}}{x} \, dx = \sqrt{x^2 - 1} - \cos^{-1} \frac{1}{x} + C$$

Based on the triangle, we could actually represent θ itself in many ways, leading to expressions for the antiderivative which seem to be different, but really aren't — such as,

$$\int \frac{\sqrt{x^2 - 1}}{x} \, dx = \sqrt{x^2 - 1} - \tan^{-1} \sqrt{x^2 - 1} + C$$

Be sure you are able to adapt between different expressions for the angle θ, such as

$$\theta = \cos^{-1} \frac{1}{x} = \tan^{-1} \sqrt{x^2 - 1}$$

If you compare different answers to the same antiderivative, you want to know if the results are truly different or just cosmetically different. ∎

You Try It

  (2) Find $\displaystyle\int \frac{\sqrt{x^2 - 9}}{x^3} \, dx$.

Integrals Containing $\sqrt{a^2 + x^2}$

Useful Fact 7.8. *When the integral contains the term $\sqrt{a^2 + x^2}$, use the substitution*

$$x = a \tan \theta \quad ; \quad dx = a \sec^2 \theta \, d\theta$$

$\boxed{\textbf{EX 3}}$ Find $\displaystyle\int_0^1 \frac{x}{\sqrt{4x^2 + 16}} \, dx$.

Although we could solve this by regular substitution, we'll treat it as a trigonometric substitution problem. First, we do a bit of factoring:

$$\int_0^1 \frac{x}{\sqrt{4x^2 + 16}} \, dx = \frac{1}{2} \int_0^1 \frac{x}{\sqrt{x^2 + 4}} \, dx$$

Then we see that the root term matches the form $\sqrt{x^2 + a^2}$ with $a = 2$, so the best substitution is

$$x = 2 \tan \theta \quad ; \quad dx = 2 \sec^2 \theta \, d\theta$$

Since this is a definite integral, we should also convert the limits at this point. When $x = 0$, we have $\tan \theta = 0$, or $\theta = 0$. When $x = 1$, we have $\tan \theta = 1/2$; this is not one of the "nice" angles, so we're stuck with

$\theta = \tan^{-1}(1/2)$. The integral is now

$$\frac{1}{2}\int_0^1 \frac{x}{\sqrt{x^2+4}}\,dx = \frac{1}{2}\int_0^{\tan^{-1}(1/2)} \frac{2\tan\theta}{\sqrt{4\tan^2\theta+4}} \cdot (2\sec^2\theta\,d\theta)$$

$$= \frac{1}{2}\int_0^{\tan^{-1}(1/2)} \frac{2\tan\theta}{2\sec\theta} \cdot (2\sec^2\theta\,d\theta)$$

$$= \int_0^{\tan^{-1}(1/2)} \tan\theta \sec\theta\,d\theta = (\sec\theta)\Big|_0^{\tan^{-1}(1/2)}$$

$$= \sec\left(\tan^{-1}\frac{1}{2}\right) - 1$$

To evaluate this final term, we use a triangle again. The expression $\sec(\tan^{-1}(1/2))$ refers to the secant of an angle whose tangent is $1/2$. If an angle's tangent is $1/2$, its opposite side has length 1, and its adjacent side has length 2. Therefore the hypotenuse has length $\sqrt{5}$ and the secant of this same angle is hyp/adj, or $\sqrt{5}/2$. We have, at last,

$$\int_0^1 \frac{x}{\sqrt{4x^2+16}}\,dx = \sec\left(\tan^{-1}\frac{1}{2}\right) - 1 = \frac{\sqrt{5}}{2} - 1 \quad \blacksquare$$

You Try It

   (3) Find $\displaystyle\int \frac{1}{\sqrt{x^2+16}}\,dx$.

Right before EX 1, there was a remark that the fancy integration techniques presented in this chapter put us in a position to derive many of the integral formulas presented in a table of integrals (such as in the back of a standard calculus textbook). If you enjoy puzzling through integration problems, then take a class in Complex Variables when you get a chance. Learning how to integrate functions of a complex variable $x + iy$ opens a whole other pipeline of integration formulas. If you want to see a very impressive collection of integral formulas, I encourage you to look up the book *Table of Integrals, Series, and Products*, by I.S. Gradshteyn and I.M. Ryzhik, originally published in 1943. It's huge! This is the sort of reference needed before the advent of computer algebra systems and similar on-line resources.

Quick Reference

- When the integral contains the term $\sqrt{a^2-x^2}$, use the substitution $x = a\sin\theta$ with $dx = a\cos\theta\,d\theta$.

- When the integral contains the term $\sqrt{x^2 - a^2}$, use the substitution $x = a \sec\theta$ with $dx = a\sec\theta\tan\theta\,d\theta$.
- When the integral contains the term $\sqrt{a^2 + x^2}$, use the substitution $x = a\tan\theta$ with $dx = a\sec^2\theta\,d\theta$.

Trigonometric Substitution — Problem List

Trigonometric Substitution — You Try It

These appeared above; solutions begin on the next page.

(1) Find $\displaystyle\int \frac{1}{x^2\sqrt{25 - x^2}}\,dx$.

(2) Find $\displaystyle\int \frac{\sqrt{x^2 - 9}}{x^3}\,dx$.

(3) Find $\displaystyle\int \frac{1}{\sqrt{x^2 + 16}}\,dx$.

Trigonometric Substitution — Practice Problems

Try these as you get the hang of the You Try It problems. Solutions to these problems are available in Sec. A.1.5.

(1) Find $\displaystyle\int \sqrt{1 - 4x^2}\,dx$.

(2) Find $\displaystyle\int \frac{x}{\sqrt{x^2 - 7}}\,dx$.

(3) Evaluate $\displaystyle\int_0^1 x\sqrt{x^2 + 4}\,dx$.

Trigonometric Substitution — Challenge Problems

Try these problems to test your skills with the ideas in this section. Solutions to these problems are available in Sec. B.1.4.

(1) Find $\displaystyle\int \frac{x^5}{\sqrt{x^2 + 2}}\,dx$.

(2) Find $\displaystyle\int \frac{x}{\sqrt{25 - x^2}}\,dx$.

(3) Evaluate $\displaystyle\int_0^1 \sqrt{x^2 + 1}\,dx$.

Trigonometric Substitution — You Try It — Solved

(1) Find $\displaystyle\int \frac{1}{x^2\sqrt{25-x^2}}\,dx$.

□ The square root term here matches the form $\sqrt{a^2-x^2}$ with $a=5$, so according to Useful Fact 7.6, we assign

$$x = 5\sin\theta \quad ; \quad dx = 5\cos\theta\,d\theta$$

which converts the integral as follows:

$$\int \frac{1}{x^2\sqrt{25-x^2}}\,dx = \int \frac{1}{25\sin^2\theta\sqrt{25-25\sin^2\theta}}5\cos\theta\,d\theta$$

$$= \int \frac{5\cos\theta}{25\sin^2\theta\,5\cos\theta}\,d\theta = \frac{1}{25}\int \csc^2\theta\,d\theta$$

$$= -\frac{1}{25}\cot\theta + C$$

(A)   (B)

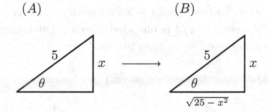

Fig. 7.4   Encoding $\sin\theta = x/5$ into a right triangle.

We reverse the substitution using the information in Fig. 7.4, where $\sin\theta = x/5$ is encoded into a right triangle to give

$$\cot\theta = \frac{\sqrt{25-x^2}}{x}$$

so that

$$-\frac{1}{25}\cot\theta + C = -\frac{1}{25}\frac{\sqrt{25-x^2}}{x} + C$$

and altogether,

$$\int \frac{1}{x^2\sqrt{25-x^2}}\,dx = -\frac{1}{25}\frac{\sqrt{25-x^2}}{x} + C \quad\blacksquare$$

(2) Find $\displaystyle\int \frac{\sqrt{x^2-9}}{x^3}\,dx$.

☐ The square root term matches the form $\sqrt{x^2-a^2}$ with $a = 3$, so according to Useful Fact 7.7, we assign

$$x = 3\sec\theta \quad ; \quad dx = 3\sec\theta\tan\theta\,d\theta$$

which converts the integral as follows:

$$\int \frac{\sqrt{x^2-9}}{x^3}\,dx = \int \frac{\sqrt{9\sec^2\theta-9}}{27\sec^3\theta}\cdot 3\sec\theta\tan\theta\,d\theta$$

$$= \int \frac{3\tan\theta}{9\sec^2\theta}\tan\theta\,d\theta = \frac{1}{3}\int \sin^2\theta\,d\theta\ldots$$

A half-angle formula is called for here; continuing,

$$\ldots = \frac{1}{3}\int \frac{1}{2}(1-\cos 2x)\,dx = \frac{1}{6}\left(\theta - \frac{1}{2}\sin 2\theta\right)+C$$

$$= \frac{1}{6}(\theta - \sin\theta\,\cos\theta)+C$$

(in the final step, we used $\sin 2\theta = 2\sin\theta\cos\theta$). To reverse the substitution, the data $\sec\theta = x/3$ is encoded into a right triangle, and then that triangle completed — as seen in Fig. 7.5:

(A)   (B)

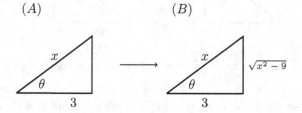

Fig. 7.5 Encoding $\sec\theta = x/3$ into a right triangle.

$$\sin\theta = \frac{\sqrt{x^2-9}}{x} \quad ; \quad \cos\theta = \frac{3}{x}$$

So that altogether,

$$\int \frac{\sqrt{x^2-9}}{x^3}\,dx = \frac{1}{6}\sec^{-1}\frac{x}{3} - \frac{1}{6}\frac{\sqrt{x^2-9}}{x}\cdot\frac{3}{x}+C$$

$$= \frac{1}{6}\sec^{-1}\frac{x}{3} - \frac{\sqrt{x^2-9}}{2x^2}+C \quad ∎$$

(3) Find $\displaystyle\int \frac{1}{\sqrt{x^2+16}}\,dx$.

□ The square root term here matches the form $\sqrt{a^2+x^2}$ with $a=4$, so according to Useful Fact 7.8, we assign

$$x = 4\tan\theta \qquad ; \qquad dx = 4\sec^2\theta\,d\theta$$

which converts the integral as follows:

$$\int \frac{1}{\sqrt{x^2+16}}\,dx = \int \frac{1}{\sqrt{16\tan^2\theta+16}} \cdot 4\sec^2\theta\,d\theta$$

$$= \int \frac{4\sec^2\theta}{4\sec\theta}\,d\theta = \int \sec\theta\,d\theta = \ln|\sec\theta+\tan\theta| + C$$

(In the last step, we used the result from Eq. (7.17). From the original substitution, we have $\tan\theta = x/4$; encoding this data into a right triangle — as shown in Fig. 7.6 — we also get

$$\sec\theta = \frac{\sqrt{x^2+16}}{4}$$

(a)   (b)

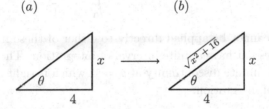

Fig. 7.6   Encoding $\tan\theta = x/4$ into a right triangle.

Therefore,

$$\int \frac{1}{\sqrt{x^2+16}}\,dx = \ln|\sec\theta+\tan\theta| + C = \ln\left|\frac{\sqrt{x^2+16}}{4} + \frac{x}{4}\right| + C$$

$$= \ln|\sqrt{x^2+16}+x| - \ln 4 + C$$

$$= \ln|\sqrt{x^2+16}+x| + C$$

(What happened to the $\ln 4$?)   ∎

7.6   Improper Integrals

Introduction

A definite integral is "proper" if it allows for use of the Fundamental Theorem of Calculus (FTOC). Recall that to assist with finding the value of a definite integral of some $f(x)$ on an interval $[a, b]$, the FTOC requires that $f(x)$ meets two criteria:

(1) The interval $[a, b]$ must be a *finite* interval.
(2) The function $f(x)$ must be continuous over the closed interval $[a, b]$.

An **improper integral** is one in which one or both of those criteria are not met. The FTOC does not apply to improper integrals. But, we can still trick the FTOC into working, if we cheat a little bit.

Discussion / Examples

Consider these two definite integrals:

$$\int_1^\infty \frac{dx}{x^3} \quad \text{and} \quad \int_0^2 \frac{dx}{x-1}$$

The FTOC cannot be applied directly to either of these integrals. The first integral does not have a finite interval of integration. The second integral contains an infinite discontinuity at $x = 1$, which is right in the middle of the interval of integration.

Now you may think that any definite integral which contains an infinite interval of integration would have result of ∞ (i.e. would *diverge*) because the area under an infinitely long curve has to be infinite. It turns out that this is not true — in fact, the left integral above ends up having a value of $1/2$. Here's what that means: imagine you are driving the Calc-mobile along the positive x-axis, starting at $x = 1$. The Calcmobile is a convertible, so we can look up through the roof and see the curve of $f(x) = 1/x^3$ above us. Suppose we stop at every power of 10, $x = 10, 100, 1000$, and so on — and find the total area under the curve from $x = 1$ out to our stopping point. What we would observe is that that total area would sneak closer and closer to $1/2$, but would never actually get to it, or exceed it. As we go farther out along the x-axis, the additional area under $f(x) = 1/x^3$ gets smaller and smaller — the end result is that the total area under the curve remains finite. The calculations we'd do along the way would be (with $f(x) = 1/x^3$):

$$\int_1^1 f(x)\,dx = 0$$

$$\int_1^{10} f(x)\,dx = \frac{99}{200}$$

$$\int_1^{100} f(x)\,dx = \frac{9999}{20000}$$

$$\vdots \qquad \vdots$$

You may recognize this as a limit process in disguise, and limits are exactly the thing we use to unlock values of improper integrals when they exist. If you're really comfortable with limits, you might note that the above sequence of integral calculations can be represented as

$$\lim_{t \to \infty} \int_1^t f(x)\,dx$$

If that makes sense to you, then you're well on your way to solving integrals over an infinitely long interval of integration. Here's our dialog with the Fundamental Theorem of Calculus:

Us: We would like to integrate $f(x) = 1/x^3$ from $x = 0$ to $x = \infty$.

FTOC: Sorry, I only work on finite integrals.

Us: OK, we'll just go out to some unspecified end value called t. Can you give us the result for $\int_1^t f(x)\,dx$?

FTOC: Sure! Here it is:

$$\int_1^t \frac{1}{x^3}\,dx = \left(-\frac{1}{2x^2}\right)\Bigg|_1^t$$

$$= -\frac{1}{2t^2} - \left(-\frac{1}{2(1)^2}\right)$$

$$\to \int_1^t \frac{1}{x^3}\,dx = \frac{1}{2} - \frac{1}{2t^2}$$

Us: Ha! We tricked you, because now we're going to find out what happens when t goes to infinity!

$$\lim_{t \to \infty} \int_1^t \frac{1}{x^3}\,dx = \lim_{t \to \infty} \left[\frac{1}{2} - \frac{1}{2t^2}\right] = \frac{1}{2}$$

And that's how we trick the Fundamental Theorem of Calculus into doing what we want.

But that's just one type of improper integral; another version arises when we have a finite interval of integration, and one of the endpoints, or a point inside the interval, locates a discontinuity of the integrand. The area under a curve with a discontinuity in it, or under a curve over an infinitely long stretch of the x-axis, can can be finite or infinite. You just never know in advance!

All in all, there are three cases when we have to either find an alternative to the Fundamental Theorem of Calculus in order to evaluate a definite integral, or at least find a sneaky way around it. These cases are:

(1) When one or both of the endpoints of integration is $+\infty$ or $-\infty$
(2) When there is an infinite discontinuity (*) in the function somewhere on the interval of integration. This includes discontinuities at the endpoints of integration themselves.
(3) A combination of the above two cases.

(*) Remember that an *infinite discontinuity* is a serious discontinuity like in $2/x$ at $x = 0$ or $\tan(x)$ at $x = \pi/2$. It is not just a boring jump discontinuity such as at the origin for $g(x) = x/|x|$.

The overall strategy for solving improper integrals is to examine the interplay between the integrand $f(x)$ and the interval of integration, and "break up" the integral so that discontinuities (or an infinity) are isolated as endpoints of integration. Here's what we mean by "break up" the integral. You should remember from Calculus 1 that, for example,

$$\int_1^4 f(x)\,dx = \int_1^2 f(x)\,dx + \int_2^3 f(x)\,dx + \int_3^4 f(x)\,dx$$

Up until now, there would be no reason to break up the integral like that, unless $f(x)$ was a routine piecewise function. But suppose $f(x)$ has a discontinuity at $x = 2$. Then we can't use the FTOC to evaluate the integral on the left in one shot. But maybe we can break up the integral to isolate $x = 2$ as an endpoint of the *subintegrals*, and take care of that location through trickery, like in the infinite integral above.

We break up overall intervals of integration into subintervals such that any subintegral has only one "bad" endpoint (a discontinuity or infinity). Here are some example scenarios presented with interval notation. In interval notation, we write things like $[a, b]$, $[a, b)$, $(a, b]$, or (a, b); in each, we are describing a stretch of the x-axis from $x = a$ to $x = b$; a square bracket means the endpoint is included, and a round parentheses means the endpoint is excluded; in the context of improper integrals, a "bad" endpoint is one that must be excluded, as it is infinity, or locates a discontinuity.

- Suppose we want to integrate $f(x)$ on $[1, 5]$, but $f(x)$ has a discontinuity at $x = 3$. We blow up the interval $[1, 5]$ as follows:

$$[1, 5] \overset{kaboom}{\longrightarrow} [1, 3) \cup (3, 5]$$

  (The symbol $\cup$ means "union", which is just mathematical scotch tape; in the given expression, we are "snipping" the interval $[1, 5]$ at $x = 3$ so that $x = 3$ is excluded.)
- Suppose we want to integrate $g(x)$ on $[0, \infty)$, and $g(x)$ has a discontinuity at $x = 10$. Is the following sufficient?

$$[0, \infty) \overset{kaboom}{\longrightarrow} [0, 10) \cup (10, \infty)$$

  No, this is not sufficient. Both endpoints of the second subinterval $(10, \infty)$ are "bad" (the discontinuity and the infinity). So we must snip the interval somewhere in between. Where? It doesn't matter; just pick your favorite number between 10 and ∞. How about 55? (That's my age as of the writing of this section.)

$$[0, \infty) \overset{kaboom}{\longrightarrow} [0, 10) \cup (10, 55] \cup [55, \infty)$$

- Suppose we want to integrate $h(x)$ on $[-4, 4]$, and $h(x)$ has discontinuities at both $x = 0$ and $x = 4$. A necessary break-up of the whole interval could be:

$$[-4, 4] \overset{kaboom}{\longrightarrow} [-4, 0) \cup (0, 2] \cup [2, 4)$$

  The endpoint of $x = 2$ is there simply to keep $x = 0$ and $x = 4$ away from each other, so that there's only one bad endpoint per subinterval.

Once the breaking up of intervals of integration is settled, then we perform the overall integration by integrating over each necessary subinterval, treating the excluded endpoints with limits, and then summing the results. (And note that limits must be left or right hand limits chosen appropriately for each subinterval — examples coming up.) With the three scenarios given above, the resulting integral would be laid out as follows.

- Integrating $f(x)$ on $[1, 5]$, where $f(x)$ has a discontinuity at $x = 3$. A first separation is:

$$\int_1^5 f(x)\, dx = \int_1^{(3)} f(x)\, dx + \int_{(3)}^5 f(x)\, dx$$

where the endpoint in parentheses cannot be included, and so we treat it with a limit:

$$\int_1^5 f(x)\, dx = \lim_{a \to 3^-} \int_1^a f(x)\, dx + \lim_{b \to 3^+} f(x)\, dx$$

- Integrating $g(x)$ on $[0, \infty)$, where $g(x)$ has a discontinuity at $x = 10$. Separating and applying limits,

$$\int_0^\infty g(x)\, dx = \lim_{a \to 10^-} \int_0^a g(x)\, dx + \lim_{b \to 10^+} \int_b^{50} g(x)\, dx \lim_{c \to \infty} \int_{50}^c g(x)\, dx$$

- Integrating $h(x)$ on $[-4, 4]$, where $h(x)$ has discontinuities at both $x = 0$ and $x = 4$.

$$\int_{-4}^4 h(x)\, dx = \lim_{a \to 0^-} \int_{-4}^a h(x)\, dx + \lim_{b \to 0^+} \int_b^2 h(x)\, dx + \lim_{c \to 4^-} \int_2^c h(x)\, dx$$

🔴 FFT: Do you understand the application of left and right hand limits in these expressions? Hint: Consider the position of the corresponding endpoint. If we have an interval $[a, b]$ and we approach the left endpoint $x = a$ from within the interval, what is the direction of that approach? 🔴

Once an improper integral is broken up into subintegrals, then if *any* of the limits / subintegrals diverges, then the whole integral does. The primary integral only has a value if each and every subintegrals converges, and the value of the primary integral is the sum of the values of the individual limits / subintegrals.

🔴 FFT: When we say that a definite integral "does not converge", that does not necessarily mean the area under the curve has to be infinite. Can you think of a function for which the total area under the curve from $x = 0$ to ∞ always remains finite, but which will not tend towards one specific value as the upper endpoint of integration gets bigger and bigger? 🔴

Here are complete examples for different varieties of improper integrals.

Useful Fact 7.9. *For integrals over $[a, \infty)$ or $(-\infty, b]$, where neither $x = a$ nor $x = b$ are locations of infinite discontinuities of the integrand: simply replace the infinite endpoint with a dummy parameter, solve the resulting integral, and then apply the appropriate limit.*

$\boxed{\textbf{EX 1}}$ Evaluate $\displaystyle\int_1^\infty \frac{1}{(3x+1)^2}\,dx$.

The integrand $f(x) = 1/(3x+1)^2$ has a discontinuity at $x = -1/3$, but that is not within the interval of integration. However, the upper endpoint of integration is ∞. We do not have to break up the interval of integration, but we do have to replace the upper endpoint with a parameter (say c) and solve, to get:

$$\int_1^\infty \frac{1}{(3x+1)^2}\,dx = \int_1^{(c)} \frac{1}{(3x+1)^2}\,dx = \left(-\frac{1}{3}\cdot\frac{1}{3x+1}\right)\Big|_1^c$$

$$= -\frac{1}{3}\left(\frac{1}{3c+1} - \frac{1}{4}\right)$$

Now, we can ask "what happens to this result as c goes to ∞?

$$\lim_{c\to\infty} -\frac{1}{3}\left(\frac{1}{3c+1} - \frac{1}{4}\right) = \frac{1}{12} \quad\blacksquare$$

Note that to speed up the process a bit, we can embed the limit operation within the sequence of calculations, rather than waiting until the end. For EX 1, this modified sequence would look like this:

$$\int_1^\infty \frac{1}{(3x+1)^2}\,dx = \lim_{c\to\infty}\int_1^c \frac{1}{(3x+1)^2}\,dx = \lim_{c\to\infty}\left(-\frac{1}{3}\cdot\frac{1}{3x+1}\right)\Big|_1^c$$

$$= -\frac{1}{3}\lim_{c\to\infty}\left(\frac{1}{3c+1} - \frac{1}{4}\right) = -\frac{1}{3}\cdot\left(-\frac{1}{4}\right) = \frac{1}{12}$$

You can take your pick as to how you want to lay out the solution. But, do not attempt the combined limit and integral notation until you know exactly what is going on, because combining the notations like that opens the door to errors.

You Try It

(1) Evaluate $\displaystyle\int_{-\infty}^0 \frac{1}{2x-5}\,dx$.

Useful Fact 7.10. *For integrals over the entire real line $(-\infty, \infty)$ and where the integrand has no discontinuities, split the integral at some real number (such as $x = 0$); replace each infinite endpoint with a dummy parameter, solve the resulting integrals, and then apply the appropriate limits.*

EX 2 Evaluate $\displaystyle\int_{-\infty}^{\infty} xe^{-x^2}\, dx$.

Here we have two trouble-maker endpoints of integration; however, we can only have one "bad" endpoint per integral. So, we have to split the interval of integration at some convenient point in between the current endpoints, such as $x = 0$. Visually, we "blow up" the interval of integration as:

$$(-\infty, \infty) \overset{kaboom}{\longrightarrow} (-\infty, 0] \cup [0, \infty)$$

Breaking the integral into subintegrals accordingly,

$$\int_{-\infty}^{\infty} xe^{-x^2}\, dx = \int_{-\infty}^{0} xe^{-x^2}\, dx + \int_{0}^{\infty} xe^{-x^2}\, dx$$

$$= \lim_{c \to -\infty} \int_{c}^{0} xe^{-x^2}\, dx + \lim_{d \to \infty} \int_{0}^{d} xe^{-x^2}\, dx$$

The whole integral converges only if both of the subintegrals converge to finite values. We can investigate them one at a time:

$$\lim_{c \to -\infty} \int_{c}^{0} xe^{-x^2}\, dx = \lim_{c \to -\infty} \left(-\frac{1}{2} e^{-x^2} \right) \Big|_{c}^{0}$$

$$= \lim_{c \to -\infty} \left(-\frac{1}{2}(1 - e^{-c^2}) \right) = -\frac{1}{2}$$

So that one converges. How about the other?

$$\lim_{d \to \infty} \int_{0}^{d} xe^{-x^2}\, dx = \lim_{d \to \infty} \left(-\frac{1}{2} e^{-x^2} \right) \Big|_{0}^{d}$$

$$= \lim_{d \to \infty} \left(-\frac{1}{2}(e^{-d^2} - 1) \right) = \frac{1}{2}$$

and that one converges, too! So the whole integral is the sum of the values of the subintegrals:

$$\int_{-\infty}^{\infty} xe^{-x^2}\, dx = -\frac{1}{2} + \frac{1}{2} = 0$$

Be sure you agree that a value of 0 for this integral is perfectly reasonable (hint: consider symmetry in the graph of the function). ∎

You Try It

(2) Evaluate $\displaystyle\int_{-\infty}^{\infty} x^2 e^{-x^3}\, dx$.

Useful Fact 7.11. *For integrals over a finite interval* (a, b), *where either* $x = a$ *or* $x = b$ *is the location of an infinite discontinuity of the integrand: replace the "bad" endpoint with a dummy parameter, solve the resulting integral, and then apply the appropriate limit.*

$\boxed{\text{EX 3}}$ Evaluate $\int_0^3 \dfrac{1}{\sqrt{x}}\, dx$.

Here, the trouble-maker endpoint is $x = 0$, and this must be replaced with a temporary parameter. We do not need to subdivide the interval of integration.

$$\int_0^3 \frac{1}{\sqrt{x}}\, dx = \lim_{c \to 0^+} \int_c^3 \frac{1}{\sqrt{x}}\, dx = \lim_{c \to 0^+} \left. (2\sqrt{x}) \right|_c^3$$
$$= \lim_{c \to 0^+} (2\sqrt{3} - 2\sqrt{c}) = 2\sqrt{3} \quad \blacksquare$$

You Try It

(3) Evaluate $\int_0^1 \dfrac{1}{\sqrt{1 - x^2}}\, dx$.

Useful Fact 7.12. *For integrals over a finite interval* (a, b), *where neither* $x = a$ *or* $x = b$ *locates an infinite discontinuity of the integrand, but there is an infinite discontinuity of the integrand inside the interval of integration, say at some* $x = d$, *then: split the integral at the location of the discontinuity (so there are now subintervals* (a, d) *and* (d, b)*); replace each "bad endpoint" at* $x = d$ *with distinct dummy parameters, solve the resulting integrals, and then apply the appropriate limits.*

$\boxed{\text{EX 4}}$ Evaluate $\int_{-2}^3 \dfrac{1}{x^4}\, dx$.

There is a discontinuity of this function at $x = 0$, which is inside the interval of integration. Therefore, we place the mathematical dynamite at that location and blow up the interval of integration as:

$$[-2, 3] \overset{kaboom}{\longrightarrow} [-2, 0) \cup (0, 3]$$

The endpoint $x = 0$ now appears in two different sub-integrals, and is taken care of with a limit in each case:

$$\int_{-2}^{3} \frac{1}{x^4}\,dx = \int_{-2}^{(0)} \frac{1}{x^4}\,dx + \int_{(0)}^{3} \frac{1}{x^4}\,dx$$

$$= \lim_{c \to 0^-} \int_{-2}^{c} \frac{1}{x^4}\,dx + \lim_{d \to 0^+} \int_{d}^{3} \frac{1}{x^4}\,dx$$

The whole integral converges only if both of the subintegrals converge to finite values. We can investigate them one at a time:

$$\lim_{c \to 0^-} \int_{-2}^{c} \frac{1}{x^4}\,dx = \lim_{c \to 0^-} \left. \frac{-1}{3x^3} \right|_{-2}^{c}$$

$$= \lim_{c \to 0^-} \left(\frac{-1}{3c^3} - \frac{1}{24} \right) \quad \leftarrow \text{ diverges}$$

Since the first of the subintegrals diverges, we're done. We don't even need to look at the second subintegral. The whole integral diverges. ∎

You Try It

(4) Evaluate $\displaystyle\int_{0}^{2} \frac{3x^2}{(x^3 - 1)^2}\,dx$.

Useful Fact 7.13. *For a combination of the above cases, split the overall integral of integration into as many subintervals as necessary such that each resulting subintegral displays only one "bad" endpoint; once the trouble-maker endpoints are isolated to one per subintegral, replace all of them with distinct dummy parameters, solve all the resulting integrals, and then apply the appropriate limits.*

$\boxed{\text{EX 5}}$ Break up the integral $\displaystyle\int_{-\infty}^{\infty} \frac{dx}{x^3}\,dx$ into appropriate subintegrals with limits, but do not solve the resulting integrals.

Not only does this integral have two infinite endpoints, but there is an infinite discontinuity at $x = 0$. Let's start breaking up the interval of integration. Is the following break-up enough?

$$(-\infty, \infty) \overset{kaboom}{\longrightarrow} (-\infty, 0) \cup (0, \infty)$$

No, this isn't enough. Each subinterval has two "bad" endpoints (remember, $x = 0$ is a discontinuity), so each of these must also be split at some convenient point in between — say, $x = -1$ and $x = 1$ — so a proper breaking of the interval is,

$$(-\infty, \infty) \overset{kaboom}{\longrightarrow} (-\infty, -1] \cup [-1, 0) \cup (0, 1] \cup [1, \infty)$$

Then, applying these subintervals along with dummy parameters to hold the places of the "bad" endpoints,

$$\int_{-\infty}^{\infty} \frac{dx}{x^3}\, dx = \lim_{c_1 \to -\infty} \int_{c_1}^{-1} \frac{dx}{x^3} + \lim_{c_2 \to 0^-} \int_{-1}^{c_2} \frac{dx}{x^3} + \lim_{c_3 \to 0^+} \int_{c_3}^{1} \frac{dx}{x^3} + \lim_{c_4 \to \infty} \int_{1}^{c_4} \frac{dx}{x^3}$$

As always, the whole integral converges only if each subintegral converges, so that we can get their sum. We should go through them one at a time, because if any one of them diverges, we don't need to waste time evaluating the others. The left-most sub-integral is (with some details omitted),

$$\int_{c_1}^{-1} \frac{dx}{x^3} = \frac{1}{2c_1^2} - \frac{1}{2}$$

$$\to \lim_{c_1 \to -\infty} \int_{c_1}^{-1} \frac{dx}{x^3} = -\frac{1}{2}$$

Then for the next one,

$$\int_{-1}^{c_2} \frac{dx}{x^3} = \frac{1}{2} - \frac{1}{2c_1^2}$$

$$\to \lim_{c_2 \to 0^-} \int_{-1}^{c_2} \frac{dx}{x^3} \quad \text{diverges}$$

Since this subintegral has diverged, the whole integral diverges. ∎

You Try It

(5) Break up the integral $\displaystyle\int_{-\infty}^{\infty} \frac{dx}{x^2 - b}\, dx$ (where b is some number, $4 < b < 100$) into appropriate subintegrals with limits, but do not solve the resulting integrals.

A final note about discontinuities: in these examples, we were worried about *infinite* discontinuities. If the integrand has jump discontinuities inside the interval of integration, we should split the integral at those locations, but no limits are required. For example, remember that $f(x) = |x|/x$ is a step function which reduces to $f(x) = -1$ for $x < 0$ and $f(x) = 1$ for

$x > 0$. The discontinuity at $x = 0$ is a jump discontinuity. We could then write

$$\int_{-2}^{2} \frac{|x|}{x} \, dx = \int_{-2}^{0} \frac{|x|}{x} \, dx + \int_{0}^{2} \frac{|x|}{x} \, dx$$

with no limits. (The final value of this integral is 0.) If there is a jump discontinuity at one of the endpoints themselves, we do not need to make any modifications to the integral there. If we were to define

$$g(x) = \begin{cases} x^2 - 4 & \text{for } x \leq 2 \\ 2x - 3 & \text{for } x > 2 \end{cases}$$

then $g(x)$ has a jump discontinuity at $x = 2$. However, a measure of the area under $g(x)$ on $[2, 5]$ could be presented as

$$\int_{2}^{5} g(x) \, dx$$

without modification.

Improper Integrals — Problem List

Improper Integrals — You Try It

These appeared above; solutions begin on the next page.

(1) Evaluate $\displaystyle\int_{-\infty}^{0} \frac{1}{2x-5}\, dx$.

(3) Evaluate $\displaystyle\int_{0}^{1} \frac{1}{\sqrt{1-x^2}}\, dx$.

(2) Evaluate $\displaystyle\int_{-\infty}^{\infty} x^2 e^{-x^3}\, dx$.

(4) Evaluate $\displaystyle\int_{0}^{2} \frac{3x^2}{(x^3-1)^2}\, dx$.

(5) Break up the integral $\displaystyle\int_{-\infty}^{\infty} \frac{dx}{x^2-b}\, dx$ (where $4 < b < 100$) into appropriate subintegrals with limits, but do not solve the resulting integrals.

Improper Integrals — Practice Problems

Try these as you get the hang of the You Try It problems. Solutions to these problems are available in Sec. A.1.6.

(1) Evaluate $\displaystyle\int_{-\infty}^{-1} e^{-2t}\, dt$.

(4) Evaluate $\displaystyle\int_{-2}^{2} \frac{1}{x^2-1}\, dx$.

(2) Evaluate $\displaystyle\int_{0}^{\pi} \tan x\, dx$.

(5) Evaluate $\displaystyle\int_{0}^{\infty} \frac{1}{(x-1)^4}\, dx$.

(3) Evaluate $\displaystyle\int_{4}^{20} \frac{1}{\sqrt{x-4}}\, dx$.

(6) Evaluate $\displaystyle\int_{-\infty}^{\infty} \frac{1}{x^2+1}\, dx$.

Improper Integrals — Challenge Problems

Try these problems to test your skills with the ideas in this section. Solutions to these problems are available in Sec. B.1.5.

(1) Evaluate $\displaystyle\int_{0}^{\infty} \frac{1}{x^2+x-6}\, dx$.

(2) Evaluate $\displaystyle\int_{-1}^{\infty} \frac{e^{1/x}}{x^2}\, dx$.

(3) If the area under $f(x) = \ln x / x^2$ on $[1, \infty)$ is finite, find the coordinate $x = A$ ($A > 1$) at which the area under $f(x)$ on $[1, A]$ is half of the total area over $[1, \infty)$. (You will not be able to solve for A exactly, so plan to estimate it.)

Improper Integrals — You Try It — Solved

---

(1) Evaluate $\int_{-\infty}^{0} \dfrac{1}{2x - 5}\, dx$.

☐ The integrand has a discontinuity at $x = 5/2$, but that's outside the interval of integration; we only have to worry about the endpoint at $-\infty$. To get started, we set up

$$\int_{-\infty}^{0} \frac{1}{2x - 5}\, dx = \lim_{c \to -\infty} \int_{c}^{0} \frac{1}{2x - 5}\, dx$$

Since

$$\int_{c}^{0} \frac{1}{2x - 5}\, dx = \left(\frac{1}{2} \ln |2x - 5| \right) \Bigg|_{c}^{0} = \ln 3 - \ln |2c - 5|$$

then

$$\lim_{c \to -\infty} \int_{c}^{0} \frac{1}{2x - 5}\, dx = \lim_{c \to -\infty} \left(\ln 3 - \ln |2c - 5| \right)$$

which diverges.   ∎

---

(2) Evaluate $\int_{-\infty}^{\infty} x^2 e^{-x^3}\, dx$.

☐ The integrand is defined and continuous for all real numbers, so we only have to deal with the two infinite endpoints. We might as well make our cut at $x = 0$, right? The overall set-up is then,

$$\int_{-\infty}^{\infty} x^2 e^{-x^3}\, dx = \lim_{c \to -\infty} \int_{c}^{0} x^2 e^{-x^3}\, dx + \lim_{d \to \infty} \int_{0}^{d} x^2 e^{-x^3}\, dx$$

The whole integral converges only if each subintegral converges. Let's check them one at a time. For the left-hand subintegral, a quick substitution $u = x^3$ yields (details omitted):

$$\int_{c}^{0} x^2 e^{-x^3}\, dx = -\frac{1}{3}(1 - e^{-c^3})$$

so that

$$\lim_{c \to -\infty} \int_{c}^{0} x^2 e^{-x^3} = \lim_{c \to -\infty} \left(-\frac{1}{3}(1 - e^{-c^3}) \right)$$

which diverges (the combination of negative signs causes e^{-c^3} to become infinite as $c \to -\infty$). Therefore, the whole integral diverges. ∎

(3) Evaluate $\displaystyle\int_0^1 \frac{1}{\sqrt{1-x^2}}\,dx$.

☐ The integrand has a discontinuity at the endpoint $x = 1$, so let's treat that with a parameter. I hope you remember your inverse trig functions and their derivatives & antiderivatives! This one is simple enought to just bundle up all together:

$$\int_0^1 \frac{1}{\sqrt{1-x^2}}\,dx = \lim_{c\to1^-}\int_0^c \frac{1}{\sqrt{1-x^2}}\,dx = \lim_{c\to1^-} \left.\sin^{-1}(x)\right|_0^c$$
$$= \lim_{c\to1^-}\left(\sin^{-1}(c) - \sin^{-1}(0)\right) = \frac{\pi}{2}$$

Hey, we finally got one to converge! ∎

(4) Evaluate $\displaystyle\int_0^2 \frac{3x^2}{(x^3-1)^2}\,dx$.

☐ The integrand has an infinite discontinuity at $x = 1$. So we can split the initial integral into two subintegrals at that location, and apply dummy parameters with limits. Hopefully by now you recognize that the antiderivative of the integrand is $-1/(x^3 - 1)$.

$$\int_0^2 \frac{3x^2}{x^3-1}\,dx = \int_0^{(1)} \frac{3x^2}{(x^3-1)^2}\,dx + \int_{(1)}^2 \frac{3x^2}{(x^3-1)^2}\,dx$$
$$= \lim_{c\to1^-}\int_0^c \frac{3x^2}{(x^3-1)^2}\,dx + \lim_{d\to1^+}\int_d^2 \frac{3x^2}{(x^3-1)^2}\,dx$$

The overall integral converges only if both of the subintegrals converge. The first one goes like this:

$$\int_0^c \frac{3x^2}{(x^3-1)^2}\,dx = \left(\frac{-1}{c^3-1} + 1\right)$$

so that

$$\lim_{c\to1^-}\int_0^c \frac{3x^2}{(x^3-1)^2}\,dx = \lim_{c\to1^-}\left(\frac{-1}{c^3-1} + 1\right) \quad \leftarrow \text{ diverges}$$

Since one of the subintegrals diverges, the whole integral does. We don't even need to look at the second subintegral, but it happens to diverge as well. ∎

(5) Break up the integral $\int_{-\infty}^{\infty} \frac{dx}{x^2 - b} \, dx$ (where $4 < b < 100$) into appropriate subintegrals with limits, but do not solve the resulting integrals.

□ The integrand has a discontinuity at $x = \sqrt{b}$. To isolate the infinite endpoints *and* the discontinuity so that there is only one of these per subinterval, we can first blow up the interval of integration as:

$$(-\infty, \infty) \overset{kaboom}{\longrightarrow} (-\infty, \sqrt{b}) \cup (\sqrt{b}, \infty)$$

But this is not sufficient, because each subinterval has *two* bad endpoints each. So they each need to be split. We know that $b > 4$, so that $\sqrt{b} > 2$. So let's split $(-\infty, \sqrt{b})$ at $x = 2$, and then $(\sqrt{b}, \infty)$ at, oh, say $x = 10$. In all, then,

$$(-\infty, \infty) \overset{kaboom}{\longrightarrow} (-\infty, 2] \cup [2, \sqrt{b}) \cup (\sqrt{b}, 10] \cup [10, \infty)$$

Building this structure into the integral, and applying limits, we have:

$$\int_{-\infty}^{\infty} \frac{dx}{x^2 - b} \, dx$$

$$= \int_{-\infty}^{2} \frac{dx}{x^2 - b} + \int_{2}^{(\sqrt{b})} \frac{dx}{x^2 - b} + \int_{(\sqrt{b})}^{10} \frac{dx}{x^2 - b} + \int_{10}^{\infty} \frac{dx}{x^2 - b}$$

$$= \lim_{c_1 \to -\infty} \int_{c_1}^{2} \frac{dx}{x^2 - b} + \lim_{c_2 \to \sqrt{b}^-} \int_{2}^{c_2} \frac{dx}{x^2 - b} + \lim_{c_3 \to \sqrt{b}^+} \int_{c_3}^{10} \frac{dx}{x^2 - b}$$

$$+ \lim_{c_4 \to -\infty} \int_{10}^{c_4} \frac{dx}{x^2 - b} \quad \blacksquare$$

Chapter 8

The Mathematics Chainsaw Massacre

Introduction to Partitioning

Introduction

One of our growing set of Calculus superpowers is that we can solve problems on a large scale by zooming in and examining the problem on a small scale. It's like having a mathematical microscope. As there are macroeconomics and microeconomics, and microbiology and macrobiology, perhaps Calculus is micro mathematics? Here's how it works: Given a large scale problem that cannot yet be solved at once in general, we break down (or partition) the problem into smaller pieces. Then we solve, or at least approximate, the solution to the problem on one small piece. Armed with the individual small-scale solutions, we assemble them into an estimate of the large scale solution. Finally, we ask: what would it take to improve this estimate?

We are about to see several examples in which we use partioning to break a problem down into small pieces, solve the problem on the small pieces, and extrapolate the results back up to the big problem. Before we dive into those, let's review the overall process and set the steps in place, using a problem you have already seen — finding the area under a curve. This will help reintroduce the associated notation. If you feel like you already have that mastered, there are no exercises in this section, and you can skip ahead to the next.

Example: Approximating the Area Under a Curve

Finding the area under a curve involves using simple and known geometry to estimate the area under irregular shapes. The general problem statement is: "Find the area under a function $f(x)$ between $x = a$ and $x = b$." It's easy to find the area of a simple geometric shape, like a rectangle, triangle, or circle. And so if $f(x)$ is linear, i.e. its graph is a straight line, this problem is easy — the area under $f(x)$ will either be a triangle or a trapezoid, and finding that area is simple geometry. But what happens when $f(x)$ is, say, a parabola like $f(x) = x^2$? The area under that function is NOT a recognizable, simple geometric shape — and this is where partitioning comes in. We are going to delineate four main steps in the partitioning process, review the notation from our known case, and prepare to extend that to upcoming, more complicated, scenarios.

The four main primary steps in the partitioning process are

(1) Design the partition — meaning, select the number of "chops", name and denote the subsequent endpoints, and choose representative points from each partition.
(2) Try to solve, or at least approximate, the solution to the problem on one small piece.
(3) Extrapolate the solution on the small pieces back up to an approximate solution to the whole problem.
(4) Consider (and implement) ways to make the approximation better.

We are going to review this process in detail to find the area under $f(x) = x^2$ on the interval from $x = 1$ to $x = 4$.

Step 1: Design of the Partition

Figure 8.1 shows the region below $f(x) = x^2$ on $[1, 4]$ partitioned into ten subintervals. (Technically, we are only partitioning the x-axis, and then extending that partitioning upwards.)

The number of partitions starts as a somewhat arbitrary choice, and so we need a general representative for the number of partitions selected in any problem. The standard parameter for this is n. Given our choice $n = 10$, we immediately know the *width* of each partition: with $[1, 4]$ split into 10 evenly sized pieces, the width of each piece is $3/10$. This can be

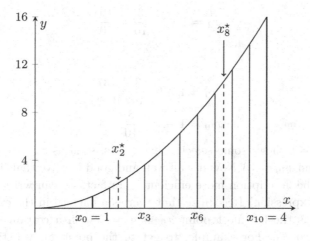

Fig. 8.1   Partitioning of $[1, 4]$ with $n = 10$ for $f(x) = x^2$.

written more generically. With a and b representing the left and right edges of our region of interest, a partition of that region into n pieces would result in partitions of width $(b - a)/n$. If we quantify "width" as Δx, then for our sample partition or any others, we can simply write

$$\Delta x = \frac{b - a}{n} \tag{8.1}$$

Our sample number of partitions $n = 10$ and resulting widths $\Delta x = 3/10$ gives us the ability to instantly roll out a list of all the endpoints created by the partitioning. First, recognize that if we take the interval from $x = a$ to $x = b$ and split it into n pieces, we actually end up with $n + 1$ endpoints. For example, take $x = 0$ to $x = 1$ and split it into 2 pieces; the resulting endpoints are $0, 0.5$, and 1 — so 2 partitions gave us 3 points. Since all these endpoints are x coordinates, it is reasonable to name them as such. Since there are $n + 1$ of them, we can either count them off as $x_1, x_2, \ldots, x_n, x_{n+1}$ or $x_0, x_1, \ldots, x_{n-1}, x_n$. The second choice is the one usually taken. The list of endpoints in our ongoing example would be (written to show a pattern):

$$x_0 = 1 + 0 \cdot \frac{3}{10} = 1$$

$$x_1 = 1 + 1 \cdot \frac{3}{10} = \frac{13}{10}$$

$$x_2 = 1 + 2 \cdot \frac{3}{10} = \frac{16}{10}$$

$$\vdots$$

$$x_9 = 1 + 9 \cdot \frac{3}{10} = \frac{37}{10}$$

$$x_{10} = 1 + 10 \cdot \frac{3}{10} = 4$$

(Figure 8.1 shows some selected endpoints, such as $x_0 = 1$, $x_3 = 1.9$, $x_6 = 2.8$, and $x_{10} = 4$.) Now it's all well and good to have that list, but we can make the description more efficient — in fact, we can write the entire list in one expression! Recognize that there's a pattern in the expressions: each point is the left-most point $x_0 = a = 1$ with a certain number of widths added on. For example, to get to the point x_4, we start at the leftmost point $a = 1$ and hop over by 4 widths, giving

$$x_4 = 1 + 4 \cdot \frac{3}{10} \quad \left(= \frac{22}{10} \right)$$

So in general, to get to *any* point x_i, we start at the leftmost point $a = 1$ and hop over by i widths, giving

$$x_i = 1 + i \cdot \frac{3}{10} \quad \left(= \frac{10 + 3i}{10} \right)$$

But it gets even better! Because remember, we can describe our widths more generally as Δx, so:

$$x_i = 1 + i \cdot \Delta x = a + i \cdot \left(\frac{b - a}{n} \right)$$

And now we've taken our example and drawn out from it a way to denote this list of endpoints for ANY interval $[a, b]$ and ANY number of partitions n. That index i only makes sense if it is a value between 0 and n, so to be more proper, a list of endpoints for a partition of the interval $[a, b]$ into n parts is:

$$x_i = 1 + i \cdot \Delta x = a + i \cdot \left(\frac{b - a}{n} \right) \quad (0 \le i \le n)$$

Having created a list of endpoints resulting from a division of the interval $[a, b]$ into n partitions, we also get the list of the resulting subintervals:

$$[a, x_1], [x_1, x_2], [x_2, x_3], \ldots, [x_{n-2}, x_{n-1}], [x_{n-1}, b]$$

(remember that $x_0 = a$ and $x_n = b$). The last step in designing the partition is to select our *representative points*. From each subinterval, we must select a point where we'll gather data related to whatever question is being solved. In our area problem, each subinterval became the base of a pretend-rectangle, and from each rectangle we had to select a representative point at which to find the "height" of each partition. The different choices for the method of selecting representative points led to the left hand rule, the right hand rule, and the midpoint rule.

In general, since there are n intervals, we can call these representative points x_i^* (the i counts the interval, and the asterisk just reminds us that its a special point selected from the subinterval). Figure 8.1 shows two such representative points, x_2^* and x_8^* selected at the midpoint of their subintervals.

Step 2: Gather Data and Solve a Mini Problem

The general notation developed for number, widths, endpoints, and representative points of a partition are components of pretty much any partitioning problem. But from there, the question becomes: what are you going to do with this partitioning? In the sample area problem, we use the representative points to determine the heights of each partition (rectangle). But in a different problem, the partitioning might be used for something else entirely. So from this point on, you need to become more comfortable with mathematical notation in general, since the data we need to gather will vary from problem to problem.

For the purpose of developing and demonstrating notation, let's take a deeper dive into the area problem. As noted, we compute the area of each partition by selecting either the left edge, right edge, or midpoint of each partition to gather the heights. Here are descriptions of how the representative points x_i^* are determined by the general endpoints x_i in different cases. Be sure these make sense to you!

- Left-Hand Rule: $x_i^* = x_{i-1}$ (because, for example, the left edge of the 2nd partition is the endpoint x_1)
- Right-Hand Rule: $x_i^* = x_i$ (because, for example, the left edge of the 2nd partition is the endpoint x_2)
- Midpoint Rule: $x_i^* = (x_{i-1} + x_i)/2$ (because the midpoint of each subinterval is the average of its endpoints)

If these make sense, then it should not be a surprise that the area of each partition / rectangle can be written as

$$A_i = \text{height} \times \text{width} = f(x_i^\star) \cdot \Delta x$$

We have now solved the mini-problem: how do we express the area of any one individual rectangle (partition)?

Step 3: Extrapolate Back to the Big Problem

Once we've solved the mini problem on one partition (i.e. here, finding the expression for each individual area, $A_i = f(x_i^\star) \cdot \Delta x$, what do we do with all the individual solutions? We add them up! Since we used 10 partitions, then the total estimated area under $f(x) = x^2$ over $[1, 4]$ is the sum of the 10 individual areas. Expressed in much (too much) detail, this is:

$$A_{tot} = A_1 + A_2 + A_3 + \ldots + A_{10}$$
$$= f(x_1^\star) \cdot \Delta x + f(x_2^\star) \cdot \Delta x + \ldots f(x_{10}^\star) \cdot \Delta x$$

That's not very efficient, though. How about we write it instead with summation notation:

$$A_{tot} = \sum_{i=1}^{10} f(x_i^\star) \cdot \Delta x$$

But wait, why restrict ourselves to 10 partitions? With any number n of partitions, this becomes

$$A_{tot} = \sum_{i=1}^{n} f(x_i^\star) \cdot \Delta x$$

And since we denote the area under $f(x)$ on $[a, b]$ with the integral symbol, we can write

$$\int_a^b f(x)dx \approx \sum_{i=1}^{n} f(x_i^\star) \cdot \Delta x \qquad (8.2)$$

Please note something important, namely that this general expression for the total area is the same for the left hand, right hand, and midpoint rules. We know in the background that we'd choose the $x_i^\star$'s, differently in each case, but that doesn't affect the general expression itself — and that's one of the main reasons to use this horrible, abstract notation. There, in one line, is the description of the ENTIRE solution process for finding an estimate to the area under a curve. It's cryptic, but it's the most efficient way to write it.

Now, the notation can meet you half-way. Equation (8.2) is wonderful because — if you understand the notation — you know the entire story of how to estimate the area under a curve. But if you to embed some details related to the individual method you choose to select the representative points for each partition, then you can certainly do that. For example, with the right hand rule in mind,

$$\int_a^b f(x)\, dx \approx \sum_{i=1}^n f\left(a + i \cdot \left(\frac{b-a}{n}\right)\right) \cdot \left(\frac{b-a}{n}\right)$$

Do you understand why it looks that way?

Step 4: The Big Payoff, Where Your Hard Work Pays Off

In any partitioning problem, the answer you generate from your partitioning is only an approximation of the true answer as long as the number of partitions is finite. Usually a small number of partitions gives a poor estimate, and that estimate improves as the number of partitions increases.

The final big payoff of the partitioning process happens when you investigate what happens as the number of partitions goes to infinity, and success at obtaining the payoff usually depends on your ability to recognize the sum / limit definition of a definite integral when you see it. That definition is this:

$$\int_a^b f(x)dx = \lim_{n \to \infty} \sum_{i=1}^n f(x_i^*)\Delta x$$

The example area problem is what led to this notation to begin with, and upcoming problems will rely on this result. The left side is the symbol used to denote area under $f(x)$ on $[a, b]$; the right side is the means by which that area is found (at least until other shortcuts present themselves).

By the end of a problem solved through partitioning, we will likely have created a sum of terms, based on our n partitions; we have to investigate the behavior of this sum as $n \to \infty$. This investigation usually pops out a definite integral — but you won't recognize it unless you know the notation.

We can't have any examples of these limits producing new integral expressions because we have not started any examples other than the area problem. Once you're ready, hop into the next section to try this out.

8.1   The Area Between Two Curves

Introduction

The simplest example of partitioning, which you encountered first, is the problem of finding the area under a curve. An immediate complication of that problem is to find the area between two curves. We can solve this quickly using the solution to the regular area problem, and we can solve it from the ground up as another application of the partitioning process.

In some previous chapters, we have saved derivations of formulas for the end of a section; in this chapter, the process of deriving a given formula is the entire point. So, the derivations (labeled as *Into the Pit!*) will appear early in a section.

Solution Based on the (Solved) Area Problem

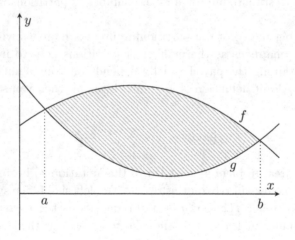

Fig. 8.2   The area between two curves.

We know that the area under $f(x)$ on the interval $[a, b]$ is given by the definite integral of $f(x)$ over that interval, and this integral can often be evaluated using the Fundamental Theorem of Calculus. Suppose, though, that we are interested in the area between two curves, $f(x)$ and $g(x)$, as shown in Fig. 8.2.

Let's name the upper function $f(x)$ and the lower function $g(x)$. Based on Fig. 8.2, it looks like we can just find the total area under $f(x)$ on $[a, b]$

and then subtract away the area under $g(x)$ on that same interval. What's left is the area between the two curves. This gives that the area between the curves is

$$A_{f,g} = \int_a^b f(x)dx - \int_a^b g(x)dx = \int_a^b (f(x) - g(x))dx \qquad (8.3)$$

🔟 FFT: It's pretty clear this holds for the curves in the figure. Would it still hold if both f and g were *below* the x-axis? Or if f was above and g was below? Or what if the region in question straddled the y-axis? What if the curves crossed each other somewhere between $x = a$ and $x = b$? Should we be cautious about leaping to too many conclusions about the applicability of (8.3)? 🔟

Area Between Two Curves, From Scratch

Into the Pit!!

The entire purpose of this section is the development of an integral formula via partitioning. Thus, we visit "the Pit" right away!

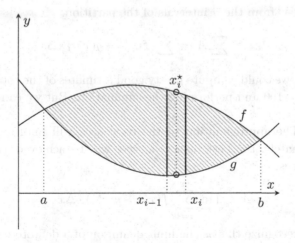

Fig. 8.3 Partitioning for the area between two curves.

Let's suppose that we never solved the original area problem (area under $f(x)$) and had to tackle this new one from scratch. This is a job for partitioning, as laid out in Chap. 8: Interlude

(1) Break down (or partition) the problem into smaller pieces.
(2) Try to solve, or at least approximate, the solution to the problem on one small piece.
(3) Extrapolate the solution on the small pieces back up to an approximate solution to the whole problem.
(4) Consider (and implement) ways to make the approximation better.

Step 1: Let's chop $[a, b]$ into n subintervals. One such subinterval, with its endpoints x_{i-1} and x_i, and representative point (midpoint) x_i^* are shown in Fig. 8.3.

Step 2: The contribution to the total area from this ith partition is approximately rectangular, so let's pretend it is a rectangle. The width of the rectangle is $\Delta x = (b - a)/n$. Selecting from the partition a representative point x_i^* (which happens to be the midpoint), we can assign the height of the rectangle to be $f(x_i^*) - g(x_i^*)$. The resulting area of this one partition is

$$A_i = \text{height} \cdot \text{width} = (f(x_i^*) - g(x_i^*)) \cdot \Delta x$$

Step 3: An estimate of the overall area of the region is the sum of the individual areas from the n intervals of the partition,

$$A_{tot} \approx \sum_{i=1}^{n} A_i = \sum_{i=1}^{n} (f(x_i^*) - g(x_i^*)) \cdot \Delta x$$

At this point, we could compute pretty good estimates of the total area, say by setting $n = 100$ and performing the summation. But we can do better.

Step 4: The approximation to the total area will improve as n gets larger. Ultimately, the best (exact) answer will be achieved as n goes to infinity. That is,

$$A = \lim_{n \to \infty} \sum_{i=1}^{n} (f(x_i^*) - g(x_i^*)) \cdot \Delta x$$

We then recognize this as the limit definition of a definite integral, and have

$$A = \int_a^b (f(x) - g(x)) \, dx \tag{8.4}$$

And so we get the same solution as we did starting from the known area problem.

Discussion and First Examples

The simplest computations of the area between two curves is merely a matter of evaluating the appropriate integral. This in itself isn't very interesting, but we gotta do it.

$\boxed{\textbf{EX 1}}$ Find the area between $y = 2\sqrt{x}$ and $y = x^3/3$ on the interval $[0, 1]$.

We don't have to do much work at all because the interval of integration is given explicitly. The curves do not intersect at $x = 1$, but that doesn't matter. The only brainpower needed now is to decide which function is above the other one, because in the integral formula for area between two curves the function on top goes first as $f(x)$. On the interval $[0, 1]$, $y = 2\sqrt{x}$ is above $y = x^3/3$ (to see why, compare their values at, say, $x = 1/4$). And so we have

$$A = \int_0^1 \left(2\sqrt{x} - \frac{x^3}{3} \right) dx = \left(\frac{4}{3}x^{3/2} - \frac{x^4}{12} \right) \Big|_0^1 = \frac{5}{4} \quad \blacksquare$$

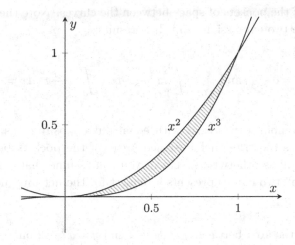

Fig. 8.4   The area between $f(x) = x^2$ and $g(x) = x^3$.

$\boxed{\textbf{EX 2}}$ Find the area between $y = x^2$ and $y = x^3$.

We are not given endpoints of an interval explicitly, so we have to assume we're just looking for any region trapped between these curves. Since these functions intersect only at $(0,0)$ and $(1,1)$ (see Fig. 8.4), our interval is from $x = 0$ to $x = 1$. On the interval $[0,1]$, $y = x^2$ is above $y = x^3$ (don't just trust the graph, compare their values at, say, $x = 1/2$). And so we have

$$A = \int_0^1 (x^2 - x^3)\, dx = \left(\frac{1}{3}x^3 - \frac{1}{4}x^4\right)\Big|_0^1 = \frac{1}{12} \quad \blacksquare$$

You Try It

(1) Find the area between $y = x$ and $y = x^2$.

One minor increase in complexity of these area problems happens when the functions intersect somewhere inside the given interval of integration. Since the integral formula specifies that the function above the other comes first, then if the functions cross and switch places, the integral must be split in two at that point. Generally, when we are seeking "area between two curves", we are asking for the "net unsigned area", meaning the true geometric area of the pockets of space between the curves. Note the difference between these two expressions and their results:

$$\int_{-1}^1 x - x^3\, dx = 0 \qquad \text{and} \qquad \int_{-1}^0 x^3 - x\, dx + \int_0^1 x - x^3\, dx = \frac{1}{4} + \frac{1}{4} = \frac{1}{2}.$$

The former combines two areas with equal but opposite signs, while the latter measures the true "net unsigned area" of the pockets between the two curves. Unless otherwise specified, you can assume that a request for "area between" two curves presents a request for the net unsigned area.

EX 3 Find the area between $y = 2x + 2$ and $y = x^3 + 2$ on $\left[-\frac{1}{2}, 1\right]$.

The curves intersect at $x = 0$. So in our interval, $y = x^3 + 2$ is above the other between $x = -1/2$ and $x = 0$, but between $x = 0$ and $x = 1$, the curve $y = 2x + 2$ is on top. Therefore,

$$A = \int_{-1/2}^{0} \left((x^3 + 2) - (2x + 2) \right) dx + \int_{0}^{1} \left((2x + 2) - (x^3 + 2) \right) dx$$

$$= \int_{-1/2}^{0} (x^3 - 2x) \, dx + \int_{0}^{1} (2x - x^3) \, dx$$

$$= \left(\frac{1}{4} x^4 - x^2 \right) \Big|_{-1/2}^{0} + \left(x^2 - \frac{1}{4} x^4 \right) \Big|_{0}^{1}$$

$$= - \left(\frac{1}{64} - \frac{1}{4} \right) + \left(1 - \frac{1}{4} \right) = \frac{63}{64} \quad \blacksquare$$

You Try It

(2) Find the area between $y = \sin x$ and $y = \cos x$ on $[0, \pi/2]$.

Other Variations

Position above / below x-axis

In this new integration scenario, we have to address again the idea that the area "under" a function $f(x)$ really means the area "between the function and the x-axis", and any part of $f(x)$ that is below the x-axis contributes a negative value to the corresponding integral. For example, the area "under" $\sin x$ on $[\pi, 2\pi]$ is actually the area of the pocket *above* $\sin x$ on that interval. The area "under" $\sin x$ on $[0, 2\pi]$ combines one pocket above the x-axis on $[0, \pi]$ with its symmetric counterpart below the axis on $[\pi, 2\pi]$. And so,

$$\int_{\pi}^{2\pi} \sin(x) \, dx > 0 \quad ; \quad \int_{\pi}^{2\pi} \sin(x) \, dx < 0 \quad ; \quad \int_{0}^{2\pi} \sin(x) dx = 0$$

By contrast, in this new problem of finding the area between two curves, we usually intend to measure the actual geometric area between the two functions, so that the final value should be positive regardless of whether the functions are above or below the axis. Fortunately, Eq. (8.4) works it out automatically: as long as we maintain the convention that when we use Eq. (8.4), we assign $f(x)$ to be the function that's *above* the other function in a north-south orientation, then the result will be positive. Be sure you see why!

EX 4 Write an integral that shows the area between $y = x$ and $y = -x^2$ on $[0, 1]$

This area would be written as

$$\int_0^1 \left(x - (-x^2) \right) dx$$

and the value of the integral, which will be positive, gives the true area between the curves. The fact that $y = x$ is above the x-axis and $y = -x^2$ is below the x-axis doesn't matter. ∎

You Try It

    (3) Write an integral that shows the net unsigned area between $y = -x$ and $y = -(e^x)$ on $[1, 2]$.

Numerical Estimates of the Points of Intersection

In the event that two curves intersect at points that cannot be found algebraically, we may need to rely on numerical estimates of those points of intersection.

EX 5 Find (approximately) the area between $y = x$, $y = \cos(x)$, and $x = 0$.

The two functions and line bound a region from $x = 0$ on the left to the intersection of $y = x$ and $y = \cos(x)$ on the right. This intersection is found by seeking solutions to $x = \cos(x)$, which cannot be solved by hand. We can use tech to determine the approximate intersection is $x \approx 0.739$, and so the area can be approximated by the integral (which is easy to evaluate)

$$A \approx \int_0^{0.739} (\cos(x) - x)\, dx = \left(\sin(x) - \frac{1}{2}x^2 \right) \Big|_0^{0.739}$$

$$= \sin(0.739) - \frac{1}{2}(0.739)^2 \approx 0.4 \quad ∎$$

You Try It

    (4) Find (approximately) the area between $y = \ln x$ and $y = 3x - 5$.

Symmetry

Sometimes symmetry in an area between curves can be used to modify limits of integration. Usually this relies on seeing that the entire area in question has repetitions of a smaller portion — for example, being symmetric across a horizontal or vertical line.

EX 6 Write an expression that relates the area between $y = x^2$ and $y = 2 - x^2$ on $[-1, 1]$ to the area between the two curves on $[0, 1]$.

These curves intersect at $x = \pm 1$; the region between them (plot it!) is symmetric about $x = 0$. So

$$\int_{-1}^{1} \left((2 - x^2) - x^2\right)\, dx = 2 \cdot \int_{0}^{1} \left((2 - x^2) - x^2\right) \quad \blacksquare$$

You Try It

(5) Write an expression that relates the area between $y = x^4 - 2x^2 + 1$ and $y = 2$ on $[-1, 1]$ to the area between the two curves on $[-1, 0]$.

Functions Described as $x = f(y)$

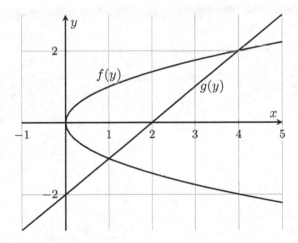

Fig. 8.5   Looking out from the y-axis (with EX 7).

Sometimes curves are given as functions of y, or perhaps as expressions not solved for either variable yet, and it may be advantageous to do the problem with respect to the y-axis instead of the x-axis. In this case, nothing changes except the variable of integration, and the fact that choosing the function "above" the other is done from the perspective of the y-axis.

$\boxed{\textbf{EX 7}}$ Write an integral that gives the area between $f(y) = y^2$ and $g(y) = y + 2$.

Figure 8.5 shows these curves. They intersect at $y = -1, 2$ (estimated visually, but confirmed algebraically, of course). The curve $x = y + 2$ is "above" $x = y^2$ from the perspective of the y-axis, and so

$$A = \int_{-1}^{2} \left((y+2) - y^2 \right) \, dy = \ldots = \frac{9}{2}$$

(details of the integration are trivial and omitted). ∎

You Try It

(6) Find the area between $x = 1 - y^2$ and $x = y^2 - 1$.

The Area Between Two Curves — Problem List

The Area Between Two Curves — You Try It

These appeared above; solutions begin on the next page.

(1) Find the area between $y = x$ and $y = x^2$.
(2) Find the area between $y = \sin x$ and $y = \cos x$ on $[0, \pi/2]$.
(3) Write an integral that shows the net unsigned area between $y = -x$ and $y = -(e^x)$ on $[1, 2]$.
(4) Find (approximately) the area between $y = \ln x$ and $y = 3x - 5$.
(5) Write an expression that relates the area between $y = x^4 - 2x^2 + 1$ and $y = 2$ on $[-1, 1]$ to the area between the two curves on $[-1, 0]$.
(6) Find the area between $x = 1 - y^2$ and $x = y^2 - 1$.

The Area Between Two Curves — Practice Problems

Try these as you get the hang of the You Try It problems. Solutions to these problems are available in Sec. A.2.1.

(1) Find the area between $y = x^2$ and $y = x^4$.
(2) Find the area between $y = 12 - x^2$ and $y = x^2 - 6$.
(3) Find the area between $y = x^3 - x$ and $y = 3x$.
(4) Find (approximately) the area between $y = e^{-x}$ and $y = 2 - x$.
(5) Write an expression that relates the net unsigned area between $\sin x$ and the x-axis on $[-12\pi, 12\pi]$ to the area under $\sin x$ on $[0, \pi]$. Then find that total net unsigned area.
(6) Find the area between $4x + y^2 = 12$ and $x = y$.
(7) Find the area between $f(x) = \cosh(x)$ and $f(x) = \sinh(x)$ on $[0, \ln 2]$ and write that value as a rational number.

The Area Between Two Curves — Challenge Problems

Try these problems to test your skills with the ideas in this section. Solutions to these problems are available in Sec. B.2.1.

(1) Find the area between $y = 1/x$, $y = 1/x^2$ and $x = 2$.
(2) Find the net unsigned area between $y = 8 - x^2$, $y = x^2$, $x = -3$ and $x = 3$.
(3) Find (approximately) the area between $y = x \cos(x^2)$ and $y = x^3$.

The Area Between Two Curves — You Try It — Solved

---

(1) Find the area between $y = x$ and $y = x^2$.

☐ These curves intersect at $(0,0)$ and $(1,1)$. The line $y = x$ is above $y = x^2$ over that interval. So, the area between them is:

$$A = \int_0^1 [(x) - (x^2)]\, dx = \left(\frac{1}{2}x^2 - \frac{1}{3}x^3 \right) \Big|_0^1 = \frac{1}{6} \quad \blacksquare$$

---

(2) Find the area between $y = \sin x$ and $y = \cos x$ on $[0, \pi/2]$.

☐ These two curves intersect at $x = \pi/4$. To the left, $\cos x$ is on top, and to the right, $\sin x$ is on top. The total area between them on the whole interval is then

$$A = \int_0^{\pi/4} (\cos(x) - \sin(x))\, dx + \int_{\pi/4}^{\pi/2} (\sin(x) - \cos(x))\, dx$$

$$= (\sin(x) + \cos(x)) \Big|_0^{\pi/4} + (-\cos(x) - \sin(x)) \Big|_{\pi/4}^{\pi/2}$$

$$= \left(\sin \frac{\pi}{4} + \cos \frac{\pi}{4} - 1 \right) + \left[\left(-\cos \frac{\pi}{2} - \sin \frac{\pi}{2} \right) - \left(-\cos \frac{\pi}{4} - \sin \frac{\pi}{4} \right) \right]$$

$$= \left(\sqrt{2} - 1 \right) + \left(-1 + \sqrt{2} \right) = 2\sqrt{2} - 2 \quad \blacksquare$$

---

(3) Write an integral that shows the net unsigned area between $y = -x$ and $y = -(e^x)$ on $[1, 2]$.

☐ Both functions are below the x-axis on this interval, and $y = -x$ is above $y = -(e^x)$ in a north-south orientation. So, the net unsigned area between the curves is

$$A = \int_1^2 (-x - (-e^x))\, dx \quad \blacksquare$$

---

(4) Find (approximately) the area between $y = \ln x$ and $y = 3x - 5$.

☐ Using tech, we can find that these curves intersect at $x \approx 0.007$ and $x \approx 1.88$. The curve $y = \ln x$ is above the curve $y = 3x - 5$ in a north-south orientation, so:

$$A \approx \int_{0.007}^{1.88} \ln x - (3x - 5)\, dx \approx 3.4 \quad \blacksquare$$

(5) Write an expression that relates the area between $y = x^4 - 2x^2 + 1$ and $y = 2$ on $[-1, 1]$ to the area between the two curves on $[-1, 0]$.

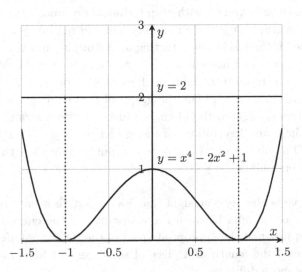

Fig. 8.6   Using symmetry for areas between curves (with YTI 5).

☐ The region between these curves is shown in Fig. 8.6. They do not intersect at $x = -1$ and $x = 1$, but that's fine. The overall region is symmetric across the y-axis, so we can write:

$$\int_{-1}^{1} 2 - (x^4 - 2x^2 + 1)\, dx = 2 \int_{0}^{1} 2 - (x^4 - 2x^2 + 1)\, dx \quad \blacksquare$$

(6) Find the area between $x = 1 - y^2$ and $x = y^2 - 1$.

☐ This region is easier to deal with from the perspective of the y-axis. The curves intersect at $y = -1$ and $y = 1$. The curve $x = 1 - y^2$ is "above" $x = y^2 - 1$ on that interval. Further, we can use the symmetry around the origin to get the area between them:

$$A = 2 \int_{0}^{1} \left[(1 - y^2) - (y^2 - 1) \right] dx = 2 \int_{0}^{1} (2 - 2y^2)\, dx$$

$$= 2 \left(2y - \frac{2}{3}y^3 \right) \Big|_{0}^{1} = \frac{8}{3} \quad \blacksquare$$

8.2   The Arc Length of a Curve

Introduction

At the risk of turning you off with deep technical terminology, I think you'll agree that the graph of $y = \sin x$ is *long and wiggly*. Imagine you fire up the point-sized Calculus-Mobile, starting at the origin, and you drive along the graph of $y = \sin x$ until you reach a stop sign at $x = 20$. What linear distance did you drive? Or, thought of another (more sane?) way, if you take your math scissors and cut the graph of $y = \sin x$ at the origin and at $x = 20$, then straighten the piece you just cut into a straight line, how long is it? These are descriptions of the arc length of $y = \sin x$ from $x = 0$ to $x = 20$. Our job is to find a way to compute the arc length of a curve between two endpoints.

If the curve is already a straight line, its arc length is easy to find: just use the distance formula between the endpoints. If the curve is part of a circle, we can just find the appropriate part of the total circumference. But what if we need the length of a piece of some general function $f(x)$, like $\sin x$? That's more difficult.

Given the placement of this topic, you may be getting suspicious about the technique we'll use to derive a general arc length formula. And you're right. We are going to implement a partitioning scheme; as a reminder, that means we will ...

(1) ... break down (or partition) the problem into smaller pieces.

(2) ... try to solve, or at least approximate, the solution to the problem on one small piece.

(3) ... extrapolate the solution on the small pieces back up to an approximate solution to the whole problem.

(4) ... consider (and implement) ways to make the approximation better.

Partitioning the Arc Length Problem

Into the Pit!!

Again, since the entire purpose of this section is the development of an integral formula via partitioning, we visit "the Pit" right away!

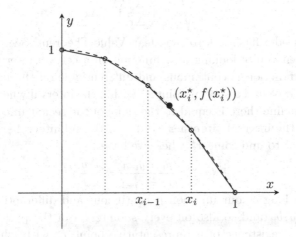

Fig. 8.7 Sample partitioning for an arc length problem.

Step 1: We must design our partition. To find the arc length of $f(x)$ on the interval $[a, b]$, we start by subdividing $[a, b]$ into n partitions. This leads to the partitioning of $f(x)$ itself into n small pieces. Figure 8.7 shows a sample partitioning. There we see a function $f(x)$ partitioned into four pieces; the curve itself is the dashed arc. We have sample endpoints x_{i-1} and x_i, and a representative point $x_i^\star$ within that subinterval. (How these representative points are chosen is shown below.) The goal of our partitioning is that on each subinterval, we will pretend the corresponding piece of the curve is a straight line — these are the four solid line segments that follow along the curve.

Step 2: We must solve our problem on one representative subinterval; the arc length of the curve in any one subinterval will be estimated by the length L_i of the straight line joining the endpoints (x_{i-1}, y_{i-1}) and (x_i, y_i).

The distance formula tells us that

$$L_i = \sqrt{(x_i - x_{i-1})^2 + (y_i - y_{i-1})^2}$$

Now $x_i - x_{i-1}$ is just the width of the ith partition, which we know to call Δx. To be consistent, let's also use Δy to represent $y_i - y_{i-1}$. The length of the ith piece of $f(x)$ is then

$$L_i = \sqrt{(\Delta x)^2 + (\Delta y)^2}$$

We can factor out the Δx to make

$$L_i = \sqrt{1 + \left(\frac{\Delta y}{\Delta x}\right)^2} \cdot \Delta x \qquad (8.5)$$

Now we need a flashback to the Mean Value Theorem (Sec. 5.5), which says that when you're looking at a function on a certain interval, then as long as your function is continuous and differentiable on the interval, you are guaranteed to find a special point $x^\star$ within the interval where the slope of the tangent line there is equal to the slope of the secant line joining the endpoints of the interval. In other words, on the subinterval $[x_{i-1}, x_i]$, we are guaranteed to find some $x_i^\star$ where we have

$$f'(x_i) = \frac{y_i - y_{i-1}}{x_i - x_{i-1}} = \frac{\Delta y}{\Delta x}$$

Therefore, as long as our function is continuous and differentiable on the big interval $[a, b]$ (and so also on every subinterval in the partition), then we are assured existence of a representative point $x_i^\star$ with which we can rewrite (8.5) as

$$L_i = \sqrt{1 + (f'(x_i^\star))^2} \cdot \Delta x \qquad (8.6)$$

(This is a subtle change in our identification of our representative points in each subinterval. In previous cases, we have found them at a specific location — such as the midpoint of the interval. In this new case, we don't know exactly where $x_i^\star$ happens to be, but we are guaranteed that they will exist — and that's the important thing.) This expression for the estimated arc length within any one partition is what we need, so it's time to put them all together.

Step 3: An estimate of the overall arc length of $f(x)$ on $[a, b]$ is the sum of the individual arc lengths from the n intervals of the partition. Using (8.6),

$$L \approx \sum_{i=1}^{n} \sqrt{1 + (f'(x_i^\star))^2} \cdot \Delta x$$

Step 4: The approximation to the total length will improve as n gets larger. Ultimately, the best (exact) answer will be achieved as n goes to infinity. That is,

$$L = \lim_{n \to \infty} \sum_{i=1}^{n} \sqrt{1 + (f'(x_i^\star))^2} \cdot \Delta x$$

We then recognize this as the limit definition of a definite integral (you recognize this is an integral, right?), and we can pose this final conclusion:

Useful Fact 8.1. *Let $f(x)$ be continuous and differentiable on the interval $[a, b]$. Then the arc length of $f(x)$ on that interval is given by*

$$L = \int_a^b \sqrt{1 + (f'(x))^2}\, dx \qquad (8.7)$$

(Technically, this result is a *theorem*, which we have proven, but we'll flag it as a Useful Fact since it's more of a utilitarian formula.)

EX 1 Find the arc length of $f(x) = \ln(\cos x)$ on $[0, \pi/4]$.

The workflow of an arc length problem is improved if we shake out the contents of the square root in Useful Fact 8.1 before loading up the integral. When $f(x) = \ln(\cos x)$, we have

$$f'(x) = \frac{1}{\cos x} \cdot (-\sin x) = -\tan x$$

and then

$$1 + [f'(x)]^2 = 1 + \tan^2 x = \sec^2 x$$

so that

$$\sqrt{1 + [f'(x)]^2} = \sqrt{\sec^2 x} = \sec x$$

(The contents of the square root, $1 + [f'(x)]^2$, are designed to be positive, so we don't have any of that $\pm$ nonsense with the square root.) Now we can assemble (8.7)

$$L = \int_a^b \sqrt{1 + [f'(x)]^2}\, dx = \int_0^{\pi/4} \sqrt{\sec^2 x}\, dx = \int_0^{\pi/4} \sec x\, dx$$

This is an integral we solved once in Sec. 7.4 and marked it as Eq. (7.17) so we would not have to go through the hassle of solving it repeatedly. The result allows us to carry on as:

$$L = \int_0^{\pi/4} \sec x \, dx = \ln|\sec x + \tan x| \Big|_0^{\pi/4}$$

$$= \ln\left|\sec\left(\frac{\pi}{4}\right) + \tan\left(\frac{\pi}{4}\right)\right| - \ln|\sec 0 + \tan 0|$$

$$= \ln|\sqrt{2} + 1| - \ln|1 + 0| = \ln|\sqrt{2} + 1| \blacksquare$$

You Try It

(1) Find the arc length of $y = 1 + 6x^{3/2}$ for $0 \le x \le 1$.
(2) Find the arc length of $y = \cosh x$ for $0 \le x \le 1$.

In most of these arc length problems, setting up the integral isn't difficult. The fun variations of these problems arise when doing the integral itself, or when you have to do a little interpretation of the function in order to generate your endpoints.

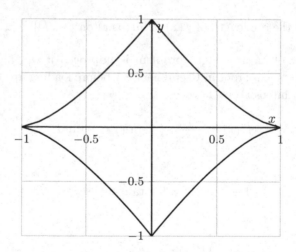

Fig. 8.8   Arc length of $x^{2/3} + y^{2/3} = 1$ (with EX 2).

EX 2 Find the arc length of $x^{2/3} + y^{2/3} = 1$.

Figure 8.8 shows this curve. By symmetry, the total arc length is four times the length of one side, so let's find the length of the upper quarter from

$x = 0$ to $x = 1$. We can prepare the integrand of (8.7) first,

$$f(x) = (1 - x^{2/3})^{3/2}$$
$$f'(x) = -x^{-1/3}(1 - x^{2/3})^{1/2}$$
$$1 + [f'(x)]^2 = 1 + x^{-2/3}(1 - x^{2/3}) = x^{-2/3}$$
$$\sqrt{1 + [f'(x)]^2} = \sqrt{x^{-2/3}} = x^{-1/3}$$

and the arc length of one quarter of the curve is:

$$L_1 = \int_a^b \sqrt{1 + [f'(x)]^2}\, dx = \int_0^1 \frac{1}{x^{1/3}}\, dx = \left(\frac{3}{2}x^{2/3}\Big|_0^1\right) = \frac{3}{2}$$

The total length of the entire curve is:

$$L_{tot} = 4L_1 = 4\left(\frac{3}{2}\right) = 6 \quad \blacksquare$$

You Try It

(3) Approximate the total arc length of the ellipse $2x^2 + 3y^2 = 6$. (This means you get to use tech to solve the integral!)

Arc Length of a Curve — Problem List

Arc Length of a Curve — You Try It

These appeared above; solutions begin on the next page.

(1) Find the arc length of $y = 1 + 6x^{3/2}$ for $0 \leq x \leq 1$.
(2) Find the arc length of $y = \cosh x$ for $0 \leq x \leq 1$.
(3) Approximate the total arc length of the ellipse $2x^2 + 3y^2 = 6$. (This means you get to use tech to solve the integral!)

Arc Length of a Curve — Practice Problems

  Try these as you get the hang of the You Try It problems. Solutions to these problems are available in Sec. A.2.2.

(1) Find the arc length of $y = 2(x + 4)^{3/2}$ for $0 \leq x \leq 2$.
(2) Find the arc length of $y = \ln \sec x$ for $0 \leq x \leq \pi/4$.
(3) Write an integral that will give the arc length of $x = y^2 - y + 3$ for $-1 \leq y \leq 1$, but do not solve the integral.
(4) Find the arc length of $y = \cosh(x)$ for $0 \leq x \leq \ln 4$ and write that value as a rational number.

Arc Length of a Curve — Challenge Problems

  Try these problems to test your skills with the ideas in this section. Solutions to these problems are available in Sec. B.2.2.

(1) Find the arc length of $y = \ln x$ for $1 \leq x \leq \sqrt{3}$.
(2) Find the arc length of $y = x^2/2 + 6\pi$ for $0 \leq x \leq 1/\sqrt{3}$.
(3) You have a skeptical friend who does not believe the circumference of a circle is $2\pi r$. You now have the tools to conclusively prove him wrong. Do it, for his own sake.

The Arc Length of a Curve — You Try It — Solved

(1) Find the arc length of $y = 1 + 6x^{3/2}$ for $0 \le x \le 1$.

☐ We'll do a little prep-work to get ready for (8.7) with $f(x) = 1 + 6x^{3/2}$:

$$f'(x) = 9x^{1/2}$$
$$1 + [f'(x)]^2 = 1 + 81x$$

So,

$$L = \int_a^b \sqrt{1 + [f'(x)]^2}\,dx = \int_0^1 \sqrt{1 + 81x}\,dx$$
$$= \frac{2}{243}(1 + 81x)^{3/2}\Big|_0^1 = \frac{2}{243}\left((82)^{3/2} - 1\right) \blacksquare$$

(2) Find the arc length of $y = \cosh x$ for $0 \le x \le 1$.

☐ We'll do a little prep-work to get ready for (8.7) with $f(x) = \cosh x$:

$$f(x) = \cosh x$$
$$f'(x) = \sinh x$$
$$1 + [f'(x)]^2 = 1 + \sinh^2 = \cosh^2 x$$
$$\sqrt{1 + [f'(x)]^2} = \sqrt{\cosh^2 x} = \cosh x$$

and so the arc length is

$$L = \int_a^b \sqrt{1 + [f'(x)]^2}\,dx = \int_0^1 \cosh x\,dx = \sinh x\Big|_0^1 = \sinh 1 \blacksquare$$

(3) Approximate the total arc length of the ellipse $2x^2 + 3y^2 = 6$. (This means you get to use tech to solve the integral!)

The complete ellipse is shown in Fig. 8.9. The left and right endpoints on the x-axis are $x = \pm\sqrt{3}$ (be sure you know why). If we say that L_1 is the arc length of the upper quarter of the ellipse over $0 < x < \sqrt{3}$, then the total arc length is $L_{tot} = 4L_1$. To find L_1, we can solve the equation of the ellipse as

$$y = \sqrt{2 - \frac{2}{3}x^2}$$

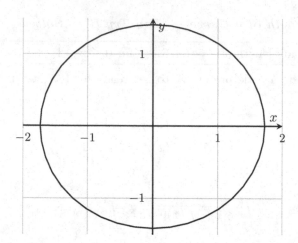

Fig. 8.9   Arc length of $2x^2 + 3y^2 = 6$ (with YTI 3).

and proceed:

$$f(x) = \sqrt{\frac{6 - 2x^2}{3}} = \frac{1}{\sqrt{3}} \left(6 - 2x^2\right)^{1/2}$$

$$f'(x) = \frac{1}{\sqrt{3}} \cdot \frac{1}{2} \left(6 - 2x^2\right)^{-1/2} (-4x) = -\frac{2x}{\sqrt{3}} \left(6 - 2x^2\right)^{-1/2}$$

$$[f'(x)]^2 = \frac{4x^2}{3} (6 - 2x^2)^{-1} = \frac{2x^2}{3(3 - x^2)}$$

$$1 + [f'(x)]^2 = 1 + \frac{2x^2}{3(3 - x^2)} = \frac{9 - 3x^2 + 2x^2}{9 - 3x^2} = \frac{9 - x^2}{3(3 - x^2)}$$

$$\sqrt{1 + [f'(x)]^2} = \sqrt{\frac{9 - x^2}{3(3 - x^2)}} = \frac{1}{\sqrt{3}} \cdot \sqrt{\frac{9 - x^2}{3 - x^2}}$$

Wolfram Alpha estimates $\displaystyle\int_0^{\sqrt{3}} \sqrt{\frac{9 - x^2}{3 - x^2}}\, dx \approx 4.29$, so: Then

$$L_{tot} = 4L_1 = \frac{4}{\sqrt{3}} \int_0^{\sqrt{3}} \sqrt{\frac{9 - x^2}{3 - x^2}}\, dx \approx \frac{4}{\sqrt{3}}(4.29) \approx 9.9$$

The shift from a circle to an ellipse makes the arc length calculation significantly harder. There is an entire catalog of mathematical tools called "elliptic functions" which arise from problems on an elliptical rather than a circular geometry. If you push ahead into some quantitatively intense STEM fields, you may encounter them! ∎

8.3   Average Value of a Function

Introduction

You already know how to find an average of a bunch of numbers. You add them all up then divide by the number of values you had. But what if you wanted to find the average of something that takes on an infinite number of values, like an average of a function $f(x)$ over an interval $[a, b]$? Well, you can estimate the average by using only a finite number of the values, then see what happens to your estimate as the number of values approaches infinity. Does that process sound familiar?

Using Partitions for the Average Value Problem

Into the Pit!!

To find the average of $f(x)$ on the interval $[a, b]$, we subdivide $[a, b]$ into n partitions. From each partition, we select a representative point $x_i^\star$ and get the function's value there, $f(x_i^\star)$. We can get an estimate of the average value by considering these n functional values:

$$f_{avg} \approx \frac{1}{n}\left(f(x_1^\star) + \cdots + f(x_n^\star)\right) = \frac{1}{n}\sum_{i=1}^{n} f(x_i^\star)$$

As usual, this estimate can be made an equality by letting the number of partitions go to infinity:

$$f_{avg} = \lim_{n\to\infty}\left(\frac{1}{n}\sum_{i=1}^{n} f(x_i^\star)\right) \tag{8.8}$$

Normally at this point, we expect to see something that looks like the definition of an integral. The current expression (8.8) isn't quite there yet. The term $1/n$ looks a bit out of place, and we're missing an instance of Δx. But the definition of Δx for such a partition,

$$\Delta x = \frac{b - a}{n}$$

can be rearranged like this:

$$\frac{1}{n} = \frac{\Delta x}{b - a}$$

And now look what happens when we continue (8.8):

$$f_{avg} = \lim_{n\to\infty} \left(\frac{1}{n}\sum_{i=1}^{n} f(x_i^*)\right) = \lim_{n\to\infty} \left(\frac{\Delta x}{b-a}\sum_{i=1}^{n} f(x_i^*)\right)$$

$$= \frac{1}{b-a}\left(\lim_{n\to\infty}\sum_{i=1}^{n} f(x_i^*)\Delta x\right)$$

And now we can see that our average really does have an integral in it! The result can be presented formally:

Useful Fact 8.2. *Let $f(x)$ be piecewise continuous on the interval $[a, b]$. Then the average value of $f(x)$ on that interval is given by*

$$f_{avg} = \frac{1}{b-a}\int_a^b f(x)\,dx \qquad (8.9)$$

| **EX 1** | Find the average value of $f(x) = \sin x$ on $[0, 2\pi]$.

$$f_{avg} = \frac{1}{b-a}\int_a^b f(x)\,dx = \frac{1}{2\pi - 0}\int_0^{2\pi} \sin x \, dx$$

$$= -\frac{1}{2\pi}\cos x\Big|_0^{2\pi} = -\frac{1}{2\pi}(1-1) = 0$$

The fact that the average of $\sin x$ on $[0, 2\pi]$ is zero should not be surprising, since the curve is perfectly symmetric across the x-axis on that interval.

∎

You Try It

    (1) Find the average of $g(x) = \cos x$ on $[0, \pi/2]$.

    (2) Find the average of the hourly temperature function $T(t) = 50 + 14\sin(\pi t/12)$ over a 12 hour period, i.e. for $t \in [0, 12]$.

The Mean Value Theorem for Integrals

Please recall the Mean Value Theorem (for Derivatives), which says: If $f(x)$ is continuous and differentiable on an interval $[a, b]$ then there's a magic point c within that interval where

$$f'(c) = \frac{f(b) - f(a)}{b - a}$$

You have been through the story of how that formula can be interpreted in more than one way.

There is another Mean Value Theorem (well, it's actually the same theorem, expressed in a different way) that we'll call the Mean Value Theorem for Integrals. It says: If $f(x)$ is continuous on an interval $[a, b]$, then there exists a magic point c in that interval where

$$f(c) = \frac{1}{b-a} \int_a^b f(x)\, dx$$

Now before you just see this as a bunch of symbols, say "Nevermind", and move on, look more closely. This theorem says something that's almost obvious: If a function is continuous between two points, then there has to be at least one point in between where the function is equal to its own average value. Your *average* height over your lifetime so far is somewhere between 0 inches and your current height. Wouldn't you think that at some point in your life, your actual height was equal to what is now your average height? Sure! Or consider EX 1: the average value of $\sin x$ on $[0, 2\pi]$ is zero. Is there a point c between $x = 0$ and $x = 2\pi$ where $\sin c = 0$? Sure, at $c = \pi$.

EX 2 Find the average of $f(x) = x^3 - 1$ on $[0, 2]$ and a value of c such that $f(c) = f_{avg}$.

First, we find the average:

$$f_{avg} = \frac{1}{b-a} \int_a^b f(x)\, dx = \frac{1}{2-0} \int_0^2 (x^3 - 1)\, dx$$

$$= \frac{1}{2}\left(\frac{1}{4}x^4 - x\right)\Big|_0^2 = \frac{1}{2}(4-2) = 1$$

Now let's find a point guaranteed by the Mean Value Theorem where the function equals its own average value on this interval. We are looking for $x = c$ such that $f(c) = f_{avg}$, or $c^3 - 1 = 1$. Solving for c, we get $c = \sqrt[3]{2}$. This value is indeed between $x = 0$ and $x = 2$. ∎

You Try It

(3) Find the average of $f(x) = (x - 3)^2$ on $[2, 5]$ and a value of c such that $f(c) = f_{avg}$.

The Average Value of a Function — Problem List

The Average Value of a Function — You Try It

These appeared above; solutions begin on the next page.

(1) Find the average of $g(x) = \cos x$ on $[0, \pi/2]$.
(2) Find the average of the hourly temperature function $T(t) = 50 + 14\sin(\pi t/12)$ over a 12 hour period, i.e. for $t \in [0, 12]$.
(3) Find the average of $f(x) = (x - 3)^2$ on $[2, 5]$ and a value of c such that $f(c) = f_{avg}$.

The Average Value of a Function — Practice Problems

Try these as you get the hang of the You Try It problems. Solutions to these problems are available in Sec. A.2.3.

(1) Find the average of $g(x) = x^2\sqrt{1 + x^3}$ on $[0, 2]$.
(2) Find the average of $f(x) = \sqrt{x}$ on $[0, 4]$ and a value of c such that $f(c) = f_{avg}$.
(3) Suppose the water level in Lake Michigan over a 30 year period starting in 1990 has followed the function

$$M(t) = 580 + 2\sin\left(\frac{\pi}{12}t\right)$$

(where M is in feet). What has been the average water level in Lake Michigan for this period, i.e. for $0 \le t \le 30$?

The Average Value of a Function — Challenge Problems

Try these problems to test your skills with the ideas in this section. Solutions to these problems are available in Sec. B.2.3.

(1) Find the average of $f(\theta) = \sec\theta\tan\theta$ on $[0, \pi/4]$.
(2) Find the average of $f(x) = 2\sin x - \sin 2x$ on $[0, \pi]$ and estimate a value of c such that $f(c) = f_{avg}$.
(3) Consider the arc of the parabola $y = x^2$ on the interval $x \in [0, 2]$. What is the average distance to the origin of all the points on this arc?

Average Value — You Try It — Solved

(1) Find the average of $g(x) = \cos x$ on $[0, \pi/2]$.

$$\square \quad g_{avg} = \frac{1}{b-a} \int_a^b g(x)dx = \frac{1}{\pi/2 - 0} \int_0^{\pi/2} \cos x \, dx$$

$$= \frac{2}{\pi}\left(\sin\frac{\pi}{2} - \sin 0\right) = \frac{2}{\pi} \quad \blacksquare$$

(2) Find the average of the hourly temperature function $T(t) = 50 + 14\sin(\pi t/12)$ over a 12 hour period, i.e. for $t \in [0, 12]$.

$$\square \quad T_{avg} = \frac{1}{b-a}\int_a^b T(t)\,dt = \frac{1}{12-0}\int_0^{12}\left(50 + 14\sin\frac{\pi t}{12}\right)dt$$

$$= \frac{1}{12}\left(50t - 14\cdot\frac{12}{\pi}\cos\frac{\pi t}{12}\right)\Bigg|_0^{12} = \frac{1}{12}\left(\frac{336}{\pi} + 600\right)$$

$$= \frac{28}{\pi} + 50 \approx 59 \quad \blacksquare$$

(3) Find the average of $f(x) = (x-3)^2$ on $[2, 5]$ and a value of c such that $f(c) = f_{avg}$.

$$\square \quad f_{avg} = \frac{1}{b-a}\int_a^b f(x)\,dx = \frac{1}{5-2}\int_2^5 (x-3)^2\,dx$$

$$= \frac{1}{3}\left(\frac{1}{3}(x-3)^3\right)\Bigg|_2^5 = \frac{1}{9}\left((2)^3 - (-1)^3\right) = 1$$

Next, we need a c such that $f(c) = f_{avg}$, i.e. $(c-3)^2 = 1$. This leads to two possibilities:

$$c - 3 = -1 \qquad \text{or} \qquad c - 3 = 1$$

which means we can pick from both $c = 2$ or $c = 4$. Since $c = 2$ is the endpoint of the interval $[2, 5]$, we should select $c = 4$, since it's in the interior of the interval. $\blacksquare$

8.4   Numerical Integration

Introduction

The ultimate use of the partitioning process is to numerically calculate values of definite integrals that are not suitable for solving by hand — whether that means they are not solved "easily" by hand, or are not solvable by hand at all. Life is great when all your integrals come with nice functions in them, but out there in the wilderness, science is often done with messy functions or raw data, not tidy algebraic functions. So let's see a few simple numerical integration schemes, to build some intuition about such processes. Like in Sec. 5.5, the use of a spreadsheet will be built into these solutions.

Left Hand, Right Hand, and Trapezoid Rules

In Sec. 6.2, we saw how to make "left hand" and "right hand" estimates of the area under a curve. Although we did not have the terminology at the time, the left hand and right hand rules are examples of numerical integration in which we estimate the value of some $\int_a^b f(x)\,dx$ using partitioning. The left and right hand rules are not very efficient, but they are a good starting place. Here's a recap:

In the left and right hand rule schemes, we divide $[a, b]$ into n partitions, with endpoints $x_0, x_1, \ldots, x_{n-1}, x_n$; note that x_0 and x_n are also known as a and b. Each partition has width $\Delta x = (b-a)/n$. At each partition point x_i, we compute the function value $f(x_i)$; this function value can serve as the "height" of a rectangle. The area of any rectangle looks like $f(x_i)\Delta x$. Flip back to Fig. 6.5 to see a diagram illustrating the left and right hand rules.

To compute the left hand rule, we add up all the values of $f(x_i)\Delta x$ except the last one:

$$\int_a^b f(x)dx \approx f(x_0)\Delta x + f(x_1)\Delta x + \ldots + f(x_{n-1})\Delta x$$

If we assume that all the endpoints in the partition are equally spaced, so that Δx is a constant $(b-a)/n$, then we can factor out that term:

$$\int_a^b f(x)dx \approx \frac{b-a}{n}\left(f(x_0) + f(x_1) + \ldots + f(x_{n-1})\right) \qquad (8.10)$$

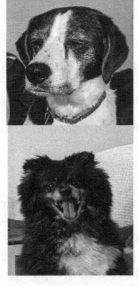

Numerical integration
by hand

Numerical integration
in a
spreadsheet

Fig. 8.10 Decisions about the process.

To compute the right hand rule, we add up all the values of $f(x_i)\Delta x$ except for the *first* one, and get — with the same assumption of constant Δx —

$$\int_a^b f(x)dx \approx \frac{b-a}{n}\left((x_1) + f(x_2) + \ldots + f(x_n)\right) \qquad (8.11)$$

We can use summation notation to tidy things up, and present the following definition:

Definition 8.1. *The left hand and right hand estimates for the definite integral of $f(x)$ over the interval $[a, b]$ are:*

$$LHR: \int_a^b f(x)\,dx \approx \frac{b-a}{n}\sum_{i=0}^{n-1} f(x_i), \quad RHR: \int_a^b f(x)\,dx \approx \frac{b-a}{n}\sum_{i=1}^{n} f(x_i)$$

EX 1 Use the left hand and right hand rules with $n = 6$ to find estimates for $\int_0^3 \sqrt{x^2 + 1}\,dx$.

In Fig. 8.10, Zoe and Tux demonstrate our philosophy about calculations. Implementation of the calculations in a spreadsheet saves time while still forcing you to really understand what's going on: you have to know what intermediate calculations to do, and how to carefully assemble the information into a final result.

Figure 8.11 shows a spreadsheet implementation (Google Sheets) of the left and right hand rules for for this integral. Here is a list of information we need to implement Def. 8.1, and how it's all placed in the spreadsheet:

- *The partitioning framework*: Cells B2–B4 show a, b, and n, and B6 has as a computed value $\Delta x = \dfrac{b-a}{n}$, with cell formula `=(B3-B2)/B4`.
- *The list of endpoints generated by the partitioning via* $x_i = a + i\Delta x$: Column C shows values of the counter i (from 1 to n); Column D contains the endpoints. The endpoint in cell D4, for example, is found by `=$B$2 + C2 * $B$6`. The dollar signs in two cell references keep the formula locked in place to those cells (B2 and B6), while C2 will get updated when the formula is copied to other cells (this is *absolute referencing* vs *relative referencing*.
- *The values of* $f(x)$ *at each endpoint.* These are held in Column E, and are found by, for example, `= SQRT(D4^2 + 1)`.
- To implement the equations in Def. 8.1, we need to find *the sum of the function values required to form the left and / or right hand estimates.*
  - The left hand estimate comes from the sum of all entries in Column E except the final one, thus forming $\displaystyle\sum_{i=0}^{5} f(x_i)$; this is held in cell G3 via `=SUM(E2:E7)`.
  - The right hand estimate comes from the sum of all entries in Column E except the first one, thus forming $\displaystyle\sum_{i=1}^{6} f(x_i)$; this is held in cell G7 via `=SUM(E3:E8)`.
- For each of the left and right hand estimates, the summation of functional values is multiplied by Δx; these are held in cells G4 and G8 respectively, with `=G3 * B6` and `=G7 * B6`.

The final left hand and right hand estimates are

$$\text{LHR: } \int_0^3 \sqrt{x^2+1}\,dx \approx 5.132 \quad ; \quad \text{RHR: } \int_0^3 \sqrt{x^2+1}\,dx \approx 6.213$$

A much more accurate estimate of the integral, via Wolfram Alpha, is 5.653. Note how the left hand estimate is too small and the right hand estimate is too large. ⌗◎⌗ FFT: Could we have used the graph of $f(x) = \sqrt{x^2 + 1}$ to predict in advance that the left hand rule would underestimate the true value of the integral, while the right hand rule would over estimate the true value? ⌗◎⌗   ■

	A	B	C	D	E	F	G
1			i	x_i	f (x_i)		
2	a = 0		0	0	1	LHR	
3	b = 3		1	0.5	1.118033989	sum:	10.26367357
4	n = 6		2	1	1.414213562	final:	5.131836785
5			3	1.5	1.802775638		
6	Delta-x =	0.5	4	2	2.236067977	RHR	
7			5	2.5	2.692582404	sum:	12.42595123
8			6	3	3.16227766	final:	6.212975615

Fig. 8.11   Spreadsheet implementation of the trapezoid rule, with EX 1.

Just for fun, go to Wolfram Alpha and enter

```
int sqrt(x^2+1) dx on [0,3] using left hand rule with n=6 .
```

and see all the information you get. Of course, this is not a substitute for knowing how to do the problem yourself, but it's nice to get confirmation that you are doing it properly!   ■

You Try It

(1) Use the left hand and right hand rules with $n = 6$ to find estimates for $\int_1^3 xe^x \, dx$.

In EX 1 (and also YTI 1), we have an integral for which the left hand rule gives an underestimate of the true value, and the right hand rule gives an overestimate. You might imagine, then, that the average of the left hand and right hand estimates would be better than either one. The average of the left hand and right hand rules is called the trapezoid rule. To form this average, we add the two estimates and divide the result by 2. So, in EX 1, the trapezoid rule estimate of the integral would be

$$\int_0^3 \sqrt{x^2+1}\, dx = \frac{1}{2}(5.132 + 6.213) = 5.6725) \tag{8.12}$$

It's important to note that since both the LHR and RHR estimates were already rounded, the final digit here is questionable. A better implementation of the trapezoid rule is to create its own direct formula. The average of the left and right hand estimates can be written as:

$$\frac{1}{2}\left[\frac{b-a}{n}\left(f(x_0) + f(x_1) + \ldots + f(x_{n-1})\right)\right.$$

$$\left. + \frac{b-a}{n}\left(f(x_1) + f(x_1) + \ldots + f(x_n)\right)\right]$$

$$= \frac{b-a}{2n}\left(f(x_0) + 2f(x_1) + \ldots + 2f(x_{n-1}) + f(x_n)\right)$$

Be sure to note the coefficients of 2 on all but the first and last terms in the sum, which occur because each endpoint except for the very first and very last shows up in both the left and right hand estimates.

Definition 8.2. *The trapezoid rule estimate for the definite integral of $f(x)$ over the interval $[a, b]$ is:*

$$\boxed{\int_a^b f(x)\, dx \approx \frac{b-a}{2n}\left(f(x_0) + 2\sum_{i=1}^{n-1} f(x_i) + f(x_n)\right)}$$

The name "trapezoid rule" comes from the geometric result of averaging the left and right hand rules. Remember that in the left hand and right hand rules, we are literally adding areas of rectangles. If we focus on one individual rectangle in the partition, when we find the average of areas determined when the height of the rectangle comes from its left edge $f(x_{i-1})$ and from its right edge $f(x_i)$, we obtain the area that would come from a trapezoid whose top edge connects $(x_{i-1}, f(x_{i-1}))$ and $(x_i, f(x_i))$. This is shown in Fig. 8.12; the left hand and right hand heights of any one partition are dashed lines, and the top edges of the resulting "average shapes" (trapezoids) are solid lines.

We can implement the trapezoid rule in a spreadsheet in two ways: (a) simply find the LHR and RHR estimates as in EX 1, then find their average, or (b) build the calculation in Def. 8.2 directly. We'll use option (b), because it creates a template that will also be useful below.

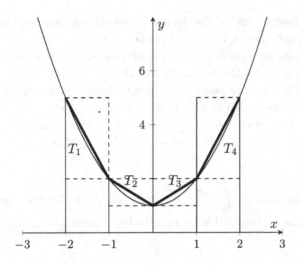

Fig. 8.12   Trapezoid rule vs LHR, RHR.

	A	B	C	D	E	F	G	H	I
fx	=sum(G2:G8)								
1			i	x_i	f(x_i)	c_i	c_i * f(x_i)		
2	a = 0		0	0	1	1	1	TRAP RULE	
3	b = 3		1	0.5	1.118033989	2	2.236067977	sum:	22.6896248
4	n = 6		2	1	1.414213562	2	2.828427125	multiple:	0.25
5			3	1.5	1.802775638	2	3.605551275	final:	5.6724062
6	Delta-x = 0.5		4	2	2.236067977	2	4.472135955		
7			5	2.5	2.692582404	2	5.385164807		
8			6	3	3.16227766	1	3.16227766		

Fig. 8.13   Spreadsheet cells for the trapezoid rule, with EX 2.

EX 2 Use the trapezoid rule with $n = 6$ to find an estimate for $\int_0^3 \sqrt{x^2 + 1}\, dx$.

We already know from Eq. (8.12) this result should come out to be about 5.6725, so we'll know if our implementation of Def. 8.2 is correct. Figure 8.13 shows a spreadsheet calculation adapted from the one in Fig. 8.11. Here are the differences:

- We have a new column (Col F) which contains the coefficients c_1 of each term $f(x_i)$ in Def. 8.2. (Remember, all but the first and last are multiplied by 2.)

- We have a list of the terms to be added together according to Def. 8.2, that is, $c_i f(x_i)$. This is Col G.
- The initial summation for the trapezoid rule is in Cell I3, and is the sum of all the entries in Col G.
- Per Def. 8.2, the summation is multiplied by $\Delta x/2$. This multiple is created in Cell I4 via =B6 / 2.
- The final trapezoid rule estimate (the sum times its multiple in Def. 8.2) is in Cell I5, and is found by = I2 * I3.

The result found here, $\displaystyle\int_0^3 \sqrt{x^2+1}\,dx \approx 5.6724$ is slightly more accurate than the average of rounded left and right hand rules found earlier as 5.6725.

∎

NOTE: When adapting the spreadsheet design in EX 1 and EX 2, it's important to carefully consider how to extend columns for more detailed partitions (i.e. larger values of n). For example, if we want to use $n = 12$, it's not sufficient to simply change Cell B4 from 6 to 12; all the columns have to be doubled in length as well. Figure 8.14 shows a spreadsheet which is intended to recreate the trapezoid rule calculation in EX 2, except using $n = 12$ instead of $n = 6$. Cell B4 has been changed from 6 to 12. Be sure you can identify what other edits are yet to be made.

	A	B	C	D	E	F	G	H	I
1			i	x_i	f (x_i)	c_i	c_i * f(x_i)		
2	a =	0	0	0	1	1	1	TRAP RULE	
3	b =	3	1	0.25	1.030776406	2	2.061552813	sum:	15.63038567
4	n =	12	2	0.5	1.118033989	2	2.236067977	multiple:	0.125
5			3	0.75	1.25	2	2.5	final:	1.953798209
6	Delta-x =	0.25	4	1	1.414213562	2	2.828427125		
7			5	1.25	1.600781059	2	3.201562119		
8			6	1.5	1.802775638	1	1.802775638		

Fig. 8.14 An unfinished spreadsheet set-up for the trapezoid rule, after EX 2.

You Try It

(2) Use the trapezoid rule with $n = 12$ to estimate the area under $f(x) = \cos(x)$ on $[0, \pi]$. How does your result compare to the true value?

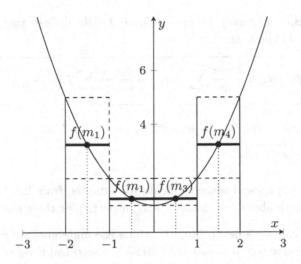

Fig. 8.15   Midpoint rule vs LHR, RHR.

The Midpoint Rule

The midpoint rule is fairly self-explanatory; we form a numerical estimate to $\int_a^b f(x)\,dx$ by forming a partition of $[a,b]$ with n rectangles, and choosing the height of each rectangle according to the midpoint of each subinterval rather than the left edge or right edge. In general, if we have n rectangles, then we have n midpoints m_i. From the usual endpoints $x_0, x_1, \ldots, x_n$, we can form the midpoint of each subinterval $[x_{i-1}, x_i]$ as $m_i = (x_{i-1} + x_i)/2$. Figure 8.15 illustrates the midpoint rule. There are four partitions in the diagram; the heights of each from the left and right hand rules are shown by dashed lines, while the height from each midpoint $(m_i, f(m_i))$ is a thick solid line. A midpoint rule estimate can be built as:

$$\int_a^b f(x)\,dx \approx \frac{b-a}{n}\left(f(m_1) + \ldots + f(m_n)\right) \tag{8.13}$$

$$= \frac{b-a}{n}\left[f\left(\frac{x_0 + x_1}{2}\right) + f\left(\frac{x_1 + x_2}{2}\right) + \ldots + f\left(\frac{x_{n-1} + x_n}{2}\right)\right] \tag{8.14}$$

In more tidy summation notation, we have:

Definition 8.3. *The midpoint rule estimate for the definite integral of* $f(x)$ *over the interval* $[a, b]$ *is:*

$$\int_a^b f(x)\,dx \approx \frac{b-a}{n} \sum_{i=1}^{n} f(m_i) = \frac{b-a}{n} \sum_{i=1}^{n} f\left(\frac{x_{i-1}+x_i}{2}\right)$$

EX 3 Use the midpoint rule with $n = 6$ to find an estimate for $\int_0^3 \sqrt{x^2+1}\,dx$.

Figure 8.16 shows a spreadsheet calculation adapted from EX 1, Fig. 8.11. I'll let you think about the changes, and in particular these questions:

- You can see a new column in which the midpoints are created. Do you know how to create the entries in (new) Col E using the entries in Col D?
- The column of function values is changed to compute $f(x)$ at the midpoints m_i rather than the endpoints x_i. Why are cells E2 and F2 empty?
- The cell formula for the final result in H4 is displayed. Be sure you know what's happening there.

The midpoint estimate with $n = 6$ is $\int_0^3 \sqrt{x^2+1}\,dx \approx 6.213$. ∎

f_X	=sqrt(E3^2+1)						
	A	B	C	D	E	F	G	H
1			i	x_i	m_i	f (m_i)		
2	a = 0		0	0			MIDPT	
3	b = 3		1	0.5	0.25	1.030776406	sum:	11.28551133
4	n = 6		2	1	0.75	1.25	final:	5.642755665
5			3	1.5	1.25	1.600781059		
6	Delta-x =	0.5	4	2	1.75	2.015564437		
7			5	2.5	2.25	2.46221445		
8			6	3	2.75	2.926174978		

Fig. 8.16   A midpoint rule implementation, with EX 3.

🔟 FFT: Is the midpoint estimate in EX 3 more or less accurate (as compared to the "true" result) than the trapezoid rule estimate from EX 2? Will this always be the case? The midpoint and trapezoid rules sound an awful lot alike: one is formed using the average of functional values, and other is formed using the average of *endpoints*. Which is which, and do you think the result from one will always be better or worse than the other, given the same number of partitions? 🔟

You Try It

   (3) Use the midpoint rule with $n = 12$ to estimate the area under $f(x) = \cos(x)$ on $[0, \pi]$. How does your result compare to the true value? How does your result compare to the trapezoid rule value in YTI 2?

Simpson's Rule[1]

All of the previous numerical techniques for estimating integrals pretend that each partition is in a shape that has a straight-line top — either a rectangle or a trapezoid. However, most functions will be curving over the top of each partition (or at the bottom, when $f(x) < 0$). Simpson's Rule is an attempt to account for that. The development of Simpson's Rule is more complicated than the others: it requires taking adjacent pairs of partitions and pretending the piece of $f(x)$ above them is parabolic in shape — thus allowing for a bit of "bending". Without details, the formula for Simpson's Rule turns out to be:

Definition 8.4. *The Simpson's rule estimate for the definite integral of* $f(x)$ *over the interval* $[a, b]$ *is (with the notation* $y_i = f(x_i)$ *to compress the recipe),*

$$\int_a^b f(x)\, dx \approx \frac{b-a}{3n} \left(y_0 + 4y_1 + 2y_2 + \ldots + 2y_{n-2} + 4y_{n-1} + y_n\right)$$

(We don't try to use a summation sign in the expression for Simpson's Rule because that becomes awkward with the alternating coefficients.) The key differences between this approximation formula and those before it are:

- The list of coefficients of the $f(x_i)$ terms starts and ends with 1, and alternates as $4, 2, 4, \ldots, 2, 4$ in between. Note that the need to stop and end the alternating coefficients with 4 means that n must be *even*. If you attempt a Simpson's Rule estimate with, say, $n = 11$, the coefficients will not work out right.

---

[1]Stop giggling.

- The multiple of Δx in front of the sum is $1/3$, not $1/2$ as in the trapezoid rule.

Otherwise, we proceed almost exactly as before. A spreadsheet template for the trapezoid rule can easily be adapted to Simpson's Rule, with only those two changes: edit the column of coefficients c_i to reflect the alternating nature of the new ones, and edit the final calculation to have the summation multiplied by $\Delta x/3$ rather than $\Delta x/2$.

In this next example, we'll continue to work with the same function as in the previous examples, which may be dull, but the function being integrated is almost secondary to the overall process. But to make things a little more interesting, let's double the number of partitions.

EX 4 Use Simpson's Rule with $n = 12$ to find an estimate for $\int_0^3 \sqrt{x^2+1}\,dx$.

f_X		=sum(G2:G14)							
	A	B	C	D	E	F	G	H	I
1			i	x_i	f (x_i)	c_i	c_i * f(x_i)		
2	a =	0	0	0	1	1	1	SIMPSON'S RULE	
3	b =	3	1	0.25	1.030776406	4	4.123105626	sum:	67.83167012
4	n =	12	2	0.5	1.118033989	2	2.236067977	multiple:	0.08333333333
5			3	0.75	1.25	4	5	final:	5.652639177
6	Delta-x =	0.25	4	1	1.414213562	2	2.828427125		
7			5	1.25	1.600781059	4	6.403124237		
8			6	1.5	1.802775638	2	3.605551275		
9			7	1.75	2.015564437	4	8.062257748		
10			8	2	2.236067977	2	4.472135955		
11			9	2.25	2.46221445	4	9.848857802		
12			10	2.5	2.692582404	2	5.385164807		
13			11	2.75	2.926174978	4	11.70469991		
14			12	3	3.16227766	1	3.16227766		

Fig. 8.17   A Simpson's rule implementation, with EX 4.

Figure 8.17 shows a spreadsheet calculation adapted from EX 2, Fig. 8.13, with the following changes:

- The column of coefficients (Col F) has been updated.
- The "multiple" value in Cell I4 represents $\Delta x/3 = 1/3$.
- Columns have been extended; the list of endpoints is accurate, and the summation of Col G is displayed, and confirmed to use all values.

The Simpson's Rule estimate with $n = 12$ is $\displaystyle\int_0^3 \sqrt{x^2 + 1}\, dx \approx 5.65264$.

■

You Try It

(4) Use Simpson's rule with $n = 12$ to find an estimate for $\displaystyle\int_1^3 xe^x\, dx$. Compare your result to the exact value, which you should know how to find!

Tabulated Data

At the start of this section was a remark that learning how to set up these numerical integration techniques using functions that have a "formula" is important to get some intuition about how they all work. Now that you've seen several examples of implementation of these methods in a spreadsheet, you will have noticed that the actual function being integrated is hidden in the background. It's not visible in the table itself; the values are visible, but the function that produced them is not. If we were to exchange a column of function values, like Column E in Fig. 8.17, with a table of some other data on the same interval $[a, b]$, then we would be integrating that new data instead. This means that the switch from integrating a function by formula to integrating a set of tabulated data is trivial. As long as we know (or can identify) the overall interval $[a, b]$ and the number of partitions n, we can use these same methods to integrate tabulated data.

$\boxed{\textbf{EX 5}}$ Use Simpson's Rule to estimate $\displaystyle\int_a^b f(x)\, dx$ for the tabulated values shown in this table:

x	0.2	0.5	0.8	1.1	1.4	1.7	2.0	2.3	2.6
$f(x)$	3.1	3.2	2.9	2.6	2.5	2.4	2.6	2.9	2.8

By counting, we can see there are 9 endpoint values x_i. Since counting of endpoints starts at x_0, we have $x_0 = 0.2$ and $x_8 = 2.6$. Each endpoint is separated by 0.3. Therefore, this data represents values of a function sampled on the interval $[0.2, 2.6]$ with $n = 8$ partitions. (Be sure you know why 9 endpoints means 8 partitions!) The integral we are estimating is $\displaystyle\int_{0.2}^{2.6} f(x)\, dx$. Just for fun, let's roll this out by hand rather than copying the values of $f(x)$ into a column of a spreadsheet implementation. Using Def. 8.4,

$$\int_{0.2}^{2.6} f(x)\,dx \approx \frac{b-a}{3n}\left(f(x_0) + 4f(x_1) + 2f(x_2) + \cdots\right.$$
$$\cdots + 2f(x_7) + 4f(x_8) + f(x_9))$$
$$\approx \frac{2.6 - 0.2}{3(8)}\left(3.1 + 4(3.2) + 2(2.9) + 4(2.6) + 2(2.5) + \cdots\right.$$
$$\cdots + 4(2.4) + 2f2.6) + 4(2.9) + 2.8) = 6.63$$

Well, OK, I'll also show the spreadsheet implementation of this: see Fig. 8.18. This gives the same result, $\int_{0.2}^{2.6} f(x)\,dx \approx 6.63$. ■

	A	B	C	D	E	F	G	H	I
1			i	x_i	f (x_i)	c_i	c_i * f(x_i)		
2	a =	0.2	0	0.2	3.1	1	3.1	SIMPSON'S RULE	
3	b =	2.6	1	0.5	3.2	4	12.8	sum:	66.3
4	n =	8	2	0.8	2.9	2	5.8	multiple:	0.1
5			3	1.1	2.6	4	10.4	final:	6.63
6	Delta-x =	0.3	4	1.4	2.5	2	5		
7			5	1.7	2.4	4	9.6		
8			6	2	2.6	2	5.2		
9			7	2.3	2.9	4	11.6		
10			8	2.6	2.8	1	2.8		

Fig. 8.18　A Simpson's rule implementation for tabulated data, with EX 5.

You Try It

(5) Use Simpson's Rule to estimate $\int_a^b f(x)\,dx$ for the tabulated values shown in this table:

x	0	0.5	1	1.5	2	2.5	3.0
$g(x)$	1	0.89	0.71	0.55	0.45	0.37	0.16

3.5	4	4.5	5	5.5	6
0.27	0.24	0.22	0.20	0.18	0.16

As a final note, the integration techniques that are programmed into computer algebra systems, or even a hand-held calculator, are more sophisticated than any of the methods seen here. But you have to start at the beginning. If the overall concepts as well as the systematic and algorithmic nature of these procedures is appealing to you, and you like the idea of actually trying to do some programming in this area, then head for your nearest Numerical Analysis course after you have satisfied whatever pre-requisites it might have (which will likely include Linear Algebra)!

Have You Learned...

- How to create a general set-up for the left and right hand rule calculations for estimating the value a definite integral?
- How to create a general set-up for trapezoid, midpoint, and Simpson's Rule calculations for estimating the value a definite integral?
- Which of the five methods you have seen so far is the most efficient — i.e. giving the most accurate result for a given value of n, or requiring the smallest value of n for a required accuracy?
- How to make the switch from integrating "formula functions" to integrating tabulated data?

Numerical Integration — Problem List

Numerical Integration — You Try It

These appeared above; solutions begin on the next page.

(1) Use the left hand and right hand rules with $n = 6$ to find estimates for
$$\int_1^3 xe^x \, dx.$$

(2) Use the trapezoid rule with $n = 12$ to estimate the area under $f(x) = \cos(x)$ on $[0, \pi]$. How does your result compare to the true value? (Hint: use =PI() to call the value of π in a spreadsheet, don't give a truncated estimate of that value like "3.14".)

(3) Use the midpoint rule with $n = 12$ to estimate the area under $f(x) = \cos(x)$ on $[0, \pi]$. How does your result compare to the true value? How does your result compare to the trapezoid rule value in YTI 2?

(4) Use Simpson's rule with $n = 12$ to find an estimate for $\int_1^3 xe^x \, dx$. Compare your result to those in YTI 1 and also to the exact value, which you should know how to find!

(5) Use Simpson's Rule to estimate $\int_a^b f(x) \, dx$ for the tabulated values shown in this table:

x	0	0.5	1	1.5	2	2.5	3.0
$g(x)$	1	0.89	0.71	0.55	0.45	0.37	0.16

3.5	4	4.5	5	5.5	6
0.27	0.24	0.22	0.20	0.18	0.16

Numerical Integration — Practice Problems

Try these as you get the hang of the You Try It problems. Solutions to these problems are available in Sec. A.2.4.

(1) Decide in advance if you would expect each of the left and right hand rules to over- or under-estimate the area under $f(x) = x^4$ on $[0, 2]$. Then perform the left and right hand estimates using $n = 12$, and confirm your guess.

(2) Find a trapezoid rule estimate for the area under $f(x) = e^{-\sqrt{x}}$ on $[0, 4]$ using $n = 6$, $n = 12$, and $n = 48$.

(3) Find a trapezoid rule estimate with $n = 12$ for $\int_0^\pi \cos(x^2) \, dx$.

(4) Find a midpoint rule estimate with $n = 12$ for $\int_0^\pi \cos(x^2)\,dx$.

(5) Find a Simpson's rule estimate with $n = 12$ for $\int_0^\pi \cos(x^2)\,dx$. Of the three results in PP 3, PP 4, and PP 5, which one do you trust the most? The least?

(6) Find a Simpson's rule estimate with $n = 6$ for $\int_0^2 e^{-(x-1)^4}\,dx$. Then do an experiment to find how many partitions are required (i.e. determine n) for the trapezoid rule to give the same result to four places after the decimal, $m.mmmm$.

(7) Use the Trapezoid Rule to estimate $\int_a^b f(x)\,dx$ for the tabulated values shown in this table (which is broken into two parts):

x	-2	-1.6	-1.2	-0.8	-0.4
$g(x)$	-0.00134	-0.05323	-0.41329	-0.70549	-0.39918

0	0.4	0.8	1.2	1.6	2.0
0	-0.39918	-0.70549	-0.41329	-0.05323	-0.00134

Numerical Integration — Challenge Problems

Try these problems to test your skills with the ideas in this section. Solutions to these problems are available in Sec. B.2.4.

(1) Do an experiment to find how big n must get for the midpoint rule to get a 5 decimal place match (within rounding) ($n.nnnnn$) to Wolfram Alpha's super-accurate answer for $\int_0^2 \dfrac{1}{\sqrt{x^3+1}}\,dx$. Use increments of 10.

(2) Do an experiment to find how big n must get for the trapezoid rule to get a 5 decimal place match (within rounding) ($n.nnnnn$) to Wolfram Alpha's super-accurate answer for $\int_0^2 \dfrac{1}{\sqrt{x^3+1}}\,dx$. Use increments of 10.

(3) Do an experiment to find how big n must get for the trapezoid rule to get a 5 decimal place match (within rounding) ($n.nnnnn$) to Wolfram Alpha's super-accurate answer for $\int_0^2 \dfrac{1}{\sqrt{x^3+1}}\,dx$. Use increments of 10. Compare to CP 1 and CP 2.

Numerical Integration — You Try It — Solved

(1) Use the left hand and right hand rules with $n = 6$ to find estimates for $\int_1^3 xe^x \, dx$.

---

fx	=D4*EXP(D4)					
	A	B	C	D	E	F	G
1			i	x_i	f (x_i)		
2	a = 1		0	1.000000	2.718281828	LHR	
3	b = 3		1	1.333333	5.05822386	sum:	93.81914739
4	n = 6		2	1.666667	8.824150084	final:	31.27304913
5			3	2.000000	14.7781122		
6	Delta-x = 0.3333		4	2.333333	24.0619365	RHR	
7			5	2.666667	38.37844292	sum:	151.3574763
8			6	3.000000	60.25661077	final:	50.45249211

Fig. 8.19   Left and right hand rule implementations, with YTI 1.

☐ Figure 8.19 shows the spreadsheet implementation, with a cell formula for the function displayed. The left and right hand estimates are:

$$\text{LHR} \int_1^3 xe^x \, dx \approx 31.273 \quad ; \quad \text{RHR} \int_1^3 xe^x \, dx \approx 50.452$$

They are very different because we are using only $n = 6$ and the value of $f(x)$ grows very quickly, so the difference beteen the left and right estimate of the area of any one partition gets more extreme as we move to the right. ∎

---

(2) Use the trapezoid rule with $n = 12$ to estimate the area under $f(x) = \cos(x)$ on $[0, \pi]$. How does your result compare to the true value? (Hint: use =PI() to call the value of π in a spreadsheet, don't give a truncated estimate of that value like "3.14".)

☐ Figure 8.20 shows the spreadsheet implementation, with the entry of π displayed. The trapezoid rule estimate ends up matching the correct value exactly: $\int_0^\pi \cos(x) \, dx \approx 0$. ∎

| | | | fx | =pi() | | | | |

	A	B	C	D	E	F	G	H	I
1				i	x_i	f(x_i)	c_i	c_i * f(x_i)		
2	a =	0		0	0.000000	1.00000	1	1.00000	TRAP RULE	
3	b =	3.14156		1	0.261799	0.96593	2	1.93185	sum:	0.00000
4	n =	12		2	0.523599	0.86603	2	1.73205	multiple:	0.1308996939
5				3	0.785398	0.70711	2	1.41421	final:	0
6	Delta-x =	0.26179		4	1.047198	0.50000	2	1.00000		
7				5	1.308997	0.25882	2	0.51764		
8				6	1.570796	0.00000	2	0.00000		
9				7	1.832596	-0.25882	2	-0.51764		
10				8	2.094395	-0.50000	2	-1.00000		
11				9	2.356194	-0.70711	2	-1.41421		
12				10	2.617994	-0.86603	2	-1.73205		
13				11	2.879793	-0.96593	2	-1.93185		
14				12	3.141593	-1.00000	1	-1.00000		

Fig. 8.20   A trapezoid rule implementation, with YTI 2.

---

(3) Use the midpoint rule with $n = 12$ to estimate the area under $f(x) = \cos(x)$ on $[0, \pi]$. How does your result compare to the true value? How does your result compare to the trapezoid rule value in YTI 2?

| | | | fx | =sum(F3:F14) | | | | |

	A	B	C	D	E	F	G	H
1				i	x_i	m_i	f(m_i)	
2	a =	0		0	0.0000000			MIDPT
3	b =	3.14159265		1	0.2617994	0.13089969	0.9914448614	sum:	0
4	n =	12		2	0.5235988	0.39269908	0.9238795325	final:	0
5				3	0.7853982	0.65449847	0.7933533403	
6	Delta-x =	0.26179938		4	1.0471976	0.91629786	0.608761429	
7				5	1.3089969	1.17809725	0.3826834324	
8				6	1.5707963	1.43989663	0.1305261922	
9				7	1.8325957	1.70169602	-0.1305261922	
10				8	2.0943951	1.96349541	-0.3826834324	
11				9	2.3561945	2.22529480	-0.608761429	
12				10	2.6179939	2.48709418	-0.7933533403	
13				11	2.8797933	2.74889357	-0.9238795325	
14				12	3.1415927	3.01069296	-0.9914448614	

Fig. 8.21   A midpoint rule implementation, with YTI 3.

☐ Figure 8.21 shows the spreadsheet implementation. The final estimate matches the trapezoid rule estimate and the exact value. This is because of the symmetry of points about $\dfrac{\pi}{2}$ where $f(x)$ is sampled.

🎞 FFT: The left and right hand rules would provide equal but opposite non-zero estimates. Without performing the calculations, could you decide which gives the positive estimate and which gives the negative estimate? 🎞 ∎

(4) Use Simpson's rule with $n = 12$ to find an estimate for $\int_1^3 xe^x \, dx$. Compare your result to those in YTI 1 and also to the exact value, which you should know how to find!

f_x	=D5*EXP(D5)							
	A	B	C	D	E	F	G	H	I
1			i	x_i	f (x_i)	c_i	c_i * f(x_i)		
2	a = 1		0	1.00000	2.718281828	1	2.718281828	SIMPSON'S RULE	
3	b = 3		1	1.16667	3.7464823	4	14.9859292	sum:	723.0877524
4	n = 12		2	1.33333	5.05822386	2	10.11644772	multiple:	0.05555555556
5			3	1.50000	6.722533606	4	26.89013442	final:	40.1715418
6	Delta-x = 0.166		4	1.66667	8.824150084	2	17.64830017		
7			5	1.83333	11.46695175	4	45.86780698		
8			6	2.00000	14.7781122	2	29.5562244		
9			7	2.16667	18.91313312	4	75.65253249		
10			8	2.33333	24.0619365	2	48.12387301		
11			9	2.50000	30.4562349	4	121.8249396		
12			10	2.66667	38.37844292	2	76.75688584		
13			11	2.83333	48.1724465	4	192.689786		
14			12	3.00000	60.25661077	1	60.25661077		

Fig. 8.22   A Simpson's rule implementation, with YTI 4.

☐ Figure 8.22 shows the spreadsheet implementation. The final Simpson's Rule estimate is $\int_1^3 xe^x \, dx \approx 40.172$. This is near the average of the left hand and right hand estimates from YTI 1. The exact value can be found with integration by parts, and it is $\int_1^3 xe^x \, dx = 2e^3 \approx 40.171$. The Simpson's Rule estimate is very close!   ∎

(5) Use Simpson's Rule to estimate $\int_a^b f(x) \, dx$ for the tabulated values shown in this table:

x	0	0.5	1	1.5	2	2.5	3.0
$g(x)$	1	0.89	0.71	0.55	0.45	0.37	0.16

3.5	4	4.5	5	5.5	6
0.27	0.24	0.22	0.20	0.18	0.16

☐ Figure 8.23 shows the spreadsheet implementation, which yields a result of $\int_0^6 f(x) \, dx \approx 2.57$. There are many different functions which

| f_x | =B6/3 |

	A	B	C	D	E	F	G	H	I
1			i	x_i	f(x_i)	c_i	c_i * f(x_i)		
2	a = 0		0	0	1.00	1	1.00	SIMPSON'S RULE	
3	b = 6		1	0.5	0.89	4	3.58	sum:	15.43
4	n = 12		2	1	0.71	2	1.41	multiple:	0.1666667
5			3	1.5	0.55	4	2.22	final:	2.57
6	Delta-x = 0.5		4	2	0.45	2	0.89		
7			5	2.5	0.37	4	1.49		
8			6	3	0.32	2	0.63		
9			7	3.5	0.27	4	1.10		
10			8	4	0.24	4	0.97		
11			9	4.5	0.22	4	0.87		
12			10	5	0.20	2	0.39		
13			11	5.5	0.18	4	0.72		
14			12	6	0.16	1	0.16		

Fig. 8.23 A Simpson's rule implementation for tabulated data, with YTI 5.

might go through these points; for example, we could design an 12-th order polynomial that goes through all of them. Perhaps the table was not created with a function at all, but rather is the result of some recording of data in an experiment. So, it's not possible to compare this result to a "true" value, to see how accurate it is. The estimate is what it is. ■

Chapter 9

Round and Round We Go: Solids of Revolution

9.1 Solids of Revolution: Surface Area

Introduction

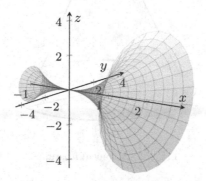

Fig. 9.1 A surface of revolution.

I hope you are ready for some excitement, because we are about to open up a third dimension to our universe. Imagine a piece of a graph of a function $f(x)$ along an interval $[a, b]$ being whirled around the x-axis like the blade of a blender. The resulting three-dimensional object is called a *surface of revolution*. Figure 9.1 shows the a surface of revolution generated by taking the piece of $f(x) = x^2$ along the interval $[-1, 2]$ and revolving it around the x-axis. Our job is to figure out the area of a surface of revolution. And just because we've left Chapter 8 behind, that doesn't mean we're done with partitioning. In case you need a reminder, we will ...

(1) ... break down (or partition) the problem into smaller pieces.

135

(2) ... try to solve, or at least approximate, the solution to the problem on one small piece.

(3) ... extrapolate the solution on the small pieces back up to an approximate solution to the whole problem.

(4) ... consider (and implement) ways to make the approximation better.

Conical Frustums

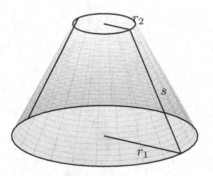

Fig. 9.2   A conical frustum.

I know that "frustum" is a silly word, but we do need to use it for a bit. Imagine a crazed samurai chopping the top and bottom off an orange traffic cone, horizontally. The resulting shape is a *conical frustum*. Figure 9.2 shows the dimensions of this shape. The top and bottom of a conical frustum are circles, with radii r_1 and r_2. The distance along the wall of the frustum between the perimeters of the bounding circles is the same all the way around, and this distance is called s. The surface area of a frustum is given by

$$A = \pi(r_1 + r_2)s \qquad (9.1)$$

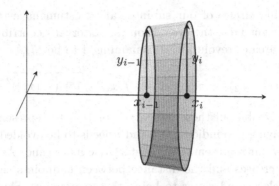

Fig. 9.3 One partition (frustum) of a surface of revolution.

Partitioning the Surface Area Problem

Into the Pit!!

Step 1: As usual, we subdivide $[a, b]$ into n partitions, with subintervals $[x_{i-1}, x_i]$. The small piece of the function connecting (x_{i-1}, y_{i-1}) and (x_i, y_i), can be approximated by a straight line segment; when it revolves around the x-axis, the resulting shape is then approximated by a conical frustum (see Fig. 9.3) with bounding circles that have radii $r_1 = y_{i-1} = f(x_{i-1})$ and $r_2 = y_i = f(x_i)$.

Step 2: The surface area of the resulting (approximate) conical frustum can be found by updating (9.1) with the data from our partition. The length s can be estimated by the arc length of this small piece of $f(x)$. Borrowing a result from Sec. 8.2, we can find in our partition (courtesy of the Mean Value Theorem) a representative point $x_i^\star$ at which we can measure

$$s = \sqrt{1 + \left(f'(x_i^\star)\right)^2} \cdot \Delta x$$

(Invoking the MVT now requires that $f(x)$ be continuous and differentiable on each subinterval, although we were likely not picturing anything else anyway.)

The bounding circles of our subinterval's frustum have radii $f(x_{i-1})$ and $f(x_i)$. So right now, the area of our subinterval's contribution to the overall surface area of revolution is (renaming A to be S_i),

$$S_i = \pi(r_1 + r_2)s = \pi(f(x_{i-1}) + f(x_i))\sqrt{1 + (f'(x_i^\star))^2} \cdot \Delta x$$

As we go to assemble all these into a summation to represent the entire surface area, having two indices $i-1$ and i needs to be avoided. So, we are going to borrow our representative point $x_i^\star$ one more time. As the number of partitions increases and the distance between endpoints shrinks, those endpoints get closer and closer to being the same point. Since this is all an approximation anyway, let's go ahead and assume the subintervals are small enough that $r_1 \approx r^2 \approx f(x_i^\star)$. With this update, we now have:

$$S_i = \pi(f(x_i^\star) + f(x_i^\star))\sqrt{1 + (f'(x_i^\star))^2} \cdot \Delta x = 2\pi f(x_i^\star)\sqrt{1 + (f'(x_i^\star))^2} \cdot \Delta x$$

Step 3: An estimate of the overall area of our surface of revolution is the sum of the individual surface areas from the n conical frustums generated by our partition,

$$S \approx \sum_{i=1}^{n} 2\pi f(x_i^\star)\left(\sqrt{1 + (f'(x_i^\star))^2} \cdot \Delta x\right)$$

Step 4: The approximation to the total surface area will improve as n gets larger. Ultimately, the best (exact) answer will be achieved as n goes to infinity. That is,

$$S = \lim_{n \to \infty} \sum_{i=1}^{n} 2\pi f(x_i^\star)\left(\sqrt{1 + (f'(x_i^\star))^2} \cdot \Delta x\right)$$

We then recognize this as the Riemann Sum representation of a definite integral (right??), and have this final result:

Useful Fact 9.1. *Let $f(x)$ be (piecewise) continuous and differentiable on the interval $[a, b]$. Then the surface formed by revolving $f(x)$ on $[a, b]$ around the x-axis has area*

$$\boxed{S = \int_a^b 2\pi f(x)\sqrt{1 + (f'(x))^2}\, dx} \qquad (9.2)$$

Examples

EX 1 Find the area of the surface formed when $y = \sin(x)$, $0 \le x \le \pi$, is revolved around the x-axis.

As with arc length problems, we can build towards the integrand of (9.2) rather than loading up the integral right away.

$$f(x) = \sin(x)$$
$$f'(x) = \cos(x)$$
$$\sqrt{1 + (f'(x))^2} = \sqrt{1 + \cos^2(x)}$$

Now we can start building the integral:

$$S = \int_a^b 2\pi f(x)\sqrt{1 + [f'(x)]^2}\,dx = \int_0^\pi 2\pi \sin(x)\sqrt{1 + \cos^2 x}\,dx$$

This can be solved with a substitution for $\cos x$:

$$\cos x = u \quad ; \quad \sin(x)\,dx = -du$$
$$x = 0 \to u = 1$$
$$x = \pi \to u = -1$$

Then,

$$S = -\int_1^{-1} 2\pi\sqrt{1 + u^2}\,du = 2\pi \int_{-1}^1 \sqrt{1 + u^2}\,du$$

Seriously? We started with a nice easy substitution and that took us to an integral needing trigonometric substitution? Good grief. Well, we've done one much like this before; it matches the form in Useful Fact 7.7, which requires

$$u = \tan\theta \quad ; \quad du = \sec^2\theta\,d\theta$$
$$u = -1 \to \theta = -\pi/4$$
$$u = 1 \to \theta = \pi/4$$

Now we have

$$S = 2\pi \int_{-\pi/4}^{\pi/4} \sqrt{1 + \tan^2\theta}\,\sec^2\theta\,d\theta = 2\pi \int_{-\pi/4}^{\pi/4} \sec^3\theta\,d\theta$$

This is a known integral, and appears as Eq. (7.18). That result leads to

$$S = 4\pi \left(\frac{1}{2} \left(\sec\theta \tan\theta + \ln|\sec\theta + \tan\theta| \right) \Big|_0^{\pi/4} \right)$$

$$= 2\pi \left(\sec\frac{\pi}{4} \tan\frac{\pi}{4} + \ln\left|\sec\frac{\pi}{4} + \tan\frac{\pi}{4}\right| \right) - 2\pi \left(\sec 0 \tan 0 + \ln|\sec 0 + \tan 0| \right)$$

$$= 2\pi \left((\sqrt{2})(1) + \ln|\sqrt{2} + 1| \right) - 2\pi \left(0 + \ln|1 + 0| \right)$$

$$= 2\pi \left((\sqrt{2})(1) + \ln|\sqrt{2} + 1| \right) - 2\pi \left(0 + \ln|1 + 0| \right)$$

$$= 2\pi \left(\sqrt{2} + \ln(\sqrt{2} + 1) \right) \quad \blacksquare$$

You might imagine from Eq. (9.2) that many integrals we build will require some technical help for evaluation.

EX 2 Find (estimate) the area of the surface formed when $y = x^4$, $0 \leq x \leq 1$, is revolved around the x-axis.

Let's get ready for the surface area integral:

$$f(x) = x^4$$
$$f'(x) = 4x^3$$
$$\sqrt{1 + (f'(x))^2} = \sqrt{1 + 16x^6}$$

Now we can start building the integral in (9.2):

$$S = \int_a^b 2\pi f(x) \sqrt{1 + [f'(x)]^2}\, dx = \int_0^1 2\pi x^4 \sqrt{1 + 16x^6}\, dx$$

We can't solve this integral by hand. So, we have to ask for some help from, say, Wolfram Alpha or Maple. The estimated final result is

$$S \int_0^1 2\pi x^4 \sqrt{1 + 16x^6}\, dx \approx 3.344 \quad \blacksquare$$

You Try It

(1) Find the surface area of the solid formed when $y = \sqrt{x}$, $4 \leq x \leq 9$, is revolved around the x-axis.

(2) Find the surface area of the solid formed when $y = \cosh x$, $0 \leq x \leq 1$, is revolved around the x-axis.

If a function is revolved around the y-axis instead of the x-axis, nothing changes except the alphabet for the function and integral. We can write an equivalent to (9.2) as

$$S = \int_c^d 2\pi g(y) \sqrt{1 + [g'(y)]^2}\, dy$$

We need the function written as a function of y, and we need limits of integration from the y-axis.

$\boxed{\text{EX 3}}$ Find the surface area of the solid formed when $x = \sqrt{a^2 - y^2}$, $0 \leq y \leq a/2$, is revolved around the y-axis.

Let's get ready for the integral:

$$g'(y) = -\frac{2y}{2\sqrt{a^2 - y^2}} = -\frac{y}{\sqrt{a^2 - y^2}}$$

$$1 + (g'(y))^2 = 1 + \frac{y^2}{a^2 - y^2} = \frac{a^2}{a^2 - y^2}$$

$$\sqrt{1 + (g'(y))^2} = \frac{a}{\sqrt{a^2 - y^2}}$$

So the surface area is

$$S = \int_c^d 2\pi g(y) \sqrt{1 + [g'(y)]^2}\, dy = \int_0^{a/2} 2\pi \sqrt{a^2 - y^2} \cdot \frac{a}{\sqrt{a^2 - y^2}}\, dy$$

$$= \int_0^{a/2} 2\pi \sqrt{a^2 - y^2 + y^2}\, dy = \int_0^{a/2} 2\pi a\, dy = \pi a^2 \quad \blacksquare$$

You Try It

(3) Find the surface area of the solid formed when $y = \sqrt[3]{x}$, $1 \leq y \leq 2$, is revolved around the y-axis.

Solids of Revolution: Surface Area — Problem List

Solids of Revolution: Surface Area — You Try It

These appeared above; solutions begin on the next page.

(1) Find the surface area of the solid formed when $y = \sqrt{x}$, $4 \le x \le 9$, is revolved around the x-axis.
(2) Find (estimate) the surface area of the solid formed when $y = \cosh x$, $0 \le x \le 1$, is revolved around the x-axis.
(3) Find the surface area of the solid formed when $y = \sqrt[3]{x}$, $1 \le y \le 2$, is revolved around the y-axis.

Solids of Revolution: Surface Area — Practice Problems

Try these as you get the hang of the You Try It problems. Solutions to these problems are available in Sec. A.3.1.

(1) Find the surface area of the solid formed when $y = \cos(2x)$, $0 \le x \le \pi/6$, is revolved around the x-axis.
(2) Find (estimate) the surface area of the solid formed when $y = x^3/6 + 1/2x$, for $1/2 \le x \le 1$, is revolved around the x-axis.
(3) Find the surface area of the solid formed when $y = 1 - x^2$, $0 \le x \le 1$, is revolved around the y-axis.
(4) Find the surface area of the solid formed when $y = \cosh(x)$, $0 \le x \le \ln 3$, is revolved around the x-axis.

Solids of Revolution: Surface Area — Challenge Problems

Try these problems to test your skills with the ideas in this section. Solutions to these problems are available in Sec. B.3.1.

(1) Find the area of the surface formed when $9x = y^2 + 18$, $2 \le x \le 6$, is revolved around the x-axis.
(2) Your skeptical friend from the last section is at it again. Now he's claiming that the surface area of a sphere isn't really $4\pi r^2$, and that quantity is just conspiracy spread by the Illuminati. Prove him wrong. (Hint: Consider how the sphere can be formed as a surface of revolution.)
(3) Show that the area of the surface formed when $y = 1/x$ is revolved around the x-axis, for $x \ge 1$, is infinite. (Hint: Is $1 + 1/x^4 > 1$ for the interval in question?).

Solids of Revolution: Surface Area — You Try It — Solved

(1) Find the surface area of the solid formed when $y = \sqrt{x}$, $4 \leq x \leq 9$, is revolved around the x-axis.

☐ Getting ready for (9.2):

$$f'(x) = \frac{1}{2\sqrt{x}}$$

$$1 + (f'(x))^2 = 1 + \frac{1}{4x} = \frac{4x+1}{4x}$$

So

$$S = \int_a^b 2\pi f(x) \sqrt{1 + [f'(x)]^2} \, dx$$

$$= \int_4^9 2\pi \sqrt{x} \cdot \sqrt{\frac{4x+1}{4x}} \, dx = \pi \int_4^9 \sqrt{4x+1} \, dx$$

A quick substitution,

$$4x + 1 = u \quad ; \quad dx = \frac{1}{4} du$$

$$x = 4 \rightarrow u = 17$$

$$x = 9 \rightarrow u = 37$$

and then,

$$S = \frac{\pi}{4} \int_{17}^{37} \sqrt{u} \, du = \frac{\pi}{4} \cdot \frac{2}{3} u^{3/2} \Big|_{17}^{37}$$

$$= \frac{\pi}{6} (37^{3/2} - 17^{3/2}) = \frac{\pi}{6} \left(37\sqrt{37} - 17\sqrt{17} \right) \quad \blacksquare$$

(2) Find (estimate) the surface area of the solid formed when $y = \cosh x$, $0 \leq x \leq 1$, is revolved around the x-axis.

☐ Getting ready for (9.2):

$$f'(x) = \sinh x$$

$$1 + (f'(x))^2 = 1 + \sinh^2 x = \cosh^2 x$$

$$\sqrt{1 + (f'(x))^2} = \cosh x$$

So

$$S = \int_a^b 2\pi f(x) \sqrt{1 + [f'(x)]^2} \, dx = \int_0^1 2\pi \cosh x \cdot \cosh x \, dx$$

$$= \int_0^1 2\pi \cosh^2 x \, dx$$

We could evaluate this by returning $\cosh x$ to its exponential form and multiplying out the integrand. Or, we can rely on tech to get

$$S = \frac{\pi}{4}(e^2 - e^{-2}) + \pi \approx 8.839 \quad \blacksquare$$

(3) Find the surface area of the solid formed when $y = \sqrt[3]{x}$, $1 \le y \le 2$, is revolved around the y-axis.

☐ Since we're revolving around the y-axis, let's rewrite the function as $x = y^3$. Then,

$$g'(y) = 3y^2$$
$$1 + (g'(y))^2 = 1 + 9y^4$$

So that

$$S = \int_a^b 2\pi g(y)\sqrt{1 + [g'(y)]^2}\, dy = \int_1^2 2\pi y^3 \sqrt{1 + 9y^4}\, dy$$

This needs a substitution,

$$1 + 9y^4 = u \quad ; \quad y^3\, dy = \frac{1}{36}\, du$$
$$x = 1 \to u = 10$$
$$x = 2 \to u = 145$$

Then,

$$S = 2\pi \int_{10}^{145} \frac{1}{36}\sqrt{u}\, du = \frac{\pi}{18} \cdot \frac{2}{3} u^{3/2} \Big|_{10}^{145}$$
$$= \frac{\pi}{27}\left(145\sqrt{145} - 10\sqrt{10}\right) \quad \blacksquare$$

9.2 Volumes I: Disc-count Mathematics

Introduction

Now we're going to fill in our surface of revolution to make a solid of revolution. Pick a stretch of a function $f(x)$ on an interval $[a, b]$ and whirl the function *and the area below it* around the x-axis. The resulting object is a filled solid, not just a surface, and is called a *solid of revolution*. Our job is to figure out the volume of that solid of revolution.

Partitioning for the New Volume Problem

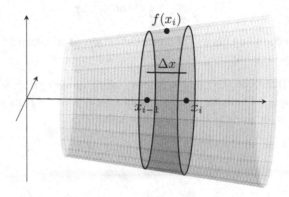

Fig. 9.4   Partitioning of a solid of revolution.

Into the Pit!!

Step 1: To find the volume of the solid formed by $f(x)$ on the interval $[a, b]$, we subdivide $[a, b]$ into n subintervals with endpoints $[x_{i-1}, x_i]$ for $i = 1 \ldots n$. When we revolve one resulting tiny piece of $f(x)$ *and the area below it* around the x-axis, we generate (approximately) a small disc, or cylinder. One such partition and resulting disc is shown in Fig. 9.4.

Step 2: The volume of a generic cylinder is $V = \pi r^2 h$. In our set-up, in which the cylinder is on its side, h is the width of the partition, so $h = x_i - x_{i-1} = \Delta x$. The radius r is the function's value $f(x_i^*)$ chosen at some representative point x_i^* within the interval. Therefore the volume of

the ith slice of this solid of revolution, associated with our ith partition, is
$V_i = \pi[f(x_i^\star)]^2 \Delta x$.

Step 3: An estimate of the overall volume of our solid is the sum of the individual volumes from the n discs / cylinders generated by our partition,

$$V \approx \sum_{i=1}^{n} \pi[f(x_i^\star)]^2 \Delta x$$

Step 4: The approximation to the total volume will improve as n gets larger. Ultimately, the best (exact) answer will be achieved as n goes to infinity. That is,

$$V = \lim_{n \to \infty} \sum_{i=1}^{n} \pi[f(x_i^\star)]^2 \Delta x$$

We then recognize this as the limit definition of a definite integral, and have the following result:

Useful Fact 9.2. *Let $f(x)$ be (piecewise) continuous and differentiable on the interval $[a, b]$. Then the volume formed by revolving $f(x)$ and the area beneath it on $[a, b]$ around the x-axis has volume*

$$\boxed{V = \int_a^b \pi[f(x)]^2 \, dx} \qquad (9.3)$$

EX 1 Find the volume of the solid formed when the region between $y = \sqrt{x}$, $x = 1$ and $x = 4$ is revolved around the x-axis.

$$V = \int_a^b \pi[f(x)]^2 dx = \int_1^4 \pi(\sqrt{x})^2 \, dx = \pi \int_1^4 x \, dx = \frac{15\pi}{2} \quad \blacksquare$$

You Try It

(1) Find the volume of the solid generated by revolving the region between $y = 1/x$, $x = 1$, and $x = 2$ around the x-axis.

What if we wanted to revolve a function of y around the y-axis? Nothing would really change, we'd use

$$\boxed{V = \int_c^d \pi[g(y)]^2 \, dy} \qquad (9.4)$$

where c and d delineate the portion of the y-axis in question.

$\boxed{\textbf{EX 2}}$ Find the volume of the solid formed when the region between $x = y^2$, $y = 0$ and $y = 2$ is revolved around the y-axis.

By (9.4), we have

$$V = \int_c^d \pi [g(y)]^2 \, dy = \int_0^2 \pi (y^2)^2 \, dy = \pi \int_0^2 y^4 \, dy = \frac{32\pi}{5} \quad \blacksquare$$

Just like we extended finding the area under one curve to finding the area between two curves, we can extend the volume problem as well. Suppose we revolve a region between two functions $f(x)$ and $g(x)$ on an interval $[a, b]$ around the x-axis? The resulting volume is just the difference between the volume of the outer solid and the volume of the inner solid,

$$V = V_{outer} - V_{inner} = \int_a^b \pi [f(x)]^2 \, dx - \int_a^b \pi [g(x)]^2 \, dx$$

or

$$\boxed{V = \int_a^b \pi \left([f(x)]^2 - [g(x)]^2 \right) \, dx} \qquad (9.5)$$

$\boxed{\textbf{EX 3}}$ Find the volume of the solid formed when the region between $y = e^x$, $y = e^{-x}$, $x = 0$ and $x = 1$ is revolved around the x-axis.

On the interval $[0, 1]$, $y = e^x$ is above $y = e^{-x}$, so we assign $f(x) = e^x$ and $g(x) = e^{-x}$, and then (9.5) gives:

$$V = \int_a^b \pi \left([f(x)]^2 - [g(x)]^2 \right) \, dx = \int_0^1 \pi \left([e^x]^2 - [e^{-x}]^2 \right) \, dx$$

$$= \pi \int_0^1 \left(e^{2x} - e^{-2x} \right) \, dx = \frac{\pi}{2} \left(e^{2x} + e^{-2x} \right) \Big|_0^1 = \frac{\pi}{2} \left(e^2 + e^{-2} - 2 \right) \quad \blacksquare$$

You Try It

(2) Find the volume of the solid generated by revolving the region between $y = x^2$ and $y^2 = x$ around the x-axis.

Ready for another wrinkle? What if we wanted to revolve a function around a horizontal line other than the x-axis? For example, what if we wanted to revolve $y = \sqrt{x}$ on $[1, 3]$ around the line $y = 2$? There are two ways to handle this. One is to try to generate a new formula to compute that volume. I guess some people might think that's fun. The other way, which we'll pursue, is to come up with an equivalent, but simpler, volume to compute; that is, figure out what function would have to get get revolved around the x-axis to produce exactly the same shape and volume as the stated problem. This is a matter of translating a function down (or up) by the correct distance: revolving $f(x)$ around $x = c$ (for $c > 0$) is equivalent to revolving the function $f(x) - c$ around the x-axis. Figures 9.5 and 9.6 show a sample region between some $f(x)$ and $y = c$ about to be revolved around $y = c$, and the translated region between $y = f(x) - c$ and $y = 0$, now ready for revolving around the x-axis. The resulting volumes will be the same, and we know how to compute the latter without needing to generate a new formula.

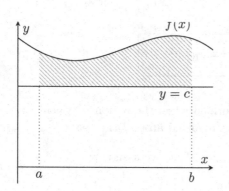

Fig. 9.5   The area between $y = f(x)$ and $y = c$, ready to be revolved around $y = c$.

Fig. 9.6   The equivalent area under $y = f(x) - c$ ready to be revolved around $y = 0$.

EX 4 Find the volume of the solid formed when the region between $y = \sqrt{x}$, $x = 1$ and $x = 3$ is revolved around the line $y = 2$.

We can find an equivalent volume by shifting everything down by 2 units. This will move $y = 2$ to the x-axis. When we move $y = \sqrt{x}$ down by 2 units, we get $y = \sqrt{x} - 2$. So the volume we want is the same as the volume obtained when the region between $y = \sqrt{x} - 2$, $x = 1$, and $x = 3$ is revolved around the x-axis. Now we don't need a new formula.

$$V = \int_a^b \pi[f(x)]^2 dx = \int_1^3 \pi(\sqrt{x} - 2)^2\, dx$$

$$= \pi \int_1^3 (x - 4\sqrt{x} + 4)\, dx = \pi \left(\frac{1}{2}x^2 - \frac{8}{3}x^{3/2} + 4x \right) \Big|_1^3$$

$$= \pi \left(\frac{9}{2} - \frac{8}{3}(3\sqrt{3}) + 12 - \left(\frac{1}{2} - \frac{8}{3} + 4 \right) \right) = \pi \left(\frac{28}{3} - 8\sqrt{3} \right) \quad \blacksquare$$

You Try It

(3) Find the volume of the solid generated by revolving the region between $y = x$ and $y = \sqrt{x}$ around the line $y = 1$.

(4) Find the volume of the solid generated by revolving the region between $y = 0$ and $y = \sin(x)$ from $x = 0$ to $x = \pi$ around the line $y = 1$.

The technique(s) discussed in this section are often known as the method of discs or the method of washers. When you generate a solid of revolution formed by revolving some $f(x)$ around the x-axis, or some $g(y)$ revolved around the y-axis, the cross sections of these regions are discs. When you revolve the region in between two functions, say, $f(x)$ and $g(x)$, around the x-axis, then the cross sections look like washers. (One cross section looks like you've cut laterally through a pineapple before coring it, and the latter is like cutting laterally through a pineapple *after* you've cored it.) Be sure to Note the match between the variable of the function and the axis of revolution, such as $f(x)$ revolved around the x-axis. In the next section, we'll create solids of revolution where some $f(x)$ on an interval $[a, b]$ is revolved around the y axis (note the mismatch between the variable of the function and the axis of revolution). This leads to a fundamentally different shape — a cake rather than a pineapple. (And if you think using food to visualize three-dimensional shapes is bad here, just wait until multivariable Calculus!)

Finally, a note about evaluation of integrals. At this point, if you are able to set up the correct integral that's needed to solve a given problem, then the solving of that integral is secondary. In the solutions, I will be presenting the integral and the final result, unless something special is to be learned in the evaluation process. If you want to use tech to evaluate integrals once they've been set up, go for it. Just be sure you have a critical eye on your results.

Solids of Revolution: Volume I — Problem List

Solids of Revolution: Volume I — You Try It

These appeared above; solutions begin on the next page.

(1) Find the volume of the solid generated by revolving the region between $y = 1/x$, $x = 1$, and $x = 2$ around the x-axis.

(2) Find the volume of the solid generated by revolving the region between $y = x^2$ and $y^2 = x$ around the x-axis.

(3) Find the volume of the solid generated by revolving the region between $y = x$ and $y = \sqrt{x}$ around the line $y = 1$.

(4) Find the volume of the solid generated by revolving the region between $y = 0$ and $y = \sin(x)$ from $x = 0$ to $x = \pi$ around the line $y = 1$.

Solids of Revolution: Volume I — Practice Problems

Try these as you get the hang of the You Try It problems. Solutions to these problems are available in Sec. A.3.2.

(1) Find the volume of the solid generated by revolving the region between $y = \sqrt{x - 1}$, $x = 2$, and $x = 5$ around the x-axis.

(2) Find the volume of the solid generated by revolving the region between $y = x^{2/3}$, $x = 1$ and $y = 0$ around the y-axis.

(3) Find the volume of the solid generated by revolving the region between $y = x^2$ and $y = 4$ around the line $y = 4$.

(4) Find the volume of the solid generated by revolving the region between $y = 0$ and $y = \sin(x)$ from $x = 0$ to $x = \pi$ around the line $y = -2$.

(5) Find the volume of the solid generated by revolving the region between $y = \sinh(x)$, $y = \cosh(x)$, $x = a$ and $x = b$, around the x-axis, for any a and b where $b > a$.

Solids of Revolution: Volume I — Challenge Problems

Try these problems to test your skills with the ideas in this section. Solutions to these problems are available in Sec. B.3.2.

(1) Find the volume of the solid generated by revolving the region between $y = \sec x$, $y = 1$, $x = -1$, and $x = 1$ around the x-axis.

(2) Find the volume of the solid generated by revolving the region between $y = 1/x$, $y = 0$, $x = 1$ and $x = 3$ around the line $y = -1$.

(3) Approximate the volume of the solid generated by revolving the region between $y = x^2$ and $y = \ln(x + 1)$ around the x-axis.

Volumes I: Disc-count Mathematics — You Try It — Solved

---

(1) Find the volume of the solid generated by revolving the region between $y = 1/x$, $x = 1$, and $x = 2$ around the x-axis.

☐ By (9.3), we have:

$$V = \int_a^b \pi [f(x)]^2 \, dx = \int_1^2 \pi \left(\frac{1}{x} \right)^2 \, dx = \frac{\pi}{2} \quad \blacksquare$$

---

(2) Find the volume of the solid generated by revolving the region between $y = x^2$ and $y^2 = x$ around the x-axis.

☐ These curves intersect at $x = 0$ and $x = 1$. The curve $y^2 = x$ is also known as $y = \sqrt{x}$ on that interval, and $y = \sqrt{x}$ is above $y = x^2$. Assigning $f(x) = \sqrt{x}$ and $g(x) = x^2$, then (9.5) gives:

$$V = \int_a^b \pi \left([f(x)]^2 - [g(x)]^2 \right) \, dx = \int_0^1 \pi \left([\sqrt{x}]^2 - [x^2]^2 \right) \, dx$$

$$= \int_0^1 \pi \left(x - x^4 \right) \, dx = \frac{3\pi}{10} \quad \blacksquare$$

---

(3) Find the volume of the solid generated by revolving the region between $y = x$ and $y = \sqrt{x}$ around the line $y = 1$.

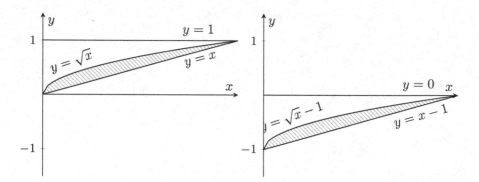

Fig. 9.7   The area between $y = x$ and $y = \sqrt{x}$, ready to be revolved around $y = 1$.

Fig. 9.8   The equivalent area between $y = x - 1$ and $y = \sqrt{x} - 1$, ready to be revolved around $y = 0$.

☐ These curves intersect at $x = 0$ and $x = 1$. When they are both revolved around the line $y = 1$, the line $y = x$ forms the outer solid and $y = \sqrt{x}$ forms the inner one. We would like to "move" the problem so that we find the equivalent volume of revolution around the x-axis, which is $y = 0$. Figures 9.7 and 9.8 show the two given curves ready to be revolved around $y = 1$, and the equivalent region between $y = \sqrt{x}-1$ and $y = x - 1$ ready for revolving around the x-axis. Following the latter, we get

$$V = \int_a^b \pi \left([f(x)]^2 - [g(x)]^2\right) dx$$

$$= \int_0^1 \pi \left([x - 1]^2 - [\sqrt{x} - 1]^2\right) dx = \frac{\pi}{6} \quad \blacksquare$$

---

(4) Find the volume of the solid generated by revolving the region between $y = 0$ and $y = \sin(x)$ from $x = 0$ to $x = \pi$ around the line $y = 1$.

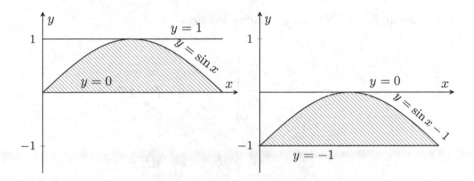

Fig. 9.9   The area between $y = \sin x$ and $y = 0$, ready to be revolved around $y = 1$.

Fig. 9.10   The equivalent area between $y = \sin x - 1$ and $y = -1$, ready to be revolved around $y = 0$.

☐ If we translate both functions and the axis of revolution down by 1 unit, we find the equivalent problem: revolve the region between $y = -1$ and $y = \sin x = 1$, for $0 \le x \le \pi$, around the x-axis. Figures 9.9 and 9.10 show the two given curves ready to be revolved around $y = 1$, and the equivalent region between $y = \sin x - 1$ and $y = -1$ ready for revolving around the x-axis. For that, we get:

$$V = \int_u^b \pi \left([f(x)]^2 - [g(x)]^2\right) dx = \int_0^\pi \pi \left([-1]^2 - [\sin x - 1]^2\right) dx$$

$$= \int_0^\pi \pi \left(1 - [\sin x - 1]^2\right) dx = \pi \left(4 + \frac{\pi}{2}\right) \quad \blacksquare$$

9.3   Volumes II: The Shell Game

Introduction

In Sec. 9.2, we created solids of revolution by revolving a piece of some $f(x)$ around the x-axis, or some piece of $g(y)$ around the y-axis. The variable of the function matches the axis of revolution, and the cross sections of the solid were either discs or washers. This time, we are going to criss cross the variable of the function and the axis of integration — such as taking a piece of some $f(x)$ on an interval $a \leq x \leq b$ and revolving it around the y-axis. Or, by taking a piece of some $g(y)$ and revolving it around the x-axis. Our job is to find a way to compute the volume of this second kind of solid of revolution.

Partitioning for the New New Volume Problem

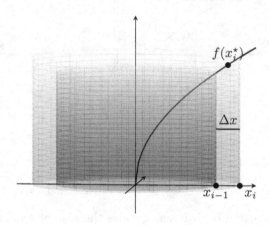

Fig. 9.11   One partition (standing cylinder) of a solid of revolution via the method of cylindrical shells.

Into the Pit!!

Step 1: To find the volume of the solid formed by revolving a piece of some $f(x)$, between $x = a$ and $x = b$, around the y-axis, we ... sigh ... subdivide $[a, b]$ into n partitions with endpoints $[x_{i-1}, x_i]$, for $i - 1 \ldots n$. and consider the contribution of each partition to the total volume.

Step 2: If you have done any baking, you have used measuring cups. Do you know how measuring cups stack inside each other, from the smallest out to the largest? You can imagine one of these new solids of revolution being a full stack of nested measuring cups; when the ith partitioned segment of $f(x)$, between $(x_{i-1}, f(x_{i-1}))$ and $(x_i, f(x_i))$, together with the area under it, gets revolved around the y-axis, the resulting solid is like one of the measuring cups nested in between the ones generated by the adjacent $(i-1)$st and $(i+1)$st subintervals in the partition. Figure 9.11 shows the scheme. When the ith partition (between x_{i-1} and x_i is revolved around the y-axis, the resulting shape is a standing (hollow) cylinder. All the nested standing cylinders generated by all n pieces of the partition together form the whole solid volume.

For a generic hollow cylinder with inner radius r_1, outer radius r_2, and height h, we'd find that the total volume is the volume of the outer cylinder minus the volume of inner cylinder. Or, with the regular formula for volume of a cylinder,

$$V = \pi r_2^2 h - \pi r_1^2 h = \pi h(r_2^2 - r_1^2) = \pi h(r_2 + r_1)(r_2 - r_1)$$

Now we can tie this to our partition. As the ith partition is revolved around the y-axis and forms a standing cylinder, the inner radius of the cylinder would be $r_1 = x_{i-1}$, and the outer radius would be $r_2 = x_i$. Then $r_2 - r_1$ is just Δx. Further, if we write

$$r_2 + r_1 = x_i + x_{i-1} = 2\left(\frac{x_i + x_{i-1}}{2}\right)$$

we can recognize the latter term as the midpoint of of the interval — which we can then name as our representative point $x_i^\star$. Thus, $r_2 + r_1 = 2x_i^\star$. And, using the representative point $x_i^\star$, we can assign the height h of the revolved cylinder to be $h = f(x_i^\star)$. Altogether, the volume generated by the revolution of the ith partition is

$$V_i = \pi h(r_2 + r_1)(r_2 - r_1) = \pi \cdot f(x_i^\star) \cdot (2x_i^\star) \cdot \Delta x = 2\pi x_i^\star f(x_i^\star)\Delta x$$

Step 3: An estimate of the overall volume of our solid is the sum of the individual volumes contributed by our n subintervals,

$$V \approx \sum_{i=1}^{n} 2\pi x_i^\star f(x_i^\star)\Delta x$$

Step 4: The approximation to the total volume will improve as n gets larger. Ultimately, the best (exact) answer will be achieved as n goes to infinity. That is,

$$V = \lim_{n \to \infty} \sum_{i=1}^{n} 2\pi x_i^* f(x_i^*) \Delta x$$

We then recognize this as the limit definition of a definite integral and have the following result:

Useful Fact 9.3. *Let $f(x)$ be (piecewise) continuous and differentiable on the interval $[a, b]$. Then the volume formed by revolving $f(x)$ and the area beneath it on $[a, b]$ around the y-axis has volume*

$$\boxed{V = \int_a^b 2\pi x f(x)\, dx} \tag{9.6}$$

This method of computing the volume of a solid of revolution is called the method of "cylindrical shells".

EX 1 Find the volume of the solid formed when the region between $y = \sqrt{x}$, $x = 1$ and $x = 4$ is revolved around the y-axis.

By (9.6), we have

$$V = \int_a^b 2\pi x f(x)\, dx = \int_1^4 2\pi x(\sqrt{x})\, dx = 2\pi \int_1^4 x^{3/2}\, dx = \frac{124\pi}{5} \quad \blacksquare$$

You Try It

(1) Find the volume of the solid generated by revolving the region between $y = x^3$, $x = 1$, and $x = 2$ around the y-axis.

What if we wanted to revolve a function of y around the x-axis? Nothing would really change, we just have to change the referencing in the integral:

$$\boxed{V = \int_c^d 2\pi y g(y)\, dy} \tag{9.7}$$

where c and d delineate the portion of the y-axis in question.

EX 2 Find the volume of the solid formed when the region between $x = y^2$, $y = 0$ and $y = 2$ is revolved around the x-axis.

By (9.7), we have

$$V \int_c^d 2\pi y g(y)\, dy = \int_0^2 2\pi y(y^2)\, dy = 8\pi \quad \blacksquare$$

You Try It

(2) Find the volume of the solid generated by revolving the region between $x = e^y$, $y = 0$, and $y = 1$ around the x-axis.

What if there are two functions involved, say $f(x)$ and $g(x)$, and we want to revolve a region between them around the y-axis? As with the method of discs, the resulting volume is just the difference between the larger and smaller contributions:

$$V = \text{(volume of larger / outer solid)} - \text{(volume of smaller / inner solid)}$$

$$= \int_a^b 2\pi x f(x) dx - \int_a^b 2\pi x g(x)\, dx = \int_a^b 2\pi x\left(f(x) - g(x)\right)\, dx$$

Let's pose both versions together for efficiency:

$$V = \int_a^b 2\pi x\left(f(x) - g(x)\right)\, dx \quad ; \quad V = \int_c^d 2\pi y\left(g(y) - h(y)\right)\, dy \qquad (9.8)$$

EX 3 Find the volume of the solid formed when the region between $y = e^{x^2}$ and $y = x$ on $x \in [0, 2]$ is revolved around the y-axis.

On the interval $[0, 2]$, $y = e^{x^2}$ generates the larger volume, so we assign $f(x) = e^{x^2}$ and $g(x) = x$. By (9.8),

$$V = \int_a^b 2\pi x\left(f(x) - g(x)\right)\, dx = \int_0^2 2\pi x\left(e^{x^2} - x\right) dx$$

$$= 2\pi \int_0^2 \left(xe^{x^2} - x^2\right) dx = 2\pi \left(\frac{1}{2}e^{x^2} - \frac{1}{3}x^3\right)\Bigg|_0^2$$

$$= \pi \left(e^{x^2} - \frac{2}{3}x^3\right)\Bigg|_0^2 = \pi \left(e^4 - \frac{16}{3}\right) \quad \blacksquare$$

You Try It

(3) Find the volume of the solid generated by revolving the region between $x = 4/y$ and $x = 5 - y$ around the x-axis.

Solids of Revolution: Volume II — Problem List

Solids of Revolution: Volume II — You Try It

These appeared above; solutions begin on the next page.

(1) Find the volume of the solid generated by revolving the region between $y = x^3$, $x = 1$, and $x = 2$ around the y-axis.
(2) Find the volume of the solid generated by revolving the region between $x = e^y$, $y = 0$, and $y = 1$ around the x-axis.
(3) Find the volume of the solid generated by revolving the region between $x = 4/y$ and $x = 5 - y$ around the x-axis.

Solids of Revolution: Volume II — Practice Problems

Try these as you get the hang of the You Try It problems. Solutions to these problems are available in Sec. A.3.3.

(1) Find the volume of the solid generated by revolving the region between $y = 1/(x^2 + 1)$, $x = 1$, and $x = 2$ around the y-axis.
(2) Find the volume of the solid generated by revolving the region between $x = \sqrt{1 - y^2}$, $y = 0$, and $y = 1$ around the x-axis.
(3) Find the volume of the solid generated by revolving the region between $y = x$ and $y = x^2$ around the y-axis.

Solids of Revolution: Volume II — Challenge Problems

Try these problems to test your skills with the ideas in this section. Solutions to these problems are available in Sec. B.3.3.

(1) Find the volume of the solid generated by revolving the region between $y = \ln x$, $x = 1$, and $x = 2$ around the y-axis.
(2) Find the volume of the solid generated by revolving the region between $x = \sin(y)$, $y = 0$, and $y = \pi$ around the x-axis.
(3) Remember your skeptical friend from the Challenge Problems of the previous two sections? He has a sister! She claims to not believe that the volume of a sphere of radius r is $V = 4\pi r^2$. Use the method of cylindrical shells to set her straight.

Volumes II: The Shell Game — You Try It — Solved

---

(1) Find the volume of the solid generated by revolving the region between $y = x^3$, $x = 1$, and $x = 2$ around the y-axis.

□ Using (9.6),

$$V = \int_a^b 2\pi x f(x)\, dx = \int_1^2 2\pi x (x^3)\, dx = \frac{62\pi}{5} \quad \blacksquare$$

---

(2) Find the volume of the solid generated by revolving the region between $x = e^y$, $y = 0$, and $y = 1$ around the x-axis.

□ Using (9.7),

$$V = \int_c^d 2\pi y g(y)\, dy = \int_0^1 2\pi y (e^y)\, dy = 2\pi \quad \blacksquare$$

---

(3) Find the volume of the solid generated by revolving the region between $x = 4/y$ and $x = 5 - y$ around the x-axis.

□ These two curves intersect at the points $(x, y) = (1, 4)$ and $(x, y) = (4, 1)$. So on the y-axis, we're using $y \in [1, 4]$. Looking out from the x-axis, we see the line $x = 5 - y$ generates the larger volume, so in preparation for Eq. (9.8), we assign $g(y) = 5 - y$ and $h(y) = 4/y$:

$$V = \int_c^d 2\pi y \left(g(y) - h(y) \right) dy = \int_1^4 2\pi y \left((5 - y) - \frac{4}{y} \right) dy = 9\pi \quad \blacksquare$$

Chapter 10

Close Only Counts in Horseshoes and Math

Introduction to Approximation

Introduction

Sometimes it is convenient or necessary to trade a function for a less complicated surrogate. This is especially true in computational settings, where it is beneficial to replace a function with a less computationally intensive version that *almost* reproduces the same values. The game of finding these surrogates (or approximations) is summed up in three parts: (1) How accurate should the approximation be? (2) Where (over what interval) is the approximation valid? (3) How can we measure error in the approximation?

You have already encountered one of the simplest types of approximation in Sec. 5.4 — local linear approximation. Here is a quick recap of that concept, so that we can borrow the notation and extend the ideas.

Linear Approximation

Let's plot a function and its tangent line at a point, say $f(x) = x^2$ at the point $x = 1$. To find the line tangent to x^2 at $x = 1$, we need (a) the point where that line attaches to $f(x)$, and (b) the slope of that tangent line. Getting the point of tangency is easy: when $x = 1$, then $y = (1)^2 = 1$, so we attach the tangent line at the point $(1, 1)$. The slope of the line tangent to $f(x)$ at $x = 1$ comes from the derivative there: $f'(1) = 2(1) = 2$. Now that we have a point on the line and the slope of the line, we use the point slope formula:

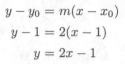

$$y - y_0 = m(x - x_0)$$
$$y - 1 = 2(x - 1)$$
$$y = 2x - 1$$

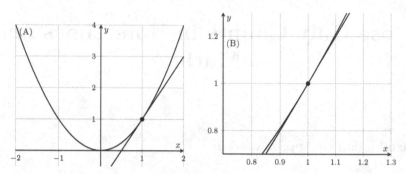

Fig. 10.1   A local linear approximation to $y = x^2$ at $(1, 1)$ on two scales.

Figure 10.1 shows $f(x) = x^2$ and its tangent line at $x = 1$, on (A) a "regular" scale, and (B) zoomed in. There is a certain interval around the point of tangency where the tangent line is fairly indistinguishable from the function itself. In that region, we have *local linear approximation*. We say "local" because the tangent line is indistinguishable from the function itself only in a small region around (local to) the point of tangency. This is a "linear" approximation because the tangent line is, well, linear!

If we are doing calculations "near" $x = 1$, we could use the tangent line (local linear approximation) as a surrogate for the function itself — as long as we recognize our results are approximate. Now granted, $y = x^2$ is not that horrible of a function to deal with, but when dealing with complicated functions, it is often preferable to work with a simpler approximation.

We need to generalize the notation here so that we can extend it. If we trade the generic notation in the point-slope formula $y - y_0 = m(x - x_0)$ to (much better) functional notation, it becomes

$$L_1(x) - f(x_0) = f'(x_0)(x - x_0)$$

The $L_1(x)$ in this expression represents that our approximation is a first order (linear) approximation. Rearranging,

$$L_1(x) = f(x_0) + f'(x_0)(x - x_0) \tag{10.1}$$

This is the expression we use to define the local linear approximation to $f(x)$ at x_0.

If you have taken physics class, you likely have fond memories of computing percent error, and will surely enjoy this next example. Remember that if you have a value v_{est} that estimates an actual value v_{act}, then the percent error E_{pct} in the estimated value is given by:

$$E_{pct} = \left| \frac{v_{est} - v_{act}}{v_{act}} \right| \cdot 100 \qquad (10.2)$$

EX 1 (a) Find the local linear approximation of $f(x) = \sqrt{x}$ at $x = 4$. (b) Use the local linear approximation to estimate the values of $\sqrt{4.1}$ and $\sqrt{4.3}$. (c) Determine the percent error between our approximations and the actual values, and compare them.

(a) Using (10.1), the local linear approximation of $f(x) = \sqrt{x}$ at $x = 4$ is found by:

$$L_1(x) = f(x_0) + f'(x_0)(x - x_0) = \sqrt{4} + \frac{1}{2\sqrt{4}}(x - 4) = 2 + \frac{1}{4}(x - 4)$$

which cleans up as

$$L_1(x) = \frac{1}{4}x + 1$$

(b) To estimate $\sqrt{4.1}$ and $\sqrt{4.3}$, with this local linear approximation $L_1(x)$:

$$\sqrt{4.1} \approx \frac{1}{4}(4.1) + 1 \approx 2.0250 \qquad \sqrt{4.3} \approx \frac{1}{4}(4.3) + 1 \approx 2.0750$$

(c) To compute percent error, we need the actual values; please ignore for a moment that knowing the exact values means we didn't need the approximations in the first place! The exact values, rounded to the same precision as the estimates, are: $\sqrt{4.1} = 2.0248$ and $\sqrt{4.3} = 2.0736$. Then by (10.2):

$$E_{4.1} = \left| \frac{2.0250 - 2.0248}{2.0248} \right| \cdot 100 = 0.00988\%$$

$$E_{4.3} = \left| \frac{2.0750 - 2.0736}{2.0736} \right| \cdot 100 = 0.06752\%$$

As should be expected, the error at 4.3 is larger, because it is farther away from the point of tangency. ∎

You Try It

    (1) (a) Find the local linear approximation of $f(x) = \sin(x)$ at $x = 0$.
        (b) Use the local linear approximation to estimate the value of $f(x)$
        at $x = \pi/12$ and $x = \pi/6$. (c) Determine the percent error in these
        estimates and compare them.

Quadratic Approximation

If a straight line attached to $f(x)$ at x_0 does an OK job as an approximation to that function near the point of tangency, imagine how much better a *parabola* attached there would do as an approximation! A parabola is able to bend with the function, at least a bit. Let's consider how we can extend the idea of a local linear approximation to a local *quadratic* approximation.

The link between the linear approximation (tangent line) and $f(x)$ is that they share the same point AND the same derivative (slope) at x_0. If we want to extend this one step farther, why not require that the approximation also share the same *second* derivative with $f(x)$ at x_0? Not only would the function and its approximation then have the same slope at the anchor point, but also the same concavity.

How can we make this happen? Well, a quadratic function usually has the general form $y = ax^2 + bx + c$, but let's tailor this[1] to mimic the form of the local linear approximation. Naming it L_2 because it's a second order approximation,

$$L_2(x) = a(x - x_0)^2 + b(x - x_0) + c \qquad (10.3)$$

Now, let's figure out $a, b,$ and c.

- If we want the vertex of our estimating parabola to sit at the anchor point $(x_0, f(x_0))$, we want $L_2(x_0) = f(x_0)$; this means we have $c = f(x_0)$.

---

[1]Look at the title of the next section to see how fabulous a pun this is!

- If we want the first derivative of $L_2(x)$ to match the first derivative of $f(x)$ at x_0, then we want

$$L_2'(x_0) = f'(x_0)$$

$$\left.(2a(x - x_0)^1 + b)\right|_{x_0} = f'(x_0)$$

$$2a(x_0 - x_0)^1 + b = f'(x_0)$$

$$b = f'(x_0)$$

- If we want the second derivative of $L_2(x)$ to match the second derivative of $f(x)$ at x_0, then we want

$$2a = f''(x_0)$$

$$a = \frac{1}{2}f''(x_0)$$

Altogether, our second order approximation looks like (with terms arranged in increasing order):

$$L_2(x) = f(x_0) + f'(x_0)(x - x_0) + \frac{1}{2}f''(x_0)(x - x_0)^2 \qquad (10.4)$$

Comparing this to the local linear approximation, we see that

$$L_2(x) = L_1(x) + \frac{1}{2}f''(x_0)(x - x_0)^2$$

So we didn't change $L_1(x)$, we got from $L_1(x)$ to $L_2(x)$ by adding a new term to $L_1(x)$.

EX 2 (a) Find the local quadratic approximation of $f(x) = \ln x$ at $x = 1$. (b) Plot both $f(x)$ and the approximation to confirm the agreement at and around $x = 1$. (c) Find both the linear and quadratic approximations of $\ln(1.1)$ and compare the percent error in each case.

(a) To find the local quadratic approximation of $f(x) = \ln x$ at $x = 1$, we'll need the values of the function and its first and second derivatives there. Since

$$f(x) = \ln(x) \quad ; \quad f'(x) = \frac{1}{x} \quad ; \quad f''(x) = -\frac{1}{x^2}$$

then

$$f(1) = \ln(1) = 0 \quad ; \quad f'(1) = 1 \quad ; \quad f''(1) = -1$$

Using (10.4),

$$L(x) = f(x_0) + f'(x_0)(x - x_0) + \frac{1}{2}f''(x_0)(x - x_0)^2$$

$$= 0 + (1)(x - 1) + \frac{1}{2}(-1)(x - 1)^2$$

$$= (x - 1) - \frac{1}{2}(x - 1)^2$$

(b) Figure 10.2 shows $f(x)$ and $L_2(x)$.

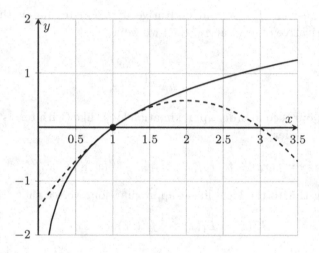

Fig. 10.2 $\ln(x)$ (solid) and its local quadratic approximation (dashed) at $x = 1$.

(c) By comparing (10.1) and (10.4), we know that $L_1(x)$ is just $L_2(x)$ without the quadratic term, so $L_1(x) = x - 1$. At $x = 1.1$, we have

$$L_1(1.1) = 1.1 - 1 = 0.1000$$

$$L_2(1.1) = (1.1 - 1) - \frac{1}{2}(1.1 - 1)^2 = 0.0.0950$$

The exact value of $\ln(1.1)$ is, rounded, 0.0953. Using (10.2),

$$E_{lin} = \left| \frac{0.1000 - 0.0953}{0.0953} \right| \cdot 100 = 4.932\%$$

$$E_{quad} = \left| \frac{0.0950 - 0.0953}{0.0953} \right| \cdot 100 = 3.148\%$$

As expected, the error from the local quadratic approximation is less than that of the local linear approximation. ∎

You Try It

> (2) (a) Find the local quadratic approximation of $f(x) = e^x$ at $x = 0$.
> (b) Plot both $f(x)$ and the approximation to confirm the agreement at and near $x = 0$. (c) Find both the linear and quadratic approximations of $e^{-0.05}$ and compare the percent error in each case.

If $L_2(x)$ was just $L_1(x)$ with a quadratic term added on, do you think that $L_3(x)$ will just be $L_2(x)$ with some cubic term added on? You get to find out! Try your own hand at developing the local cubic approximation to a function. The template for a local cubic approximation should look like this:

$$y = a(x - x_0)^3 + b(x - x_0)^2 + c(x - x_0) + d$$

Your job is to determine what the coefficients a, b, c, d should be if we want our approximation to share the same value, first, second, *and third* derivatives with $f(x)$ at x_0. This is posed as ...

You Try It

> (3) (a) Develop a formula for the local cubic approximation to any $f(x)$ at x_0. (b) Find the local cubic approximation to $f(x) = 1/x$ at $x = 2$ and plot it with $f(x)$.

Introduction to Approximation — Problem List

Intro to Approximation — You Try It

These appeared above; solutions begin on the next page.

(1) (a) Find the local linear approximation of $f(x) = \sin(x)$ at $x = 0$.
(b) Use the local linear approximation to estimate the value of $f(x)$ at $x = \pi/12$ and $x = \pi/6$. (c) Determine the percent error in these estimates and compare them.

(2) (a) Find the local quadratic approximation of $f(x) = e^x$ at $x = 0$. (b) Plot both $f(x)$ and the approximation to confirm the agreement at and near $x = 0$. (c) Find both the linear and quadratic approximations of $e^{-0.05}$ and compare the percent error in each case.

(3) (a) Develop a general formula for a local cubic approximation to any $f(x)$ at an anchor point x_0. (b) Find the local cubic approximation to $f(x) = 1/x$ at $x = 2$ and plot it with $f(x)$.

Introduction to Approximation — You Try It — Solved

---

(1) (a) Find the local linear approximation of $f(x) = \sin(x)$ at $x = 0$.
(b) Use the local linear approximation to estimate the value of $f(x)$
at $x = \pi/12$ and $x = \pi/6$. (c) Determine the percent error in these
estimates and compare them.

☐ (a) To find the local linear approximation of $f(x) = \sin(x)$ at $x = 0$,
we need the values $f(0) = \sin(0) = 0$ and $f'(0) = \cos(0) = 1$, so that
by (10.1),

$$L_1(x) = f(x_0) + f'(x_0)(x - x_0) = 0 + (1)(x - 0) = x$$

So $L_1(x) = x$ is the local linear approximation to $\sin x$ at $x = 0$.

(b) By this local linear approximation, we estimate

$$\sin\left(\frac{\pi}{12}\right) = \frac{\pi}{12} \approx 0.2618 \qquad \text{and} \qquad \sin\left(\frac{\pi}{6}\right) = \frac{\pi}{6} \approx 0.5326$$

(c) The exact values of $f(x)$ at these locations are (with rounding),

$$\sin\left(\frac{\pi}{12}\right) = 0.2588 \qquad \text{and} \qquad \sin\left(\frac{\pi}{12}\right) = \frac{1}{2} = 0.5000$$

The percent error in each approximation is, then,

$$E_{\pi/12} = \left| \frac{0.2618 - 0.2588}{0.2588} \right| \cdot 100 = 0.077\%$$

$$E_{\pi/6} = \left| \frac{0.5326 - 0.5000}{0.5000} \right| \cdot 100 = 1.63\%$$

Error at $\pi/6$ is larger, because it is farther away from the anchor point
of $x_0 - 0$. ∎

---

(2) (a) Find the local quadratic approximation of $f(x) = e^x$ at $x = 0$. (b)
Plot both $f(x)$ and the approximation to confirm the agreement at and
near $x = 0$. (c) Find both the linear and quadratic approximations of
$e^{-0.05}$ and compare the percent error in each case.

☐ (a) To find the local quadratic approximation of $f(x) = e^x$ at $x = 0$,
we'll need $f(0), f'(0)$, and $f''(0)$. Since

$$f(x) = e^x \quad ; \quad f'(x) = e^x \quad ; \quad f''(x) = e^x$$

then

$$f(0) = 1 \quad ; \quad f'(0) = 1 \quad ; \quad f''(0) = 1$$

so that by (10.4), the local quadratic approximation is

$$L_2(x) = f(x_0) + f'(x_0)(x - x_0) + \frac{1}{2}f''(x_0)(x - x_0)^2$$

$$= 1 + (1)(x - 0) + \frac{1}{2}(1)(x - 0)^2$$

$$= 1 + x + \frac{x^2}{2}$$

(b) A plot of $f(x)$ and its local quadratic approximation are shown in Fig. 10.3.

(c) At $x = -0.05$, we have

$$L_2(-0.005) = 1 + (-0.05) + \frac{1}{2}(-0.05)^2 = 0.95125$$

We can also pull right out of our recipe for $L_2(x)$ that $L_1(x) = 1 + x$, and so the local linear approximation at $x = -0.05$ is $L_1(-0.05) = 1 + (-0.05) - 0.95$. The exact value of $e^{-0.05}$ is (rounded) 0.95123. The percent errors are then

$$E_{L1} = \left| \frac{0.95000 - 0.95123}{0.95123} \right| \cdot 100 = 0.129\%$$

$$E_{L2} = \left| \frac{0.95125 - 0.95123}{0.95123} \right| \cdot 100 = 0.002\%$$

The error in the local quadratic approximation is much less than in the linear approximation, as it should be.   ∎

---

(3) (a) Develop a general formula for a local cubic approximation to any $f(x)$ at an anchor point x_0. (b) Find the local cubic approximation to $f(x) = 1/x$ at $x = 2$ and plot it with $f(x)$.

☐ We start with a generic template for $L_3(x)$: and its derivatives

$$L_3(x) = a(x - x_0)^3 + b(x - x_0)^2 + c(x - x_0) + d$$

The requirement that our local cubic approximation match $f(x_0)$, $f'(x_0)$, and $f''(x_0)$ will lead to b, c, d being the same coefficients as we have for $L_2(x)$ in (10.4), so

$$L_3(x) = a(x - x_0)^3 + \frac{f''(x_0)}{x^2}(x - x_0)^2 + f'(x_0)(x - x_0) + f(x_0)$$

The extra requirement for $L_3(x)$ is that we also match the *third* derivative of $f(x)$ at x_0. With the generic formula above, we get that $L_3'''(x_0)$,

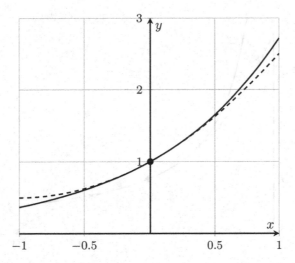

Fig. 10.3 $f(x) = e^x$ (solid) and its local quadratic approximation (dashed) at $x = 0$.

and so we need

$$L_3'''(x_0) = f'''(x_0)$$
$$6a = f'''(x_0)$$
$$a = \frac{1}{6}f'''(x_0)$$

Altogether, then, (written with ascending terms),

$$L_3(x) = f(x_0) + f'(x_0)(x - x_0) + \frac{1}{2}f''(x_0)(x - x_0)^2 + \frac{1}{6}f'''(x_0)(x - x_0)^3$$
(10.5)

We can also note that

$$L_3(x) = L_2(x) + \frac{1}{6}f'''(x_0)(x - x_0)^3$$

(b) To find the local quadratic approximation of $f(x) = 1/x$ at $x = 2$, we need $f(2)$ and the values of the first three derivatives of $f(x)$ at $x = 2$. Since

$$f(x) = \frac{1}{x} \quad ; \quad f'(x) = -\frac{1}{x^2} \quad ; \quad f''(x) - \frac{2}{x^3} \quad ; \quad f'''(x) - \frac{6}{x^4}$$

then

$$f(2) = \frac{1}{2} \quad ; \quad f'(2) = -\frac{1}{4} \quad ; \quad f''(2) = \frac{1}{4} \quad ; \quad f'''(2) = -\frac{3}{8}$$

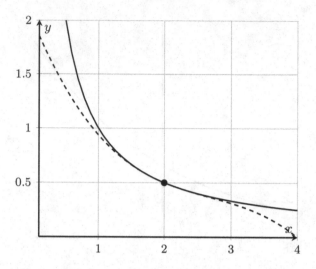

Fig. 10.4 $f(x) = 1/x$ (solid) and its local cubic approximation (dashed) at $x = 2$.

so that the local cubic approximation is

$$L_3(x) = f(x_0) + f'(x_0)(x - x_0) + \frac{1}{2}f''(x_0)(x - x_0)^2 + \frac{1}{6}f'''(x_0)(x - x_0)^3$$

$$= \frac{1}{2} - \frac{1}{4}(x - 2) + \frac{1}{2} \cdot \frac{1}{4}(x - 2)^2 - \frac{1}{6} \cdot \frac{3}{8}(x - 2)^3$$

$$= \frac{1}{2} - \frac{1}{4}(x - 2) + \frac{1}{8}(x - 2)^2 - \frac{1}{16}(x - 2)^3$$

A plot of $f(x)$ and its local cubic approximation at $x_0 = 2$ are in Fig. 10.4.

■

10.1 Taylor Polynomials

Introduction

In the previous section, Chap. 10: Interlude, we developed local linear, quadratic and cubic approximations to a function $f(x)$ at a point x_0. We found a notable pattern in the general formulas for the approximation of each degree, and this pattern allows us to extend the order of our approximations relatively easily. Let's follow this thread and see how to develop approximations of any order that we'd like.

As a word of caution, the content here is being presented in a slightly different order than in many other calculus texts. We are breaking out and seeing early a piece of a topic called "Taylor Series". Most books will hold off on that full topic until more general and theoretical background has been laid. But we are taking a peek sooner than that because this section and the next flow nicely out of the idea of approximations. Hopefully, it makes the topics a bit more intuitive.

Some Notation

In generalizing linear, quadratic, and cubic approximations upward to higher degrees, we are going to use factorials. You have probably seen these before. A factorial is an extended product of nonpositive integers. Given a positive integer n, we denote "n factorial" as $n!$,[2] and it is computed as:

$$n! = n \cdot (n-1) \cdot (n-2) \cdots 2 \cdot 1 \tag{10.6}$$

So, $1! = 1$, $2! = 2$, $3! = 6$, $4! = 24$, etc. And we also define $0! = 1$.

We're also going to need some tidy derivative notation. After two or three derivatives, the prime notation, such as $f''(x)$, gets clunky. When we need to swap notation, an exponent in parentheses will denote a derivative of that order. For example, $f^{(4)}(x)$ means the 4th derivative of $f(x)$. Also note that the 0th derivative of a function is the function itself, $f^{(0)}(x) = f(x)$.

Generalzing Approximations

Let's sweep back through the local linear, quadratic and cubic approximations we've already seen, and update the notation (which will mean

---

[2]No, this does not mean "n dammit!"

introducing some factorials and the above adjusted derivative notation). When we do this, we are going to see a very distinct pattern that allows us to generalize our approximations to any order.

We're going to work our way backwards. If we start with the local cubic approximation, we know from (10.5) in You Try It 3 of Chap. 10: Interlude:

$$L_3(x) = f(x_0) + f'(x_0)(x - x_0) + \frac{1}{2}f''(x_0)(x - x_0)^2 + \frac{1}{6}f'''(x_0)(x - x_0)^3$$

Let's standardize the use of derivative notation,

$$L_3(x) = f^{(0)}(x_0)(x - x_0)^0 + f^{(1)}(x_0)(x - x_0)^1$$
$$+ \frac{f^{(2)}(x_0)}{2}(x - x_0)^2 + \frac{f^{(3)}(x_0)}{6}(x - x_0)^6$$

Now let's look at that denominator of 6 in the final term. This was generated by the third derivative of $(x - x_0)^3$, which can be written explicitly as $3 \cdot 2 \cdot 1 \cdot (x - x_0)$, or $3! \cdot (x - x_0)$. In other words, the 6 is there because the repeated derivatives built up a 3-factorial. The coefficients of the other terms also come about the same way, but it's most obvious with the 6. Let's build that notation into $L_3(x)$:

$$L_3(x) = \frac{f^{(0)}(x_0)}{0!}(x - x_0)^0 + \frac{f^{(1)}(x_0)}{1!}(x - x_0)^1$$
$$+ \frac{f^{(2)}(x_0)}{2!}(x - x_0)^2 + \frac{f^{(3)}(x_0)}{3!}(x - x_0)^3$$

Finally, let's now recognize that we are adding four terms which all have exactly the same structre, and we can bundle it up into a summation:

$$L_3(x) = \sum_{n=0}^{3} \frac{f^{(n)}(x_0)}{n!}(x - x_0)^n \tag{10.7}$$

Now, let's go backwards to $L_2(x)$ and see if we can build up the same sort of general expression. The extended version of $L_2(x)$ comes from Eq. (10.4),

$$L_2(x) = f(x_0) + f'(x_0)(x - x_0) + \frac{1}{2}f''(x_0)(x - x_0)^2$$

If we update the derivative notation, we have

$$L_2(x) = f^{(0)}(x_0)(x - x_0)^0 + f^{(1)}(x_0)(x - x_0)^1 + \frac{f^{(2)}(x_0)}{2}(x - x_0)^2$$

Then tossing in the factorials,

$$L_2(x) = \frac{f^{(0)}(x_0)}{0!}(x - x_0)^0 + \frac{f^{(1)}(x_0)}{1!}(x - x_0)^1 + \frac{f^{(2)}(x_0)}{2!}(x - x_0)^2$$

Like with $L_3(x)$, we have terms with exactly the same structure, so we can bundle it all as:

$$L_2(x) = \sum_{n=0}^{2} \frac{f^{(n)}(x_0)}{n!}(x - x_0)^n \tag{10.8}$$

You know, we also could have just remembered that we found in the last section that $L_3(x)$ is $L_2(x)$ with the cubic term added on, and simply removed the $n = 3$ term from (10.7) to get (10.8). Oh well, it's good to see it verified. And finally, for $L_1(x)$, we can pick either one of these options:

- generate a "better" formula for $L_1(x)$ by updating the notation in
$$L_1(x) = f(x_0) + f'(x_0)(x - x_0)$$

- Remember that $L_2(x)$ is just $L_1(x)$ plus the quadratic term, and so simply drop the $n = 2$ case from (10.8) to get a version for $L_1(x)$

Either way, we end up at:

$$L_1(x) = \sum_{n=0}^{1} \frac{f^{(n)}(x_0)}{n!}(x - x_0)^n \tag{10.9}$$

Now, go back and compare (10.7), (10.8), and (10.9). The general formula for each approximation is based on exactly the same summation recipe; the only difference is how many terms we use. For $L_1(x)$, our sum goes from $n = 0$ to $n = 1$. For $L_2(x)$, our sum goes from $n = 0$ to $n = 2$. For $L_3(x)$, our sum goes from $n = 0$ to $n = 3$. Can you take a crazy guess what the local linear fourth order approximation would look like? This is how we launch into generalizing our approximations to any order.

■⊙■ FFT: If we go *backwards* with the pattern noted among (10.7), (10.8), and (10.9), and simply take one term of the sum — for $n = 0$ only. Does the result have any meaning? ■⊙■

Taylor and Maclaurin Polynomials

In the early 1700s, a very smart man named Brook Taylor built on work done by a lot of people before him and "locked in" the idea of alternate representation of functions in the form of a series. Such representations have become known as Taylor Series. You know for yourself from our experience with Riemann Sums that series can have finite or infinite number of terms. When an approximate representation of function near a certain anchor point has a finite number of terms, such as our polynomials $L_1(x), L_2(x)$, and $L_3(x)$, it is called a Taylor polynomial.[3] Therefore, although it helped us for a while, we are going to dump the L notation and switch to T. The equations (10.7), (10.8), and (10.9) will now be said to represent $T_1(x)$, $T_2(x)$, and $T_3(x)$. But the whole idea here is to generalize to higher order approximations, and that is done like so:

Definition 10.1. *A Taylor polynomial is a polynomial that approximates a function $f(x)$ at and around a point x_0. A Taylor polynomial can be of any order; if its order is N, we name it $T_N(x)$. From the above examples it's pretty clear that the Taylor polynomial of order N of $f(x)$ at x_0 is*

$$T_N(x) = \sum_{n=0}^{N} \frac{f^{(n)}(x_0)}{n!}(x - x_0)^n \tag{10.10}$$

In the special case that $x_0 = 0$, the polynomial is called a Maclaurin polynomial, because another very smart person, Colin Maclaurin, focused on this particular case in his own work. These will be denoted $M_N(x)$.

EX 1 Find the Taylor polynomial of order 4 for $f(x) = e^{-2x}$ at $x = 1/2$.

Recognize that even though the Taylor polynomial formula is written in summation notation to save space, we still need to compute it term by term; we can't do it all at once. In this example, we want to build:

$$T_4(x) = \sum_{n=0}^{4} \frac{f^{(n)}(1/2)}{n!}\left(x - \frac{1}{2}\right)^n$$

and so we need $f(x)$ and its first four derivatives evaluated at $x = 1/2$. These are:

$$f(x) = e^{-2x} \rightarrow f(1/2) = e^{-1}$$
$$f'(x) = (-2)e^{-2x} \rightarrow f'(1/2) = -2e^{-1}$$
$$f''(x) = (-2)^2 e^{-2x} \rightarrow f''(1/2) = (-2)^2 e^{-1}$$
$$f^{(3)}(x) = (-2)^3 e^{-2x} \rightarrow f^{(3)}(1/2) = (-2)^3 e^{-1}$$
$$f^{(4)}(x) = (-2)^4 e^{-2x} \rightarrow f^{(4)}(1/2) = (-2)^4 e^{-1}$$

Now let's "open up" the Taylor polynomial formula for $n = 4$ (which will give five terms), plugging in all this information, and simplifying the terms

---

[3]I'll let you break into a sweat at the idea of the infinite series version, which is coming up in the next section!

as much as we can. Starting with

$$T_4(x) = \frac{f^{(0)}(1/2)}{0!}\left(x - \frac{1}{2}\right)^0 + \frac{f^{(1)}(1/2)}{1!}\left(x - \frac{1}{2}\right)^1 + \frac{f^{(2)}(1/2)}{2!}\left(x - \frac{1}{2}\right)^2$$
$$+ \frac{f^{(3)}(1/2)}{3!}\left(x - \frac{1}{2}\right)^3 + \frac{f^{(4)}(1/2)}{4!}\left(x - \frac{1}{2}\right)^4$$

we can plug in all the derivative information to get:

$$T_4(x) = \frac{e^{-1}}{0!}\left(x - \frac{1}{2}\right)^0 + \frac{-2e^{-1}}{1!}\left(x - \frac{1}{2}\right)^1 + \frac{(-2)^2 e^{-1}}{2!}\left(x - \frac{1}{2}\right)^2$$
$$+ \frac{(-2)^3 e^{-1}}{3!}\left(x - \frac{1}{2}\right)^3 + \frac{(-2)^4 e^{-1}}{4!}\left(x - \frac{1}{2}\right)^4$$

And now we can start moving some numbers around and simplifying (lots of cancellation happens with the factorials):

$$T_4(x) = \frac{1}{e}\left(x - \frac{1}{2}\right)^0 - \frac{2}{e}\left(x - \frac{1}{2}\right)^1 + \frac{2^2}{2!e}\left(x - \frac{1}{2}\right)^2$$
$$- \frac{8}{3!e}\left(x - \frac{1}{2}\right)^3 + \frac{16}{4!e}\left(x - \frac{1}{2}\right)^4$$
$$= \frac{1}{e} - \frac{2}{e}\left(x - \frac{1}{2}\right)^1 + \frac{2}{e}\left(x - \frac{1}{2}\right)^2 - \frac{4}{3e}\left(x - \frac{1}{2}\right)^3 + \frac{2}{3e}\left(x - \frac{1}{2}\right)^4 \quad \blacksquare$$

You Try It

(1) Find the Taylor polynomial of order 3 for $f(x) = x^{2/3}$ at $x = 1$.

EX 2 Find the Maclaurin polynomial of order 3 for $f(x) = \sec x$.

Remember that a Maclaurin polynomial is a Taylor polynomial for $x_0 = 0$, so we need to build

$$M_3(x) = \sum_{n=0}^{3} \frac{f^{(n)}(0)}{n!}(x - 0)^n = \sum_{n=0}^{3} \frac{f^{(n)}(0)}{n!}x^n$$

We'll need $f(x)$ and its first three derivatives evaluated at $x = 0$. These are:

$$f(x) = \sec x \rightarrow f(0) = 1$$
$$f'(x) = \sec x \tan x \rightarrow f'(0) = 0$$
$$f''(x) = \sec^3 x + \sec^2 x \tan x \rightarrow f''(0) = 1$$
$$f^{(3)}(x) = 3\sec^3 x \tan x + 2\sec^2 x \tan^2 x + \sec^4 x \rightarrow f^{(3)}(0) = 4$$

When we open up the sum for M_3 (which gives *four* terms), we see

$$M_3(x) = \frac{f^{(0)}(0)}{0!}x^0 + \frac{f^{(1)}(0)}{1!}x^1 + \frac{f^{(2)}(0)}{2!}x^2 + \frac{f^{(3)}(0)}{3!}x^3$$

Putting in the derivative information and simplfying, we get:

$$M_3(x) = \frac{1}{0!}x^0 + \frac{0}{1!}x^1 + \frac{1}{2!}x^2 + \frac{4}{3!}x^3 = 1 + \frac{1}{2}x^2 + \frac{2}{3}x^3 \quad \blacksquare$$

You will learn that when you're asked to lay out a Maclaurin polynomial, you will say "oh, good!", but when you're asked to generate a Taylor polynomial, you will say, "oh, no!".

You Try It

(2) Find the Maclaurin polynomial of order 3 for for $f(x) = xe^{-x}$.

Error Estimation

Now we know about Taylor polynomials, which are like mathematical chameleons[4] which mimic other functions. But this agreement isn't necessarily global. When a Taylor polynomial of order N, $T_N(x)$, is used to approximate a function $f(x)$ at and around the point x_0, we know that $f(x)$ and $T_N(x)$ will be equal at x_0; they should be "close" nearby x_0, and they should drift farther apart as we move away from x_0. How do we measure any of this?

Let's borrow a word from elementary school and call the difference between $f(x)$ and its Taylor polynomial $T_N(x)$ the *remainder $R_N(x)$*. That is,

$$f(x) = T_N(x) + R_N(x) \qquad \text{or} \qquad R_N(x) = f(x) - T_N(x)$$

We should expect a couple of things to be true right away about $R_N(x)$:

- $R_N(x_0) = 0$; that is, since x_0 is the one point where we know $f(x)$ and $T_N(x)$ have the same value, the error (remainder) there should be 0.
- At any point other than x_0, but near x_0, we should see $R_N(x)$ get smaller as N gets bigger; increasing the order of the approximating polynomial should decrease the error.

---

[4]Or Replicants, if you've seen Blade Runner!

An analysis of how "close" $T_N(x)$ is to $f(x)$ is easy to do on a point by point basis: at any one location $x = c$, we can compute $R_N(c) = f(c) - T_N(c)$. More often, we're interested in some sort of overall analysis of the accuracy of $T_N(x)$ in the vicinity of the anchor point x_0. Let's specifically describe this "vicinity" as an interval $[x_0 - d, x_0 + d]$. In other words, we'll be looking at an interval centered at x_0, and going out a distance d to either side. For example, if $x_0 = 2$, the interval $[1.5, 2.5]$ corresponds to $d = 0.5$. Or if we say we have $x_0 = 4$ and $d = 3$, then the interval of interest is $[1, 7]$.

Once we have the interval $[x_0 - d, x_0 + d]$ defined, we can ask the following question: What's the worst case scenario for $R_N(x)$ anywhere in this interval? That is, how far off could $T_N(x)$ be from $f(x)$ anywhere in this interval? This information is found by an expression called Taylor's Inequality, and it requires one more piece of information (apart from $f(x)$, x_0, and d); this extra piece of information will sound awful, but it's really not that bad.

If I said we needed "an upper bound of the absolute value of the $(N+1)$st derivative of $f(x)$ over the interval $[x_0 - d, x_0 + d]$", would you scream and run away? I hope not, because this is what we need. First, don't be intimidated by the phrase "upper bound". An upper bound of a set of values is just another value that is larger than anything in the set you have. Upper bounds are not unique. For example, you know that the function $\sin(x)$ only takes up values from 0 to 1 on the interval $x \in [0, \pi]$. And so 1 is an "upper bound of $f(x) = \sin(x)$ on the interval $[0, \pi]$". But also, 2 is an upper bound. So is 44. All of them are equal to or larger than any value of $\sin(x)$ in that interval $[0, \pi]$. Now, if I asked you which is the "best" or "most efficient" upper bound for $\sin(x)$ on $[0, \pi]$, you might say 1, and I wouldn't argue. But the nice thing about upper bounds is that, since they're not unique, you can pick nice numbers to use. For example, if we have some set of values and happen to know that 2.97664 is greater than anything in the set, we might decide to just report 3 as an upper bound of the set.

So, if the idea of an interval $[x_0 - d, x_0 + d]$ doesn't bother you, and the idea of finding some higher order derivative of $f(x)$ doesn't bother you, and you're no longer intimidated by the idea of an "upper bound", then you should not sweat over finding "an upper bound of the absolute value of the $(N + 1)$st derivative of $f(x)$ over the interval $[x_0 - d, x_0 + d]$". In fact, given $f(x)$, x_0, d, and N, here's how to find such a thing.

(1) First, let's give this upper bound a name, M.
(2) Next, compute the $(N+1)$st derivative $f^{(N+1)}(x)$ and slap absolute value bars around it.
(3) Then plot the result, which is $|f^{(N+1)}(x)|$, on the interval $[x_0 - d, x_0 + d]$.
(4) Finally, look at the plot of $|f^{(N+1)}(x)|$ over the interval $[x_0-d, x_0 + d]$. Examine the y axis and select some number from the y axis that is above all values of $|f^{(N+1)}(x)|$ over that interval. Call this value M.

In that last step, make it easy on yourself by selecting a "nice" value of M. If you build a graph of some $|f^{(N+1)}(x)|$ over an interval $[x_0-d, x_0+d]$ and it sure looks to your eyes like maybe the highest up the graph gets is about 28.75, then heck, just report $M = 30$ as the upper bound. Everything we're doing is an estimate anyway, so why overcomplicate it?

With all of that being said, we have the following error estimate:

Useful Fact 10.1. *Suppose we have the Taylor polynomial of order N for $f(x)$ at and around some x_0. Let d be the width of an interval with center x_0 in which we are going to examine the remainder $R_N(x)$ between $f(x)$ and $T_N(x)$. Let M be an upper bound of the absolute value of the $(N+1)$st derivative of $f(x)$ over the interval $[x_0 - d, x_0 + d]$. Then,*

$$R_N(x) \leq \frac{M}{(N + 1)!} \cdot d^{N+1}$$

*This is called **Taylor's Inequality,** and it puts a cap on the largest largest error (remainder) between $f(x)$ and $T_N(x)$ that we could observe anywhere in the given interval.*

Useful Fact 10.1 is an **error bound**, and it's important to note that it's still very possible that all the error inside the interval could still be significantly less than $|R_N(x)|$, but certainly we'd never measure error greater than $|R_N(x)|$.

EX 3   Go get the Taylor polynomial of order 3 for $f(x) = x^{2/3}$ around $x_0 = 1$ from YTI 2, and then find the maximum possible error between $T_3(x)$ and $f(x)$ on the interval $[0.25, 1.75]$.

Hopefully attempted You Try It 2, and found this Taylor polynomial to be
$$T_3(x) = 1 + \frac{2}{3}(x-1) - \frac{1}{9}(x-1)^2 + \frac{4}{81}(x-1)^3$$
(Fig. 10.5 shows $f(x)$ and $T_3(x)$ on the interval $[0,2]$.) Now let's look for the biggest possible error between $T_3(x)$ and $f(x)$ on the interval $[0.25, 1.75]$. To use Taylor's Inequality, we need the following ingredients: N, d, and the $(N+1)$st derivative of $f(x)$.

- From the problem statement, we have $N = 3$.
- From the given interval, which is centered at $x_0 = 1$, we have $[0.25, 1.75] = [1 - 0.75, 1 + 0.75]$, so that $d = 0.75$.
- The $(N+1)$st derivative of $f(x)$ is the 4th derivative, which is $f^{(3+1)}(x) = -56/(81x^{10/3})$.
- The absolute value of this derivative is $\left| 56/(81x^{10/3}) \right|$.

Now we need a plot of this absolute value function on the interval $[0.25, 1.75]$. This is shown in Fig. 10.6. Don't be shy about using tech to generate plots like this. We'll make a rough estimate that $M = 80$ is larger than any value of the 4th derivative over the given interval. (Other values might work, but we don't want to get ridiculous about it and report, say, 200. Similarly, we may be able to zoom in to distinguish perhaps 78.77 as an upper bound, too, but that would have minimal payoff. All in all, $M = 80$ is good!)

And now we have all we need for Taylor's Inequality:
$$R_3(x) \leq \frac{M}{(3+1)!} \cdot d^{3+1} = \frac{40}{4!} \cdot (0.75)^4 \approx 0.527$$
This means that at any point in the interval $[0.25, 1.75]$, the value of the Taylor polynomial cannot be more than 0.53 away from the true value of $f(x)$. We can observe this at a test point, say $x = 1.5$. The true value of $f(x)$ is $1.5^{2/3} \approx 1.310$. The Taylor polynomial at $x = 1.25$ is $T_3(1.5) = 1.311$. The error at this point is 0.001, which is far less than the estimated maximum error of 0.527. ∎

You Try It

(3) Find the Maclaurin polynomial of order 6 for for $f(x) = \cosh(x)$; determine the maximum possible error on the interval $[-1, 1]$ and test your result at $c = 0.5$.

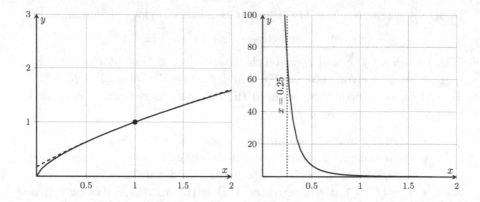

Fig. 10.5 $f(x) = x^{2/3}$ and $T_3(x)$ at $x_0 = 1$ (dashed) w/ EX 3.   Fig. 10.6   Looking for an upper bound of $|f^{(iv)}n + 1(x)|$ (w/ EX 3).

As a final note, the discussion of Taylor's Inequality is another place where I'm really leaning into this text as a work where utilitarian purposes may trump theoretical underpinnings. If we were not broaching topics until we had the all background for their theoretical basis, this could not come until the end of the next chapter. But in terms of putting it where its use fits nicely with the surrounding content, right here is pretty good. Just don't tell anyone we did it this way.

Taylor and Maclaurin Polynomials — Problem List

Taylor and Maclaurin Polynomials — You Try It

These appeared above; solutions begin on the next page.

(1) Find the Taylor polynomial of order 3 for $f(x) = x^{2/3}$ at $x = 1$.
(2) Find the Maclaurin polynomial of order 3 for for $f(x) = xe^{-x}$.
(3) Find the Maclaurin polynomial of order 6 for for $f(x) = \cosh(x)$; determine the maximum possible error on the interval $[-1, 1]$ and confirm your result using $c = 0.5$ as a test point.

Taylor and Maclaurin Polynomials — Practice Problems

Try these as you get the hang of the You Try It problems. Solutions to these problems are available in Sec. A.4.1.

(1) Find the Taylor polynomial of order 3 for $f(x) = \ln(1 + 2x)$ at $x = 1$. Determine $R_3(x)$ for an interval of $d = 0.5$ centered at $x = 1$.
(2) Find the Maclaurin polynomial of order 4 for for $f(x) = x + e^{-x/2}$. Determine $R_4(x)$ for the appropriate interval with $d = 0.5$.
(3) Find the Taylor polynomial of order 3 for for $f(x) = x + e^{-x/2}$ at $x = 2$. Determine $R_3(x)$ for an interval of $d = 0.5$ centered at $x = 2$ and confirm this $R_3(x)$ holds at the test point $c = 2.3$.
(4) Find the Maclaurin polynomial of order 3 for for $f(x) = \sin(2x)$. Determine $R_3(x)$ for the appropriate interval of $d = 0.25$ and confirm this $R_3(x)$ holds at the test point $c = 0.1$.

Taylor and Maclaurin Polynomials — Challenge Problems

Try these problems to test your skills with the ideas in this section. Solutions to these problems are available in Sec. B.4.1.

(1) Find the Maclaurin polynomial of order 3 for $f(x) = \sqrt{1 - x^2}$.
(2) Find the Taylor polynomial of order 2 for $\tan^{-1}(x)$ at $x = 1$. Determine $R_2(x)$ for an interval of $d = 0.5$ centered at $x = 1$ and confirm this $R_2(x)$ holds at the test point $c = 0.8$.
(3) Find the Taylor polynomial of order 3 for $f(x) = \ln(1 + x^2)$ at $x = 1$.

Taylor Polynomials — You Try It — Solved

---

(1) Find the Taylor polynomial of order 3 for $f(x) = x^{2/3}$ at $x = 1$.

☐ We need $f(x)$ and its first three derivatives evaluated at $x = 1$. These are:

$$f(x) = x^{2/3} \rightarrow f(1) = 1$$

$$f'(x) = \frac{2}{3}x^{-1/3} \rightarrow f'(1) = \frac{2}{3}$$

$$f''(x) = -\frac{2}{9}x^{-4/3} \rightarrow f''(1) = -\frac{2}{9}$$

$$f^{(3)}(x) = \frac{8}{27}x^{-7/3} \rightarrow f^{(3)}(1) = \frac{8}{27}$$

so

$$T_3(x) = \sum_{n=0}^{3} \frac{f^{(n)}(x_0)}{n!}(x - x_0)^n$$

$$= \frac{f^{(0)}(1)}{0!}(x - 1)^0 + \frac{f^{(1)}(1)}{1!}(x - 1)^1 + +\frac{f^{(2)}(1)}{2!}(x - 1)^2$$

$$+ \frac{f^{(3)}(1)}{3!}(x - 1)^3$$

$$= \frac{1}{0!}(x - 1)^0 + \frac{2/3}{1!}(x - 1)^1 + \frac{(-2/9)}{2!}(x - 1)^2 + \frac{8/27}{3!}(x - 1)^3$$

$$= 1 + \frac{2}{3}(x - 1) - \frac{1}{9}(x - 1)^2 + \frac{4}{81}(x - 1)^3 \quad \blacksquare$$

---

(2) Find the Maclaurin polynomial of order 3 for for $f(x) = xe^{-x}$.

☐ We need $f(x)$ and its first three derivatives evaluated at $x = 0$. These are:

$$f(x) = xe^{-x} \rightarrow f(0) = 0$$

$$f'(x) = e^{-x} - xe^{-x} \rightarrow f'(0) = 1$$

$$f''(x) = -2e^{-x} + xe^{-x} \rightarrow f''(0) = -2$$

$$f^{(3)}(x) = 3e^{-x} - xe^{-x} \rightarrow f^{(3)}(0) = 3$$

so

$$M_3(x) = \sum_{n=0}^{3} \frac{f^{(n)}(x_0)}{n!}(x-0)^n$$

$$= \frac{f^{(0)}(0)}{0!}(x)^0 + \frac{f^{(1)}(0)}{1!}(x)^1 + \frac{f^{(2)}(0)}{2!}(x)^2 + \frac{f^{(3)}(0)}{3!}(x)^3$$

$$= \frac{0}{0!}(x)^0 + \frac{1}{1!}(x)^1 + \frac{-2}{2!}(x)^2 + \frac{3}{3!}(x)^3$$

$$= x - x^2 + \frac{1}{2}x^3 \quad \blacksquare$$

---

(3) Find the Maclaurin polynomial of order 6 for for $f(x) = \cosh(x)$; determine the maximum possible error on the interval $[-1, 1]$ and confirm your result using $c = 0.5$ as a test point.

☐ We need a whole bunch of derivatives; thankfully. they are very simple and predictable. For $f(x) = \cosh(x)$,

$$f'(x) = f^{(3)}(x) = f^{(5)}(x) = \sinh(x)$$

$$f''(x) = f^{(4)}(x) = f^{(6)}(x) = \cosh(x)$$

so that

$$f'(0) = f^{(3)}(0) = f^{(5)}(0) = 0$$

$$f''(0) = f^{(4)}(0) = f^{(6)}(0) = 1$$

so with the zeroes in place,

$$M_6(x) = \sum_{n=0}^{6} \frac{f^{(n)}(x_0)}{n!}(x-0)^n$$

$$= \frac{f^{(0)}(0)}{0!}(x)^0 + \frac{f^{(2)}(0)}{2!}(x)^2 + \frac{f^{(4)}(0)}{4!}(x)^4 + \frac{f^{(6)}(0)}{6!}(x)^6$$

$$= 1 + \frac{x^2}{2!} + \frac{x^4}{4!} + \frac{x^6}{6!}$$

For the error estimate, we have $d = 1$. As $N = 6$, we need M to be an upper bound of $|f^{(7)}(x)| = |\sinh(x)|$ on $[-1.1]$. Figure 10.7 shows the graph of $|\sinh(x)|$ on that interval, and the maximum is at the endpoints, where $|\sinh(1)| = \dfrac{e^2 - 1}{2e} \approx 1.175$. So let's choose $M = 2$ as our upper bound. Then, by Taylor's Inequality,

$$R_6(x) \leq \frac{M}{(N+1)!} \cdot d^{N+1} = \frac{2}{7!} \cdot (1)^7 = \frac{2}{7!} \approx 0.0004$$

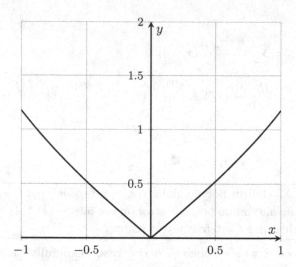

Fig. 10.7   Looking for an upper bound of $|\sinh(x)|$ on $[-1, 1]$ (w/ YTI 3).

At the test point $c = 0.5$, we have

$$M_6(0.5) = 1 + \frac{0.5^2}{2!} + \frac{0.5^4}{4!} + \frac{0.5^6}{6!} \approx 1.12762587$$

$$f(0.5) = \cosh(0.5) \approx 1.12762596$$

so that $|f(0.5) - M_6(0.5)|$ is certainly less than 0.0004. ■

10.2 Taylor Series Part 1

Introduction

Here, we'll combine two ideas:

- You may or may not have noticed, but in some of the Taylor and Maclaurin polynomials we've seen so far, the terms follow patterns. That is, once we use the defining equation (10.10), unroll several terms, fill in the required information, and simplify those terms, the results often follow a notable pattern that can then be continued without direct reference to the definition anymore. If we can identify these patterns, we can write the Taylor polynomials in compact notation.
- If 2nd order approximations to functions are good, and 3rd order are better, then how do we get to the best? And, can we get to the point where the approximation ceases to be just an approximation and actually becomes *equal* to the function itself?

Patterns in Taylor and Maclaurin Polynomials

Here are some of the Taylor and Maclaurin polynomials we've seen so far, in the notes or in problem solutions. In each case, we're going to try to predict the next term in the polynomial.

- The Maclaurin polynomial of order 2 for $f(x) = e^x$ was

$$1 + x + \frac{x^2}{2}$$

(We didn't call it a Maclaurin polynomial at the time, but when you found the linear quadratic approximation for $f(x) = e^x$ at $x = 0$ in You Try It 2 of Chap. 10: Interlude, that's exactly what you found.) Would you guess that if we wanted to extent this polynomial one more term, the next term is probably $x^3/3$? Well, don't be so sure. Remember that the defining Eq. (10.10) uses a factorial in the denominator of each coefficient of x. And, the denominators of the first 3 terms seen there are not just $1, 1$, and 2, but also $0!, 1!$, and $2!$. In fact, the latter makes more sense as a pattern. So it's also a reasonable guess that the next term would be $x^3/3!$. The way to settle this for sure would be to go back to (10.10), build one more term, and see if it's $x^3/6$. And, that's what happens. But now we're ready to roll along — the term after that one would be $x^4/4!$, and then $x^5/5!$, and so on. Now we're generating these without any direct reference to (10.10):

$$\frac{1}{0!} + \frac{x}{1!} + \frac{x^2}{2!} + \frac{x^3}{3!} + \frac{x^4}{4!} + \cdots$$

- The Taylor polynomial of order 3 for $f(x) = 1/x$ at $x = 2$ was

$$\frac{1}{2} - \frac{1}{4}(x-2) + \frac{1}{8}(x-2)^2 - \frac{1}{16}(x-2)^3$$

(We didn't call it a Taylor polynomial at the time, but when you found the linear cubic approximation for $f(x) = e^x$ at $x = 0$ in You Try It 3 of Chap. 10: Interlude, that's exactly what you found.) We need to note the alternating signs of the terms, and the rest is fairly predictable. The next term ought to be $+(x-2)^4/32$, and then $-(x-2)^5/64$, and so on. In fact, we can note each denominator is a power of 2, and see them as:

$$\frac{1}{2^1} - \frac{1}{2^2}(x-2) + \frac{1}{2^3}(x-2)^2 - \frac{1}{2^4}(x-2)^3 + \frac{1}{2^5}(x-2)^4 + \cdots$$

- The Maclaurin polynomial of order 3 for $f(x) = xe^{-x}$ was:

$$x - x^2 + \frac{x^3}{2}$$

(from You Try It 2 of Sec. 10.1). The next term is *mostly* predictable, but we have the same uncertainty as in the other Maclaurin series above regarding the denominator. We can either consider the denominators as $1, 1, 2$, or as $0!, 1!, 2!$. Let's go with the latter (again, confirmation would come from designing the next term according to (10.10)). Also note the alternating signs; the next term should be negative. Altogther, it seems like the next term will be $-x^4/3!$, and the extended polynomial is:

$$\frac{x}{0!} - \frac{x^2}{1!} + \frac{x^3}{2!} - \frac{x^4}{3!} + \cdots$$

- The Maclaurin polynomial of order 4 for for $f(x) = x + e^{-x/2}$ was presented as:

$$1 + \frac{1}{2}x + \frac{1}{2^2 2!}x^2 - \frac{1}{2^3 3!}x^3 + \frac{1}{2^4 4!}x^4$$

(from Practice Problem 3 of Sec. 10.1). These coefficients are written clearly enough to immediately extend the pattern (note the alternating signs):

$$1 + \frac{1}{2}x + \frac{1}{2^2 2!}x^2 - \frac{1}{2^3 3!}x^3 + \frac{1}{2^4 4!}x^4 - \frac{1}{2^5 5!}x^5 + \cdots$$

Now that we've recognized the patterns in each of these, we can collapse them back into summations. However, these summations would not be achieved directly from the basic definition. It's necessary to open up several of the terms, simplify them, note the patterns in the results, and *then* present a new, compact summation form. In order to do this, we have to keep a couple of things in mind:

- In Taylor and Maclaurin polynomials, we count the terms with an index n starting at 0.
- In three of the above examples, we have alternating signs. The terms $(-1)^n$ or $(-1)^{n+1}$ will create alternating signs as n increases as $0, 1, 2, \ldots$. We can use either one as needed to get the right signs on the right terms.

Here are the above examples collapsed back into summation form :

- The Maclaurin polynomial of order 3 for $f(x) = e^x$:

$$M_4(x) = \frac{1}{0!} + \frac{x}{1!} + \frac{x^2}{2!} + \frac{x^3}{3!} = \sum_{n=0}^{3} \frac{1}{n!} x^n$$

- The Taylor polynomial of order 4 for $f(x) = 1/x$ at $x = 2$:

$$T_4(x) = \frac{1}{2^1} - \frac{1}{2^2}(x-2) + \frac{1}{2^3}(x-2)^2 - \frac{1}{2^4}(x-2)^3 + \frac{1}{2^5}(x-2)^4$$

$$= \frac{(-1)^0}{2^1} + \frac{(-1)^1}{2^2}(x-2) + \frac{(-1)^2}{2^3}(x-2)^2$$

$$+ \frac{(-1)^3}{2^4}(x-2)^3 + \frac{(-1)^4}{2^5}(x-2)^4$$

$$= \sum_{n=0}^{4} \frac{1}{2^{n+1}}(x-2)^n$$

- The Maclaurin polynomial of order 4 for $f(x) = xe^{-x}$:

$$M_4(x) = x - x^2 + \frac{1}{2}x^3 - \frac{1}{3!}x^4$$

$$= \frac{(-1)^0}{0!}x^1 + \frac{(-1)^1}{1!}x^2 + \frac{(-1)^2}{2!}x^3 + \frac{(-1)^3}{3!}x^4$$

$$= \sum_{n=0}^{4} \frac{(-1)^n}{n!}x^{n+1}$$

- The Maclaurin polynomial of order 5 for for $f(x) = x + e^{-x/2}$ is a bit of a puzzler:

$$M_5(x) = 1 + \frac{1}{2}x + \frac{1}{2^2 2!}x^2 - \frac{1}{2^3 3!}x^3 + \frac{1}{2^4 4!}x^4 - \frac{1}{2^5 5!}x^5$$

Pay close attention to the pattern of signs: the signs do not actually alternate by each term in the first few terms. There are three positive terms before the first negative term, and so we don't have a direct pattern from the first term:

$$M_5(x) = \frac{(-1)^0}{2^0 0!}x^0 + \frac{(-1)^{???}}{2^1 1!}x^1 + \frac{(-1)^2}{2^2 2!}x^2 + \frac{(-1)^3}{2^3 3!}x^3$$
$$+ \frac{(-1)^4}{2^4 4!}x^4 + \frac{(-1)^5}{2^5 5!}x^5$$

The best we can do is write the first two terms by themselves, and then the rest (which do follow the pattern) in a summation:

$$M_5(x) = 1 + \frac{1}{2}x + \sum_{n=2}^{5} \frac{(-1)^n}{2^n n!}x^n$$

The reason for collapsing the Taylor and Maclaurin polynomials back into a tidy summation is that now it is trvial to generate any case. For example, we can now write the Maclaurin polynomial of order 172 for $f(x) = xe^{-x}$ as

$$M_{172}(x) = \sum_{n=0}^{172} \frac{(-1)^n}{n!}x^{n+1}$$

The Approximations Become Equalities

If you recall your first experience with finding the area under the curve, the game was to split the area into "rectangles", find the area of the individual rectangles, then add them up. And then, we realized, well, if 10 rectangles gave a pretty good approximation to the actual area, imagine what 20 would do! And then 30! The same idea arises here. If we can get a pretty good approximation to a function using a Taylor polynomial of 5 terms, imagine what 50 would do! And then 500! Just like Riemann Sums gave estimates that collapsed onto the exact area as the number of rectangles went to infinity, so here our approximating polynomials will collapse exactly onto the given function as the number of terms in the polynomial goes to infinity. The difference is that this agreement may not span the entire real line, but often just a certain segment of it.

A Taylor (or Maclaurin) polynomial in which the order (i.e. the number of terms) is allowed to go to infinity, is called a Taylor or Maclaurin **series**.

To develop the Taylor series[5] of $f(x)$ around x_0, we do exactly the same thing as in the above examples:

- Compute the first several terms of an n-th order Taylor polynomial for $f(x)$ at x_0 using (10.10).
- Identify the patterns among the terms.
- Write the polynomial in summation notation.
- Set the upper number of terms in the summation to ∞.

The number of terms you need in your initial expansion is the number of terms you need to clearly identify the patterns.

$\boxed{\textbf{EX 1}}$ Find the Taylor series for $f(x) = 1/(1-x)$ around $x_0 = 3$.

Let's write out the first few terms of $T_N(x)$ and see if we can find a pattern. For $x_0 - 3$, we have to create

$$T_N(x) = \sum_{n=0}^{N} \frac{f^{(n)}(3)}{n!}(x-3)^n \tag{10.11}$$

So we need $f(x)$ and its first few derivatives evaluated at $x = 3$. These are:

$$f(x) = \frac{1}{1-x} \rightarrow f(3) = -\frac{1}{2}$$

$$f'(x) = \frac{1}{(1-x)^2} \rightarrow f'(3) = \frac{1}{2^2}$$

$$f''(x) = \frac{2}{(1-x)^3} \rightarrow f''(3) = -\frac{2}{2^3}$$

$$f^{(3)}(x) = \frac{3!}{(1-x)^4} \rightarrow f^{(3)}(3) = \frac{3!}{2^4}$$

Even before we lay these out into their specific terms, we can identify a distinct pattern (note the alternating signs):

$$f^{(n)}(3) = (-1)^{n+1}\frac{n!}{2^{n+1}}$$

---

[5] A Maclaurin series is a special Taylor series, so we don't have to always specify Maclaurin series whenever we mention Taylor series.

Note that the $n!$ in this recipe is ready to cancel the $n!$ in the denominator of the defining sum (10.11):

$$T_N(x) = \sum_{n=0}^{N} \frac{f^{(n)}(3)}{n!}(x-3)^n = \sum_{n=0}^{N}(-1)^{n+1}\frac{n!}{2^{n+1}(n!)}(x-3)^n$$

$$= \sum_{n=0}^{N}(-1)^{n+1}\frac{1}{2^{n+1}}(x-3)^n$$

This represents the Taylor polynomial of any order N for $f(x) = \dfrac{1}{1-x}$ around $x_0 = 3$. This becomes the Taylor *series* when we let N become ∞:

$$T(x) = \sum_{n=0}^{\infty} \frac{(-1)^{n+1}}{2^{n+1}}(x-3)^n \quad \blacksquare$$

You Try It

(1) Find the Taylor series for $f(x) = e^x$ around $x_0 = 3$.

EX 2 Find the Maclaurin series for $f(x) = \tan^{-1}(x)$.

Now don't get all freaked out by the inverse tangent, remember that it goes away in its first derivative. Also, note again that we're looking for a Maclaurin series, which is a Taylor series around $x_0 = 0$. Let's write out the first few terms of $M_N(x)$ and see if we can find a pattern. We'll have

$$M_N(x) = \sum_{n=0}^{N} \frac{f^{(n)}(0)}{n!}x^n$$

So we need $f(x)$ and its first few derivatives evaluated at $x = 0$. These are (with some help from technology to compute and simplify them):

$$f(x) = \tan-1(x) \rightarrow f(0) = 0$$

$$f'(x) = \frac{1}{1+x^2} \rightarrow f'(0) = 1$$

$$f''(x) = -\frac{2x}{(1+x^2)^2} \rightarrow f''(0) = 0$$

$$f^{(3)}(x) = \frac{2(3x^2-1)}{(1+x^2)^3} \rightarrow f^{(3)}(0) = -2 = -(2!)$$

I don't think there are enough values yet to indicate a pattern, so let's get a few more:

$$f^{(4)}(x) = -\frac{24x(-1+x^2)}{(1+x^2)^4} \rightarrow f^{(4)}(0) = 0$$

$$f^{(5)}(x) = \frac{24(1+5x^4-10x^2)}{(1+x^2)^5} \rightarrow f^{(5)}(0) = 24 = 4!$$

Now, in EX 1, we were able to use the list of these derivatives themselves to establish a pattern. Here, especially because of the alternating zeros, I think we should see everything put together term by term:

$$M_N(x) = \frac{f^{(1)}(0)}{1!}(x)^1 + \frac{f^{(3)}(0)}{3!}(x)^3 + \frac{f^{(5)}(0)}{5!}(x)^5 - \cdots$$

$$= \frac{1}{1!}(x)^1 - \frac{2!}{3!}(x)^3 + \frac{4!}{5!}(x)^5 - \cdots$$

$$= x - \frac{1}{3}x^3 + \frac{1}{5}x^5 - \cdots$$

Now we can certainly find the pattern, but there's a catch. The powers of x are going up by 2 at each step. We still have to count off our terms using $n = 0, 1, 2, 3, \ldots$, that never changes. So how can we turn the counter $n = 0, 1, 2, 3, \ldots$ into exponents of $1, 3, 5, 7, \ldots$? Like this:

- We can turn $n = 0, 1, 2, 3, \ldots$ into $1, 3, 5, 7, \ldots$ using $2n + 1$.
- We can turn $n = 0, 1, 2, 3, \ldots$ into $0, 2, 4, 6, \ldots$ using $2n$.

Since we need odd exponents only, we'll use $2n + 1$, and so:

$$M(x) = \sum_{n=0}^{\infty} \frac{(-1)^n}{2n+1} x^{2n+1} \quad \blacksquare$$

You Try It

   (2) Find the Maclaurin series for $f(x) = \cos x$.

   (3) Find the Maclaurin series for $f(x) = \ln(1 + x)$.

We Need to Hit "Pause" on This for Now...

It is not possible to graph an actual full Taylor series, because you can't actually generate an infinite number of terms. But plotting a Taylor polynomial with at least several terms will show you something good. Figure 10.8 shows plot of $f(x) = e^x$ and its Maclaurin polynomial of order 10. There really are two curves in that figure, but you can't tell them apart. On the scale of the graph, the function e^x is visually indistinguishable Maclaurin polynomial / series.

Figure 10.9 shows a plot of $f(x) = 1/(1 - x)$ and its Taylor polynomial of order 6 around $x_0 = 3$. In this figure, you can see that there is an interval around $x_0 = 3$ where $f(x)$ and its Taylor polynomial $T(x)$ agree very well. As we increase the number of terms in the Taylor polynomial,

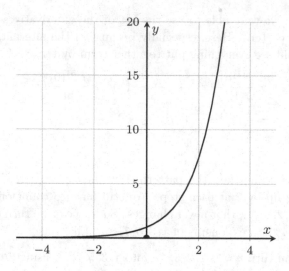

Fig. 10.8　$f(x) = e^x$ (solid) and its Maclaurin polynomial (dashed, yes it's there) for $n = 10$.

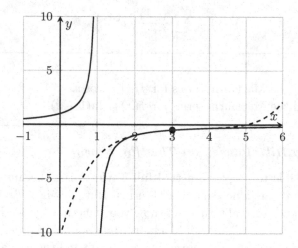

Fig. 10.9　$f(x) = 1/(1 - x)$ (solid) and its Taylor polynomial around $x_0 = 3$ for $n = 6$.

the agreement would improve, and the interval of agreement would widen ... somewhat. In this case, the asymptote at $x = 1$ provides a "wall" past which the Taylor polynomial (or series) could not maintain agreement with the function. Because we're trying to get a polynomial (albeit a longer and longer one) to mimic a function, and there is no way for a polynomial to display the same discontinuity as the function, the agreement can hold up only so far away from the anchor point. And, the the interval around $x_0 = 3$ along which we can attain agreement between the Taylor series and the function must have the same length to both sides of 3; since the "wall" of the asymptote is 2 units to the left of $x_0 = 3$, then the agreement will also fizzle at 2 units to the right of $x_0 = 3$. You can observe this behavior in the Fig. 10.9; $x = 5$ is about where the Taylor polymomial starts breaking off from the function.

An interval where a Taylor series converges to the function $f(x)$ for which it is designed is called the *interval of convergence* for that Taylor series. These intervals can be $(-\infty, \infty)$ or $[-1, 1]$ or (as seen in Fig. 10.9) the interval $[1, 5]$, or any other interval depending on the situation. It is important to know the interval of convergence of a Taylor series, since you wouldn't want to use a Taylor series or polynomial to estimate a function at a point that's not inside this interval.

In order to investigate intervals of convergence for Taylor series, we're going to have to take a detour and consider the idea of infinite series, and their convergence, in general.

Taylor and Maclaurin Series Part 1 — Problem List

Taylor and Maclaurin Series Part 1 — You Try It

These appeared above; solutions begin on the next page.

(1) Find the Taylor series for $f(x) = e^x$ around $x_0 = 3$.
(2) Find the Maclaurin series for $f(x) = \cos x$.
(3) Find the Maclaurin series for $f(x) = \ln(1 + x)$.

Taylor and Maclaurin Series Part 1 — Practice Problems

Try these as you get the hang of the You Try It problems. Solutions to these problems are available in Sec. A.4.2.

(1) Find the Maclaurin series for $f(x) = \cos \pi x$.
(2) Find the Taylor series for $f(x) = \ln x$ around $x_0 = 2$.
(3) Find the Maclaurin series for $f(x) = 1/(1 + x)^2$.

Taylor and Maclaurin Series Part 1 — Challenge Problems

Try these problems to test your skills with the ideas in this section. Solutions to these problems are available in Sec. B.4.2.

(1) Find the Maclaurin series for $f(x) = \cosh x$.
(2) Find the Taylor series for $f(x) = \sin x$ around $x_0 = \pi/2$.
(3) Find the Taylor series for $f(x) = e^{-2x}$ around $x_0 = -1/2$.

Taylor Series Part 1 — You Try It — Solved

---

(1) Find the Taylor series for $f(x) = e^x$ around $x_0 = 3$.

☐ The derivatives are simple enough we don't need to list them in advance. The definition of Taylor series gives us

$$T(x) = \frac{f(3)}{0!}(x-3)^0 + \frac{f'(3)}{1!}(x-3)^1 + \frac{f''(3)}{2!}(x-3)^2$$
$$+ \frac{f^{(3)}(3)}{3!}(x-3)^3 + \cdots$$
$$= \frac{e^3}{0!}(x-3)^0 + \frac{e^3}{1!}(x-3) + \frac{e^3}{2!}x^2 + \frac{e^3}{3!}(x-3)^3 + \cdots$$
$$= \sum_{n=0}^{\infty} \frac{e^3}{n!}(x-3)^n \quad \blacksquare$$

---

(2) Find the Maclaurin series for $f(x) = \cos x$.

☐ The definition of Maclaurin series gives us:

$$M(x) = \frac{f(0)}{0!}x^0 + \frac{f'(0)}{1!}x^1 + \frac{f''(0)}{2!}x^2 + \frac{f^{(3)}(0)}{3!}x^3 + \cdots$$

But

$$f(x) = \cos x \quad f'(x) = -\sin x \quad f''(x) = -\cos x \quad f^{(3)}(x) = \sin x \quad \cdots$$

so

$$f(0) = 1 \quad f'(0) = 0 \quad f''(0) = -1 \quad f^{(3)}(0) = 0 \quad \cdots$$

and

$$M(x) = \frac{1}{0!}x^0 + 0 + \frac{-1}{2!}x^2 + 0 + \cdots$$
$$= x^0 - \frac{1}{2!}x^2 + \frac{1}{4!}x^4 \cdots$$
$$= \sum_{n=0}^{\infty} \frac{(-1)^n}{(2n)!}x^{2n} \quad \blacksquare$$

---

(3) Find the Maclaurin series for $f(x) = \ln(1+x)$.

☐ The definition of Maclaurin series gives us:

$$M(x) = \frac{f(0)}{0!}x^0 + \frac{f'(0)}{1!}x^1 + \frac{f''(0)}{2!}x^2 + \frac{f^{(3)}(0)}{3!}x^3 + \cdots$$

But

$$f(x) = \ln(1+x) \quad , \quad f'(x) = \frac{1}{1+x}$$

$$f''(x) = -\frac{1}{(1+x)^2} \quad , f^{(3)}(x) = \frac{2!}{(1+x)^3} \quad \cdots$$

so

$$f(0) = 0 \,,\, f'(0) = 1 \,,\, f''(0) = -1 \,,\, f^{(3)}(0) = 2! \quad \cdots$$

and

$$M(x) = \frac{0}{0!}x^0 + \frac{1}{1!}x^1 + \frac{-1!}{2!}x^2 + \frac{2!}{3!}x^3 + \cdots$$

$$= x - \frac{1}{2}x^2 + \frac{1}{3}x^3 + \cdots$$

$$= \sum_{n=1}^{\infty} (-1)^{n+1}\frac{x^n}{n} \quad \blacksquare$$

10.3   Infinite Sequences

Introduction

A **sequence** is a list of numbers. I'll bet you can guess what an *infinite sequence* is. Each entry in a sequence is a **term**. Examples are

- $1, 2, 3, 4, 5, \ldots$
- $1, \dfrac{1}{4}, \dfrac{1}{9}, \dfrac{1}{16}, \dfrac{1}{25}, \ldots$
- $1, \pi, 16, -2, \sqrt{7}, \ldots$

Some questions we can ask about an infinite sequence are:

- Is there a pattern to the sequence? That is, can you predict the next value in a sequence based on the values before it?
- Are the terms in the sequence getting closer and closer to some predictable value? (It would be nice to call this a "final" value, but since the sequence is infinitely long, there is no last, final number.)

If you like number games, then answers to these questions may be fun to consider for their own sake. But also, knowing how sequences behave is the key to knowing how infinite series behave — and that's our priority.

Notation and Definitions

There are lots of ways to represent a sequence. One way is to merely write out enough terms to establish any pattern in the sequence. We'll call this *list notation*. Another method is called *bracket notation*. This notation is much more brief than list notation, but it requires that you provide a "recipe" for constructing the terms in the sequence. Bracket notation looks like this:

$$\{a_n\}_{n=c}^{\infty}$$

where n is a counter (or index), c and ∞ are the starting and ending values of the counter which is incremented by 1, and a_n is the recipe for constructing the terms of the sequence using n. For example, the first sequence written above in the Introduction can be expressed using bracket notation as:

$$1, 2, 3, 4, 5, \ldots = \{n\}_{n=1}^{\infty}$$

It could also be written as

$$1, 2, 3, 4, 5, \ldots = \{n + 1\}_{n=0}^{\infty}$$

(Note the change in the starting value of the index.) So, a sequence can be written in many different ways using bracket notation — you can adjust either the starting counter value or the recipe as long as you adjust the other accordingly to be sure you're generating the same list of numbers.

As another example, the second sequence above in the Introduction can be written

$$1, \frac{1}{4}, \frac{1}{9}, \frac{1}{16}, \frac{1}{25}, \ldots = \left\{ \frac{1}{n^2} \right\}_{n=1}^{\infty}$$

The third sequence above ($1, \pi, 16, -2, \sqrt{7}, \ldots$) has no apparent pattern, and cannot be described using bracket notation.

Another notation that's sort of a hybrid of bracket and list notation looks like this:

$$a_n = \frac{n+1}{n!} \text{ for } n = 0, 1, 2, \ldots$$

Hopefully it's self explanatory.

Here are a couple of (helpful?) reminders for when it's time to write the recipe for a sequence:

- Many sequences contain terms which alternate between positive and negative. Including a factor of

$$(-1)^n \qquad \text{or } (-1)^{n+1}$$

  as part of the recipe will easily introduce alternating $+$ or $-$ signs in a sequence.
- Factorials come in handy when writing the recipe for many sequences. Remember that the definition of $n!$ is

$$n! = n \cdot (n-1) \cdot (n-2) \cdots 2 \cdot 1$$

$\boxed{\textbf{EX 1}}$ Give three equivalent formulas for the terms in the sequence
$$\frac{1}{2}, \frac{2}{3}, \frac{3}{4}, \frac{4}{5}, \ldots$$

By adjusting the recipe according to the starting value of the index, we can write this as:

- $a_n = \dfrac{n}{n+1}$ for $n = 1, 2, 3 \ldots$
- $a_n = \dfrac{n+1}{n+2}$ for $n = 0, 1, 2 \ldots$
- $a_n = \dfrac{n-1}{n}$ for $n = 2, 3, 4 \ldots$ ■

EX 2 Give three equivalent formulas for the terms in the sequence
$$-\frac{1}{2}, +\frac{1}{4}, -\frac{1}{8}, +\frac{1}{16}, \ldots$$

The denominators are powers of 2, and we have to create the alternating signs. So, we can write this in several ways depending on how we select the starting value of the index:

- $a_n = \dfrac{(-1)^n}{2^n}$ for $n = 1, 2, 3 \ldots$
- $a_n = \dfrac{(-1)^{n+1}}{2^{n+1}}$ for $n = 0, 1, 2, \ldots$
- $a_n = \dfrac{(-1)^{n+1}}{2^{n-1}}$ for $n = 2, 3, 4, \ldots$ ■

You Try It

(1) Find a formula for the terms in the sequence $2, 7, 12, 17, \ldots$.
(2) Find a formula for the terms in the sequence $3, \dfrac{9}{2}, \dfrac{27}{6}, \dfrac{81}{24}, \ldots$.

The Limit of a Sequence

The **limit** L of a sequence is (informally) the number that the terms of the sequence are tending towards. That is,

$$L = \lim_{n \to \infty} \{a_n\}$$

A sequence is said to **converge** if it has a finite limit. If the sequence has no limit, or its limit is $\pm\infty$, it **diverges**. 📺 FFT: How can a sequence have no limit, yet not be trending towards $-\infty$ or ∞? 📺

Determining the limit of a sequence may or may not be simple. The limit L of the sequence

$$1, \frac{1}{4}, \frac{1}{9}, \frac{1}{16}, \frac{1}{25}, \ldots$$

is pretty clearly $L = 0$. But, sometimes it's not so obvious, and we'll need help. Fortunately, Calculus I will come to the rescue!

Consider the function $f(x) = 1/x$ and suppose we go along the function, starting at $x = 1$, and yank out the value of the function at each integer — that is, at $x = 1$, $x = 2$, $x = 3$, etc. We have just generated the sequence

$$\left\{ \frac{1}{n} \right\}_{n=1}^{\infty}$$

This correspondence between a sequence and a function to which it is synchronized can always be found. If you have the recipe a_n for a sequence, you also have a function that will create all the terms in the sequence: simply replace n by x.

Because sequences can be associated with functions, all the tools we have available to deal with limits of functions can be used to deal with limits of sequences. These include the usual limit laws and L-Hopital's rule. We can state this tie between functions and sequences formally, with this Useful Fact:

Useful Fact 10.2. *Let $f(x)$ be a function that produces a sequence $\{a_n\}$, meaning that $a_n = f(n)$ for integer values of n. If $\lim\limits_{x \to \infty} f(x) = L$, then $\lim\limits_{n \to \infty} \{a_n\} = L$.*

$\boxed{\text{EX 3}}$ If it exists, determine the limit of the sequence $a_n = \dfrac{n^3 + n^2}{3n^3 - 2}$.

This sequence is synchronized to the function $f(x) = (x^2 + x)/(3x^2 - 2)$, since $a_n = f(n)$ for $n = 0, 1, 2 \ldots$. (Actually, we don't specify a starting value for n; but since we are seeking the limit as $n \to \infty$, it doesn't matter if n starts at 0 or 1 or 50.) A quick run through L-Hopital's Rule will establish the limit of the function as x goes to ∞ is $1/3$. We could also play the game of factoring out the highest power of n from the numerator and denominator, shown here as a reminder:

$$\lim_{n \to \infty} \frac{n^3 + n^2}{3n^3 - 2} = \lim_{n \to \infty} \frac{n^3(1 + 1/n)}{n^3(3 - 2/n^3)} = \lim_{n \to \infty} \frac{1 + 1/n}{3 - 2/n^3} = \frac{1}{3} \quad \blacksquare$$

EX 4 If it exists, determine the limit of the sequence $a_n = \dfrac{n+1}{e^n}$.

This sequence is synchronized to the function $f(x) = (x+1)/e^x$. The limit of that function as $x \to \infty$ would be found by L-Hopital's rule. We have

$$\lim_{x \to \infty} \frac{x+1}{e^x} = \lim_{x \to \infty} \frac{1}{e^x} = 0$$

Since the limit of $f(x)$ is 0 as $x \to \infty$, we also know that the limit of the sequence $\{a_n\}$ is 0 as $n \to \infty$. ∎

You Try It

(3) If it exists, determine the limit of the sequence $a_n = \dfrac{3 + 5n^2}{n + n^2}$.

(4) If it exists, determine the limit of the sequence $a_n = \dfrac{e^n + e^{-n}}{e^{2n} - 1}$.

(5) If it exists, determine the limit of the sequence $a_n = \cos \dfrac{n\pi}{2}$.

🔟 FFT: Useful Fact 10.2 tells us that once we establish a sequence is generated by a function via $a_n = f(n)$, then $\lim_{x \to \infty} f(x) = L$ also means $\lim_{n \to \infty} \{a_n\} = L$. Can we reverse this statement? Meaning, is the following statement true or false?

$$\text{If } \lim_{n \to \infty} \{a_n\} = L, \text{ then } \lim_{x \to \infty} f(x) = L$$

Monotone and Bounded Sequences

Monotone sequences are sequences whose terms display a steady trend.

- A sequence is **increasing** if any term is equal to or greater than the preceding term.
- A sequence is **decreasing** if any term is equal to or less than the preceding term.

We add the word **strictly** to either descriptor if equality between terms is not allowed. For example, $1, 1, 2, 2, 3, 3, \ldots$ is increasing, but not strictly increasing. On the other hand, $1, 2, 3, \ldots$ is strictly increasing. Another complication is that monotomic trends might not begin with the first term — we may have to "pass by" several terms before monotonicity appears. Sequences that behave like this are said to **eventually** increase or decrease.

Sequences that are increasing or decreasing may also be **bounded**. The definition of boundedness is fairly self-explanatory. *Bounded sequences* can be bounded above and/or below.

- If a sequence is bounded from above, its terms are always smaller in magnitude than some number M, i.e. $|a_n| \leq M$ for all n.
- If a sequence is bounded from below, its terms are always larger in magnitude than some number N, i.e. $|a_n| \geq N$ for all n.

For example, the sequence $a_n = 1/n$ is decreasing for $n \geq 1$; it is bounded below by 0 and above by 1.

We investigate the trend of a sequence by checking explicitly whether a_{n+1} is always greater than or less than a_n, using the recipe for a_n: build a representation of a_{n+1} and set it against a_n to test if one is always larger or smaller than the other.

EX 5 Determine whether the sequence $a_n = \dfrac{n+3}{n-1}$, for $n \geq 2$, is increasing or decreasing, and if it is bounded.

The first few terms in the sequence are $5, 3, 7/3, 2, \ldots$. It *seems* to be decreasing, but let's show that for sure. We need to see if $a_n > a_{n+1}$ for all $n \geq 2$ by building a_{n+1} and setting it against a_n in a contest of "left side versus right side" to see which is greater. Since all the statements in this sequence are equivalent, then once we know which way the inequality goes, that relation is established for all.

$$a_n \text{ vs } a_{n+1}$$
$$\frac{n+3}{n-1} \text{ vs } \frac{(n+1)+3}{(n+1)-1}$$
$$\frac{n+3}{n-1} \text{ vs } \frac{n+4}{n}$$
$$(n+3)(n) \text{ vs } (n+4)(n-1)$$
$$n^2 + 3n \text{ vs } n^2 + 3n - 4$$
$$0 \text{ vs } -4$$

In the last line, we know $0 > -4$, so that $>$ will cascade up the stack of equivalent expressions and back to the top row: $a_n > a_{n+1}$ for all n. This means every entry in the sequence is less than the one that preceded it; the sequence is (strictly) decreasing. Also, since the first term (which is 5) is the largest term in the sequence, then the sequence is bounded above by 5.

To see if it is bounded below as it decreases, we should check if there is a limit:

$$\lim_{n\to\infty} \frac{n+3}{n-1} = \lim_{n\to\infty} \frac{n(1+3/n)}{n(1-1/n)} = \lim_{n\to\infty} \frac{1+3/n}{1-1/n} = 1$$

The sequence is decreasing and has a limit of 1. This means it will never fall below 1. Altogether, this strictly decreasing sequence is bounded above by 5 and below by 1. We have now diagnosed the heck out of this sequence. ∎

You Try It

(6) Determine whether the sequence $a_n = \dfrac{n}{n^2+1}$ is increasing or decreasing, and if it is bounded.

Infinite Sequences — Problem List

Infinite Sequences — You Try It

These appeared above; solutions begin on the next page.

(1) Find a formula for the terms in the sequence $2, 7, 12, 17, \ldots$.

(2) Find a formula for the terms in the sequence $3, \dfrac{9}{2}, \dfrac{27}{6}, \dfrac{81}{24}, \ldots$.

(3) If it exists, determine the limit of the sequence $a_n = \dfrac{3 + 5n^2}{n + n^2}$.

(4) If it exists, determine the limit of the sequence $a_n = \dfrac{e^n + e^{-n}}{e^{2n} - 1}$.

(5) If it exists, determine the limit of the sequence $a_n = \cos\dfrac{n\pi}{2}$.

(6) Determine whether the sequence $a_n = \dfrac{n}{n^2 + 1}$ is increasing or decreasing, and if it is bounded.

Infinite Sequences — Practice Problems

Try these as you get the hang of the You Try It problems. Solutions to these problems are available in Sec. A.4.3.

(1) Find a formula for the terms in the sequence $-\dfrac{1}{4}, \dfrac{2}{9}, -\dfrac{3}{16}, \dfrac{4}{25}, \ldots$.

(2) If it exists, determine the limit of the sequence $a_n = \dfrac{\sqrt{n}}{1 + \sqrt{n}}$.

(3) If it exists, determine the limit of the sequence $a_n = \cos\left(\dfrac{2}{n}\right)$.

(4) If it exists, determine the limit of the sequence $a_n = \dfrac{\ln n}{\ln 2n}$.

(5) Determine whether the sequence $a_n = ne^{-n}$ is increasing or decreasing, and if it is bounded.

Infinite Sequences — Challenge Problems

Try these problems to test your skills with the ideas in this section. Solutions to these problems are available in Sec. B.4.3.

(1) If it exists, determine the limit of the sequence $a_n = \dfrac{(-1)^n n^3}{n^3 + 2n^2 + 1}$.

(2) If it exists, determine the limit of the sequence $a_n = \ln(n + 1) - \ln(n)$.

(3) Determine whether the sequence $a_n = n + \dfrac{1}{n}$ is increasing or decreasing, and if it is bounded.

Infinite Sequences — You Try It — Solved

---

(1) Find a formula for the terms in the sequence $2, 7, 12, 17, \ldots$.

☐ Notice that each number in the sequence is two more than a multiple of 5, that is

$$\{2, 7, 12, 17, \ldots\} = \{0 + 2, 5 + 2, 10 + 2, 15 + 2, \ldots\}$$

so that $a_n = 5n + 2$ for $n \geq 0$. ∎

---

(2) Find a formula for the terms in the sequence $3, \dfrac{9}{2}, \dfrac{27}{6}, \dfrac{81}{24}, \ldots$.

☐ The numerators are powers of 3. Denominators, explicitly, are $1, 2, 6, 24, \ldots$ which should be recognizable as factorials $1!, 2!, 3!, 4!, \ldots$. So, how about

$$a_n = \frac{3^n}{n!} \quad ; \quad n \geq 1 \quad ∎$$

---

(3) If it exists, determine the limit of the sequence $a_n = \dfrac{3 + 5n^2}{n + n^2}$.

☐ We can rewrite the given sequence as

$$a_n = \frac{3 + 5n^2}{n + n^2} = \frac{n^2(\frac{3}{n^2} + 5)}{n^2(\frac{1}{n} + 1)} = \frac{\frac{3}{n^2} + 5}{\frac{1}{n} + 1}$$

so

$$\lim_{n \to \infty} a_n = 5$$

and the sequence converges to 5. ∎

---

(4) If it exists, determine the limit of the sequence $a_n = \dfrac{e^n + e^{-n}}{e^{2n} - 1}$.

☐ For the sequence

$$a_n = \frac{e^n + e^{-n}}{e^{2n} - 1}$$

we can use L-Hopital's rule, and then simplify. After replacing the numerator and denominator with their derivatives (L-Hopital's rule), we get:

$$\lim_{n \to \infty} \frac{e^n - e^{-n}}{2e^{2n}} = \lim_{n \to \infty} \frac{e^{2n} - 1}{2e^{3n}} = \lim_{n \to \infty} \left(\frac{1}{2e^n} - \frac{1}{2e^{3n}} \right) = 0$$

The sequence converges to 0. ∎

---

(5) If it exists, determine the limit of the sequence $a_n = \cos\dfrac{n\pi}{2}$.

☐ Because n is an integer, $a_n = \cos(n\pi/2)$ will oscillate among the values -1, 0, and 1, never converging to one specific value. The sequence diverges.   ∎

---

(6) Determine whether the sequence $a_n = \dfrac{n}{n^2 + 1}$ is increasing or decreasing, and if it is bounded.

☐ I have a suspicion it decreases since the sequence starts as $1/2, 2/5, 3/10, \ldots$. To show this, I need to show that $a_n > a_{n+1}$ for all $n > 0$.

$$a_n >? \ a_{n+1}$$
$$\frac{n}{n^2 + 1} >? \ \frac{(n+1)}{(n+1)^2 + 1}$$
$$(n)(n^2 + 2n + 2) >? \ (n+1)(n^2 + 1)$$
$$n^3 + 2n^2 + 2n >? \ n^3 + n^2 + n + 1$$
$$2n^2 + 2n >? \ n^2 + n + 1$$
$$n^2 + n >? \ 1$$

This is surely true for all $n > 0$, so the sequence is indeed decreasing. It is bounded above by its first term of $1/2$ and below by 0.   ∎

10.4   Infinite Series

Introduction

An infinite sequence is an infinite list of numbers. An **infinite series** is, loosely speaking, a *sum* of an infinite sequence. If $\{a_n\}$ is an infinite sequence, we can create the infinite series

$$\sum_{n=0}^{\infty} a_n = a_0 + a_1 + a_2 + \cdots$$

This should bring to mind a couple of questions:

(1) Is it possible for a summation of an infinite number of numbers to have a finite result?

(2) How do you find the sum of an infinite list of numbers?

The simple answer to question 1 is: YES. You can add an infinite number of numbers together and get a finite result. If we couldn't, Riemann Sums wouldn't work, and Calculus would be doomed! To get a handle on # 2, we need a new concept, called a partial sum.

Partial Sums

It is not actually possible to add up an infinite number of items in the logistical sense because we would never finish the job. Instead, we have to do a sort of numerical target practice. Let's take an infinite sequence $\{a_n\}$ and start making a list of numbers derived from this sequence:

- Add up the first two terms $a_1 + a_2$ and note the result. Call it s_2.
- Add up the first three terms $a_1 + a_2 + a_3$ and note the result. Call it s_3.
- Add up the first four terms $a_1 + a_2 + a_3 + a_4$ and note the result. Call it s_4. And so on. Do you get the idea?

Or, in summation notation,

$$s_2 = \sum_{i=1}^{2} a_i \quad ; \quad s_3 = \sum_{i=1}^{3} a_i \quad ; \quad s_4 = \sum_{i=4}^{2} a_i \quad \cdots$$

The values $s_2, s_3, s_4, \ldots$ are called partial sums. In general, the nth partial sum is

$$s_n = \sum_{i=0}^{n} a_i = a_0 + a_1 + a_2 + \cdots + a_n$$

And while we are not evaluating the entire series $\sum\limits_{i=0}^{\infty} a_i$, we are making very, very slow progress towards it. Each partial sum is a "shot" at the target, and the farther out we go into the list (sequence) of partial sums, the closer we get to the value of the series ... if it has one. The list of numbers $s_2, s_3, s_4, \ldots$ is itself a sequence, and if this sequence of partial sums converges, then we say the entire infinite series converges. We can put this together completely as follows:

Useful Fact 10.3. *Given an infinite series* $\sum\limits_{i=0}^{\infty} a_i$*, we define the sequence of partial sums* $\{s_n\}$*, where each* s_n *is given as* $s_n = \sum\limits_{i=0}^{n} a_i$*. If this sequence of partial sums has a limit* S*, then* S *is defined to be the value of the infinite series:*

$$\sum_{i=0}^{\infty} a_i = \lim_{n\to\infty} s_n = S$$

$\boxed{\text{EX 1}}$ Write the first few partial sums for the series $\sum\limits_{i=1}^{\infty} \dfrac{1}{2i}$ and use them to guess if the series converges or diverges.

The first few partial sums of the series with $a_i = 1/(2i)$ for $i \geq 1$ are:

$$s_1 = \frac{1}{2} \quad ; \quad s_2 = \frac{1}{2} + \frac{1}{4} = \frac{3}{4} \quad ; \quad s_3 = \frac{1}{2} + \frac{1}{4} + \frac{1}{6} = \frac{11}{12}$$

We now have a sequence of values $1/2, 3/4, 11/12, \ldots$. The fate of this sequence and the fate of the given infinite series for this problem are the same. You might think from those first entries that the sequence of partial sums converges to 1. But you'd be wrong! The next partial sum is

$$s_4 = \frac{1}{2} + \frac{1}{4} + \frac{1}{6} + \frac{1}{8} = \frac{25}{24}$$

which is bigger than 1. It turns out that this sequence of partial sums does not converge, and the sum

$$\sum_{i=1}^{\infty} \frac{1}{2i} = \frac{1}{2} + \frac{1}{4} + \frac{1}{6} + \cdots$$

does not have a finite value. ∎

You Try It

(1) Find the 5th partial sum of the series $\sum_{n=0}^{\infty} \dfrac{1}{2^n}$.

The Divergence Test

In general, the task of determining if an infinite series converges (and what it converges to) is not simple. The concept of partial sums gives us the fundamental meaning of convergence, but we hardly ever want to actually rely on partial sums to determine convergence or divergence. In fact, the next several topics after this introduction are devoted to "convergence tests", which help us determine if a series converges.

Since determining convergence is often not simple, let's start by going about it another way — can we check quickly to see if a series *diverges*? If we can determine that a series diverges, we don't have to spend any time worrying about convergence. Here's they key: if a series is going to converge, then the terms in the series *must* be approaching 0. If the terms are not approaching 0, and we're adding up an infinite number of those terms, then the series certainly cannot converge to a finite value. Let's put it more formally:

Useful Fact 10.4. *Let a_n represent the terms in an infinite series. If* $\lim\limits_{n \to \infty} |a_n| \neq 0$, *then the series* $\sum\limits_{n=1}^{\infty} a_n$ *must diverge. This is called the **divergence test**.*

$\boxed{\textbf{EX 2}}$ Investigate the convergence or divergence of $\sum\limits_{n=1}^{\infty} \dfrac{n}{n+1}$.

Since $\lim\limits_{n \to \infty} \dfrac{n}{n+1} = 1$, the limit of the terms in this series is not 0, and by the divergence test, we know this series cannot converge. It is a divergent series. ∎

You Try It

    (2) Investigate the convergence or divergence of $\displaystyle\sum_{n=2}^{\infty} \frac{n^2}{n^2-1}$.

    NOTE: There is an important thing to remember about the divergence test. The divergence test says that

$$\text{if } \lim_{n\to\infty} |a_n| \neq 0 \text{ then the series diverges}$$

This if-then only works in one direction. In other words, this statement:

$$\text{if } \lim_{n\to\infty} |a_n| = 0 \text{ then the series converges}$$

is FALSE! If you have a series in which the terms approach 0 as $n \to \infty$, you still do not know if it converges or not. For example, the series

$$\sum_{n=1}^{\infty} \frac{1}{n}$$

is called the **harmonic series** and it diverges[6] even though the terms approach 0 as n approaches ∞. All you can discover from the divergence test is its direct implication: a non-zero limit of the terms in the series means the series diverges. That's all. You cannot try to use it to conclude anything else.

The Geometric Series

One of the most important types of series is the geometric series. This will show up over and over, so learn it and love it. A geometric series is of the form

$$\sum_{n=0}^{\infty} ar^n$$

These are geometric series:

$$\sum_{n=0}^{\infty} \frac{1}{4}\left(\frac{1}{2}\right)^n \text{ and } \sum_{n=0}^{\infty} \left(\frac{1}{5}\right)^n \tag{10.12}$$

but this is not:

$$\sum_{n=0}^{\infty} \frac{1}{4}(n!)^n$$

The convergence or divergence of a geometric series is easy to confirm:

---

[6]You'll just have to take my word for it at this point!

Useful Fact 10.5. *A **geometric series** of the form* $\displaystyle\sum_{n=0}^{\infty} ar^n$ *will converge if* $|r| < 1$. *A convergent geometric series converges to the value* $\dfrac{a}{1-r}$.

A discussion of why this Useful Fact is true is at the end of this section.

Note that the convergence formula $\dfrac{a}{1-r}$ is dependent on the counter for the geometric series starting at $n = 0$, as shown in Useful Fact 10.5. When the geometric series is already presented in summation form, with the counter starting at $n = 0$, then finding the value of convergence is pretty easy. For example, for the two sample geometric series given in (10.12) follow like this:

- In $\displaystyle\sum_{n=0}^{\infty} \frac{1}{4}\left(\frac{1}{2}\right)^n$, we see $a = \dfrac{1}{4}$ and $r = \dfrac{1}{2}$. Since $|r| < 1$, the series converges to:

$$\sum_{n=0}^{\infty} \frac{1}{4}\left(\frac{1}{2}\right)^n = \frac{a}{1-r} = \frac{1/4}{1-1/2} = \frac{1}{2}$$

- In $\displaystyle\sum_{n=0}^{\infty} \left(\frac{1}{5}\right)^n$, we see $a = 1$ and $r = \dfrac{1}{5}$. Since $|r| < 1$, the series converges to:

$$\sum_{n=0}^{\infty} \left(\frac{1}{5}\right)^n = \frac{a}{1-r} = \frac{1}{1-1/5} = \frac{5}{4}$$

When series that turn out to be geometric are presented in a term by term fashion, then we have to generate the summation structure ourselves, so we can decide on the values of a and r.

EX 3 Investigate the convergence of the series $3 + \dfrac{3}{4} + \dfrac{3}{16} + \dfrac{3}{64} + \cdots$.

Writing this series in summation notation, we have

$$\sum_{n=0}^{\infty} \frac{3}{4^n} = \sum_{n=0}^{\infty} 3\left(\frac{1}{4}\right)^n$$

so it's a geometric series with $a = 3$ and $r = \dfrac{1}{4}$. Since $|r| < 1$, this series converges to

$$\frac{a}{1-r} = \frac{3}{1-1/4} = 4 \quad \blacksquare$$

It isn't enough for a series to kind of, sort of, look like the standard geometric series template in Useful Fact 10.5. Rather, we have to be sure series can be put precisely into the form specified there. This can involve some adjusting of the book-keeping relative to the counter or the exponent in play.

EX 4 Investigate the convergence of the series $\displaystyle\sum_{n=1}^{\infty} \frac{2}{3^{n+1}}$.

The series in this example has two problems with it: (1) The starting value of the index is 1, not 0. (2) The exponent in the term is $n + 1$, not n. Let's see if we can account for these issues. First, let's reindex the series to change the starting value to 0:

$$\sum_{n=1}^{\infty} \frac{2}{3^{n+1}} = \frac{2}{3^2} + \frac{2}{3^3} + \frac{2}{3^4} + \cdots = \frac{2}{3^{0+2}} + \frac{2}{3^{1+2}} + \frac{2}{3^{2+2}} + \cdots = \sum_{n=0}^{\infty} \frac{2}{3^{n+2}}$$

Reindexing is a standard trick that can be applied to any series. Note that the net result here was a trade-off between bumping down the starting value of the index by 1 (from $n = 1$ to $n = 0$) and bumping up each instance of the index in the recipe — the exponent of $n + 1$ changed to $(n + 1) + 1 = n + 2$. This ensured we still got the same numbers in the same places.

Next, we have to fix the exponent issue. We need to change the exponent in play from $n + 2$ to n; this can be done by peeling off a couple of 3's:

$$\sum_{n=0}^{\infty} \frac{2}{3^{n+2}} = \sum_{n=0}^{\infty} \frac{2}{3^2 3^n} = \sum_{n=0}^{\infty} \frac{2}{9} \left(\frac{1}{3}\right)^n$$

This is now in precisely the correct geometric series form, with $a = 2/9$ and $r = 1/3$. Since $|r| < 1$, the series converges to

$$\sum_{n=0}^{\infty} \frac{2}{9} \left(\frac{1}{3}\right)^n = \frac{a}{1 - r} = \frac{2/9}{1 - 1/3} = \frac{2}{9} \cdot \frac{3}{2} = \frac{1}{3} \quad \blacksquare$$

You Try It

(3) Investigate the convergence or divergence of $3 + 2 + \dfrac{4}{3} + \dfrac{8}{9} + \cdots$.

(4) Investigate the convergence or divergence of $\displaystyle\sum_{n=1}^{\infty} \frac{(-3)^{n-1}}{4^n}$.

Applications of Geometric Series I: Writing Repeating Decimals as Fractions

As a sort of oddball application of geometric series, consider this scenario: You have undoubtedly gone through the exercise of taking a fraction and dividing it out to see its decimal format. Often, those decimal formats containg repeating decimals. But have you ever done it the other way? That is, have you ever taken a repeating decimal and figured out the fraction it came from? We can do this with geometric series.

$\boxed{\textbf{EX 5}}$ Write $1.26262626\ldots$ as a fraction.

First, we can extract the repeated grouping in fractional form:

$$1.26262626\ldots = 1.0 + 0.26 + 0.0026 + 0.000026 + \cdots$$
$$= 1 + \frac{26}{100} + \frac{26}{10000} + \frac{26}{1000000} + \cdots$$
$$= 1 + \frac{26}{10^2} + \frac{26}{10^4} + \frac{26}{10^6} + \cdots$$

Believe it or not, there's a geometric series hiding in there:

$$1.26262626\ldots = 1 + \frac{26}{10^2} + \frac{26}{10^4} + \frac{26}{10^6} + \cdots$$
$$= 1 + \frac{26}{10^2}\left(1 + \frac{1}{10^2} + \frac{1}{10^4} + \frac{1}{10^6} + \cdots\right)$$
$$= 1 + \frac{26}{10^2}\left(1 + \frac{1}{10^2} + \frac{1}{(10^2)^2} + \frac{1}{(10^2)^3}\cdots\right)$$
$$= 1 + \sum_{n=0}^{\infty} \frac{26}{100}\left(\frac{1}{100}\right)^n$$

The series that showed up is geometric, with $a = 26/100$ and $r = 1/100$. Since $|r| < 1$, the series converges, and the whole expression converges to

$$1.26262626\ldots = 1 + \frac{26/100}{1 - 1/100} = 1 + \frac{26}{100}\cdot\frac{100}{99} = 1 + \frac{26}{99} = \frac{125}{99} \quad\blacksquare$$

You Try It

(5) Write $3.417417417\ldots$ as a fraction.

Applications of Geometric Series II: Infinite Series as Functions

Here is an infinite series; in fact, it's a convergent geometric series:

$$\sum_{n=0}^{\infty} 3\left(\frac{1}{2}\right)^n \qquad (10.13)$$

But what is this next thing?

$$\sum_{n=0}^{\infty} 3\left(\frac{x}{2}\right)^n \qquad (10.14)$$

All I did was change the numerator of the fraction from 1 to x. But in doing so, I made a huge change to the meaning of the expression. The original expression was equivalent to the value 6 (be sure you know why!). The new expression is actually a *function* because it contains a variable, x. You might imagine that some values, when plugged in for x, will lead to convergence of the series while others might lead to divergence. For example, the first expression (10.13) is simply the value of the function (10.14) for $x = 1$; we know that this value is 6.

If we plugged in 4 for x in (10.14), would the resulting series be convergent? Nope. It would be a geometric series with $r = 2$, and that series won't converge. But if we plugged in $x = 1/4$, the resulting series would converge. There is an entire domain of x values that will lead to (10.14) being a convergent series. And the nifty thing is that we can actually find the algebraic form of the function that the series (10.14) represents (when it converges) using the geometric series formula. The following example shows this.

| **EX 6** | Find the values of x for which the series $\displaystyle\sum_{n=1}^{\infty} 3\left(\frac{x}{2}\right)^n$ converges, and determine the algebraic form of the function the convergent series defines.

This is in the form of a geometric series, with $a = 3$ and $r = x/2$. In order for a geometric series to converge, we need $|r| < 1$. So for this series to converge, we need $|x/2| < 1$, i.e. $|x| < 2$. This is the interval $-2 < x < 2$. Any x value selected from that interval will give a convergent series. And further, since this is a geometric series, we know that when it converges it will converge to

$$\frac{a}{1-r} = \frac{3}{1-x/2} = \frac{6}{2-x}$$

So for values of x in the interval $-2 < x < 2$, we now know that

$$\sum_{n=1}^{\infty} 3\left(\frac{x}{2}\right)^n \text{ is the series form of the function } \frac{6}{2-x} \quad \blacksquare$$

You Try It

(6) Find the values of x for which the series $\sum_{n=1}^{\infty}\left(\frac{x}{3}\right)^n$ converges, and determine the algebraic form of the function the convergent series defines.

Into the Pit!!

Demonstration of Geometric Series Convergence Formula

Here is *why* geometric series work the way they do, as shown in Useful Fact 10.5. It's an example of how the concept of partial sums is useful as gives us the fundamental meaning of convergence, even though we may not use them directly when investigating convergence. Remember that convergence of a geometric series will be dictated by its sequence of partial sums. If we write out the nth partial sum of a geometric series, we get:

$$s_n = a + ar + ar^2 + ar^3 + \cdots + ar^n$$

Now let's multiply that expression by r (don't ask why yet!):

$$rs_n = ar + ar^2 + ar^3 + ar^4 + \cdots + ar^{n+1}$$

Now let's subtract those two expressions. Most of the terms will cancel out (hooray!) and we're left with

$$s_n - rs_n = a - ar^{n+1}$$
$$s_n(1 - r) = a(1 - r^{n+1})$$
$$s_n = \frac{a(1 - r^{n+1})}{1 - r}$$

OK, so what was the point of all that? Well, we know that a geometric series will converge if its sequence of partial sums converges. That is, we need

$$\lim_{n\to\infty} s_n = \lim_{n\to\infty} \frac{a(1 - r^{n+1})}{1 - r}$$

to be finite. The only way this will remain finite as n goes to ∞ is if $|r| < 1$, so that r^{n+1} goes to 0. When $|r| < 1$, the partial sums (and so the geometric series) converges to $a/(1 - r)$. Voila!

Infinite Series — Problem List

Infinite Series — You Try It

These appeared above; solutions begin on the next page.

(1) Find the 5th partial sum of the series $\sum_{n=0}^{\infty} \dfrac{1}{2^n}$.

(2) Investigate the convergence or divergence of $\sum_{n=2}^{\infty} \dfrac{n^2}{n^2 - 1}$.

(3) Investigate the convergence or divergence of $3 + 2 + \dfrac{4}{3} + \dfrac{8}{9} + \cdots$.

(4) Investigate the convergence or divergence of $\sum_{n=1}^{\infty} \dfrac{(-3)^{n-1}}{4^n}$.

(5) Write $3.417417417\ldots$ as a fraction.

(6) Find the values of x for which the series $\sum_{n=1}^{\infty} \left(\dfrac{x}{3}\right)^n$ converges, and determine the algebraic form of the function the convergent series defines.

Infinite Series — Practice Problems

Try these as you get the hang of the You Try It problems. Solutions to these problems are available in Sec. A.4.4.

(1) Investigate the convergence or divergence of $\sum_{n=1}^{\infty} \ln\left(\dfrac{n}{2n + 5}\right)$.

(2) Investigate the convergence or divergence of $\dfrac{1}{8} - \dfrac{1}{4} + \dfrac{1}{2} - 1 + \cdots$.

(3) Investigate the convergence or divergence of $\sum_{n=0}^{\infty} \dfrac{\pi^n}{3^{n+1}}$.

(4) Write $6.2545454\ldots$ as a fraction.

(5) Find the values of x for which the series $\sum_{n=1}^{\infty} (x - 4)^n$ converges, and determine the algebraic form of the function the convergent series defines.

Infinite Series — Challenge Problems

Try these problems to test your skills with the ideas in this section. Solutions to these problems are available in Sec. B.4.4.

(1) Investigate the convergence or divergence of $\sum_{n=1}^{\infty} \frac{e^n}{3^{n-1}}$.

(2) Write $0.123456456\ldots$ as a fraction.

(3) Find the values of x for which the series $\sum_{n=0}^{\infty} \frac{(x+3)^n}{2^n}$ converges, and determine the algebraic form of the function the convergent series defines.

Infinite Series — You Try It — Solved

---

(1) Find the 5th partial sum of the series $\sum\limits_{n=0}^{\infty} \dfrac{1}{2^n}$.

□ We add up the first five terms in the series:

$$s_5 = \frac{1}{2^0} + \frac{1}{2^1} + \frac{1}{2^2} + \frac{1}{2^3} + \frac{1}{2^4} = 1 + \frac{1}{2} + \frac{1}{4} + \frac{1}{8} + \frac{1}{16}$$

$$= \frac{16}{16} + \frac{8}{16} + \frac{4}{16} + \frac{2}{16} + \frac{1}{16} = \frac{31}{16}$$

NOTE: It turns out that this series converges to 2. The 5th partial sum is already only 1/16 away from 2. The rest of the terms in the infinite series do not add up to more than 1/16, even though there are still an infinite number of terms left to go. That's pretty cool. ∎

---

(2) Investigate the convergence or divergence of $\sum\limits_{n=2}^{\infty} \dfrac{n^2}{n^2 - 1}$.

□ Since $a_n = n^2/(n^2 - 1)$, we see that $a_n \to 1$ as $n \to \infty$. So by the divergence test, this series cannot converge. The series diverges. (Convergence requires $a_n \to 0$.) ∎

---

(3) Investigate the convergence or divergence of $3 + 2 + \dfrac{4}{3} + \dfrac{8}{9} + \cdots$.

□ Let's find the series formula, see if it's a geometric series, and determine convergence that way.

$$3 + 2 + \frac{4}{3} + \frac{8}{9} + \cdots = 3\left(1 + \frac{2}{3} + \frac{4}{9} + \frac{8}{27} + \cdots\right)$$

$$= 3\left(1 + \left(\frac{2}{3}\right)^1 + \left(\frac{2}{3}\right)^2 + \left(\frac{2}{3}\right)^3 + \cdots\right) = \sum_{n=0}^{\infty} 3\left(\frac{2}{3}\right)^n$$

This is a geometric series with $a = 3$ and $r = 2/3$. Since $|r| < 1$, the series converges, and it converges to

$$\sum_{n=0}^{\infty} 3\left(\frac{2}{3}\right)^n = \frac{a}{1 - r} = \frac{3}{1 - 2/3} = \frac{3}{1/3} = 9 \quad ∎$$

(4) Investigate the convergence or divergence of $\displaystyle\sum_{n=1}^{\infty} \frac{(-3)^{n-1}}{4^n}$.

□ This series looks geometric-ish, let's pursue that, by adjusting the starting counter to zero, and fixing the exponents to match the form in Useful Fact 10.5:

$$\sum_{n=1}^{\infty} \frac{(-3)^{n-1}}{4^n} = \sum_{n=0}^{\infty} \frac{(-3)^n}{4^{n+1}} = \sum_{n=0}^{\infty} \frac{1}{4} \cdot \left(-\frac{3}{4}\right)^n$$

This is a geometric series with $a = 1/4$ and $r = -3/4$. Since $|r| < 1$, the series converges, and it converges to

$$\sum_{n=1}^{\infty} \frac{(-3)^{n-1}}{4^n} = \sum_{n=0}^{\infty} \frac{1}{4} \cdot \left(-\frac{3}{4}\right)^n = \frac{a}{1-r} = \frac{1/4}{1+3/4} = \frac{1}{7} \quad ■$$

(5) Write $3.417417417\ldots$ as a fraction.

$$□\, 3.417417417\ldots = 3 + \frac{417}{10^3} + \frac{417}{10^6} + \frac{417}{10^9} + \cdots$$

$$= 3 + \frac{417}{10^3}\left(1 + \frac{1}{10^3} + \frac{1}{10^6} + \cdots\right)$$

$$= 3 + \sum_{n=0}^{\infty} \frac{417}{10^3}\left(\frac{1}{10^3}\right)^n$$

This is a geometric series with $a = 417/10^3$ and $r = 1/10^3$. Since $|r| < 1$, the series converges, and the whole expression converges to

$$3.417417417\ldots = 3 + \sum_{n=0}^{\infty} \frac{417}{10^3}\left(\frac{1}{10^3}\right)^n = 3 + \frac{417/10^3}{1 - 1/10^3}$$

$$= 3 + \frac{417}{999} = 3 + \frac{139}{333} = \frac{1138}{333} \quad ■$$

(6) Find the values of x for which the series $\displaystyle\sum_{n=1}^{\infty} \left(\frac{x}{3}\right)^n$ converges, and determine the algebraic form of the function the convergent series defines.

□ This is a geometric series that's almost in standard form, so let's mess with it:

$$\sum_{n=1}^{\infty} \left(\frac{x}{3}\right)^n = \sum_{n=0}^{\infty} \left(\frac{x}{3}\right)^{n+1} = \sum_{n=0}^{\infty} \frac{x}{3} \cdot \left(\frac{x}{3}\right)^n$$

Now it is a geometric series in the standard form from Useful Fact 10.5, with $a = x/3$ and $r = x/3$. It will converge when $|r| < 1$, i.e. when $|x/3| < 1$, i.e. when $|x| < 3$. When it does converge, it converges to

$$\frac{a}{1-r} = \frac{x/3}{1-x/3} = \frac{x}{3-x}$$

So, the series $\displaystyle\sum_{n=1}^{\infty}\left(\frac{x}{3}\right)^n$ is equivalent to the function $\dfrac{x}{3-x}$ on $-3 < x < 3$. ∎

10.5   Alternating Series

Introduction

Many infinite series have terms which alternate in sign. For example, did you notice that the series presented in You Try It 4 in Sec. 10.4 had that behavior? It just sort of snuck in there without warning, and maybe it didn't bother you. If so, that's good, because then you won't find *alternating series* anything to worry about!

Alternating Series

Definition 10.2. *An alternating series is an infinite series in which not all terms are positive. Usually, the terms alternate in sign like +/-/+/-,... Thus the name. The general form of an alternating series is*

$$\sum_{n=1}^{\infty}(-1)^{n+1}a_n \quad or \quad \sum_{n=1}^{\infty}(-1)^n a_n$$

For example, these are examples of alternating series:

$$(A) \quad \sum_{n=0}^{\infty}\frac{(-1)^n}{n+1} \quad and \quad (B) \quad \sum_{n=0}^{\infty}(-1)^n\frac{e^{-n}}{n^2+1}$$

We would describe the term formulas as $a_n = 1/(n+1)$ and $a_n = e^{-n}/(n^2+1)$ respectively. The sign generator, $(-1)^n$, is not included in what we call the term formula a_n.

Finding whether an alternating series converges is fairly easy.

Useful Fact 10.6. *The **alternating series test** says that for an alternating series to converge, these two things need to happen:*

   1 *Each successive term is no bigger than the one before it:*

$$a_{n+1} \le a_n \ for \ n \ge 1$$

   2 *The terms converge to zero:* $\lim_{n\to\infty} a_n = 0$

At first glance, you may think these two items are redundant, but they are not. A series could, for example, satisfy condition (1) but be converging to 1 instead of 0. For any given series, you must show that *both* conditions hold to demonstrate convergence, but you only need to show that ONE of the conditions *fails* to demonstrate divergence. Although its wording is phrased casually, the alternating series test is inherently an "if and only if" statement, or equivalently that the two conditions are "necessary and sufficient" for convergence. This means that not only do we say, "IF the conditions are met, then the series converges", but also, "if one of the conditions is not met, then the series does not converge."

$\boxed{\textbf{EX 1}}$ Investigate the convergence of $\displaystyle\sum_{n=0}^{\infty} \frac{(-1)^n}{n+1}$ and $\displaystyle\sum_{n=0}^{\infty}(-1)^n \frac{1+n}{2+n}$

The first series is called the **alternating harmonic series**. We can investigate its convergence by checking the two conditions of the alternating series test.

(1) Do the terms get progressively smaller? That is, can we show that $a_{n+1} \leq a_n$ for all n? Since $a_n = 1/(n+1)$, then

$$a_{n+1} = \frac{1}{(n+1)+1} = \frac{1}{n+2}$$

and the condition $a_{n+1} \leq a_n$ becomes

$$\frac{1}{n+2} \leq \frac{1}{n+1}$$

which is equivalent to $n+2 \geq n+1$. Since this is always true, then yes, $a_{n+1} \leq a_n$ for all n. The first condition is met.

(2) Do the sequence of (signless) terms approach zero? Well, since $a_n = 1/(n+1)$, then

$$\lim_{n\to\infty} a_n = \lim_{n\to\infty} \frac{1}{n+1} = 0$$

The second condition is also met.

Since both conditions of the alternating series test are met, the alternating harmonic series converges. (This is in contrast to the regular harmonic series, which diverges.) Figure 10.10 shows the values of the first 8 partials sums of the alternating harmonic series (n on the x-axis, and s_n on the y-axis), and you can see the partial sums closing in on some ultimate value of convergence.

In the second series, with $a_n = (1+n)/(2+n)$, we can note that $\lim_{n\to\infty} a_n = 1$. The second condition of the alternating series test is not met, and this series will diverge. $\blacksquare$

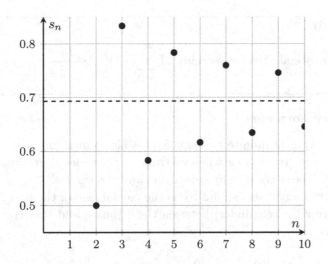

Fig. 10.10 Partial sums $S_2 \ldots S_{10}$ for the alternating harmonic series.

You Try It

(1) One of these alternating series converges and one diverges. Which is which, and how do you know?

$$(A) \sum_{n=1}^{\infty} (-1)^n \frac{n^2 + 2n + 100}{n^3} \quad ; \quad (B) \sum_{n=0}^{\infty} (-1)^n \frac{e^n}{n^2}$$

(2) Investigate the convergence of $\displaystyle\sum_{n=0}^{\infty} \frac{(-1)^n}{n^2 + 1}$.

$\boxed{\textbf{EX 2}}$ Investigate the convergence of $\displaystyle\sum_{n=1}^{\infty} (-1)^n \frac{2n + 1}{3n - 1}$.

Here, we have $a_n = (2n+1)/(3n-1)$. Since $\lim\limits_{n\to\infty} a_n = 2/3$, then the second condition of the alternating series test is not met; the terms in this series do not approach zero. So this series will *diverge*. ∎

You Try It

(3) Investigate the convergence of $\displaystyle\sum_{n=1}^{\infty}(-1)^{n+1}\frac{3n}{n^2+1}$.

Remainder Formulas

With some types of infinite series, it is possible to measure how far off we can be from the "true" value when we truncate the series after some number n of terms. These are usually called *remainder* formulas. If S is the true value of a convergent series, and S_N is the partial sum of the first N terms, then the error (or remainder) between the estimate and the true value of the series is

$$R_N = |s - s_N| = \left|\sum_{n=1}^{\infty} a_n - \sum_{n=1}^{N} a_n\right|$$

Note that this expression does not *compute* R_N, it merely defines it. If wanted to use this expression to compute R_N, we'd have to already know the true value of the series — in which case, discussion of error is rather silly.

The general scenario where remainder formulas are useful is to give this information: "If we truncate the series after N terms, we will be no more than this amount $\to$ $\leftarrow$ from the true value." We won't know the actual error, but we will know the worst case scenario for the error. This worst case scenario is R_N.

Alternating series come with a very simple remainder formula. Since alternating series have terms which alternate in sign, then as we add more and more terms to the series, we are bouncing back and forth around the true value. This means that if we stop adding after, say, 25 terms, then we are no farther away from the true value than the magnitude of the 26th term. A bit more generally, if we compute the partial sum $\displaystyle\sum_{n=0}^{N}(-1)^n a_n$, then the error is no bigger than the magnitude of the first ununsed term, a_{N+1}. Slightly more formally yet,

Definition 10.3. *Given a convergent alternating series, let S be the value of the series, i.e. the limit of the partial sums of the series. Then if the series is truncated after N terms, the remainder is bounded by $R_N < |a_{N+1}|$.*

There are lots of computer algebra systems out there that will report values of partial sums, or even give the exact value of some series. When first learning the concepts, using such tools exclusively isn't what's best, because there's little to no thinking involved. On the other hand, computing large sums by hand is tedious and not a good use of time. The middle ground is, like with Sec. 8.4, a spreadsheet. While using a computer algebra system (like Wolfram Alpha) to compute partial sums, you can demonstrate your understanding of the process by constructing them in a spreadsheet.

EX 3 Use a spreadsheet to find the 50th partial sum S_{50} of the alternating harmonic series (see EX 1). Use a computer algebra system to find the "proper" value S of the complete alternating harmonic series. Confirm that the error between S_{50} and S is no larger than a_{51}.

Figure 10.10 shows a portion of the list of terms of the alternating harmonic series, with the sum for the 50th partial sum displayed. This gives $S_{50} = 0.702855$. Wolfram Alpha reports the true value of the alternating harmonic series to be $S = \ln 2$ (how about that?). The command used to generate this result is `sum (-1)^(n) / (n+1), n = 1 to infinity`. We can write $\ln 2 \approx 0.693147$. So, $|S_n - S| = 0.0097$. The 51st term in the alternating harmonic series is $a_{51} = \dfrac{(-1)^{51}}{52} = 0.01923$, so sure enough, $|S_{50} - S| < a_{51}$. ∎

You Try It

(4) Find the 25th partial sum S_{25} of the series $\displaystyle\sum_{n=1}^{\infty} (-1)^{n+1} \frac{3n}{n^2 + 1}$ and determine the maximum possible error between S_{25} and the true value of the series.

We can reverse the questions we just examined and use remainder formulas to ask: How many terms are required to find a partial sum that is within some set error (or *tolerance*) from the unknown true value?

EX 4 How many terms are needed to find a partial sum of the alternating harmonic series that falls within 0.001 of the true value of the series?

We know from the remainder formula in Def. 10.3 that at the Nth partial sum, we are within a_{N+1} of the true value. So to ensure an error of no more than 0.001, we want to find N such that $a_{N+1} < 0.001$, i.e.

$$a_{(N+1)+1} < 0.001$$
$$\frac{1}{N+2} < \frac{1}{1000}$$
$$N+2 > 1000$$
$$N > 998$$

So we must generate a partial sum of the alternating harmonic series with at least 998 terms to get within 0.001 of the true value.   ■

You Try It

(5) How many terms are needed to find a partial sum of the series $\sum_{n=1}^{\infty} (-1)^{n+1} \dfrac{3n}{n^2+1}$ that falls within 0.005 of the true value of the series?

Alternating Series — Problem List

Alternating Series — You Try It

These appeared above; solutions begin on the next page.

(1) One of these alternating series converges and one diverges. Which is which, and how do you know?

$$(A) \sum_{n=1}^{\infty} (-1)^n \frac{n^2 + 2n + 100}{n^3} \quad ; \quad (B) \sum_{n=0}^{\infty} (-1)^n \frac{e^n}{n^2}$$

(2) Investigate the convergence of $\sum_{n=0}^{\infty} \frac{(-1)^n}{n^2 + 1}$.

(3) Investigate the convergence of $\sum_{n=1}^{\infty} (-1)^{n+1} \frac{3n}{n^2 + 1}$.

(4) Find the 25th partial sum S_{25} of the series $\sum_{n=1}^{\infty} (-1)^{n+1} \frac{3n}{n^2 + 1}$ and determine the maximum possible error between S_{25} and the true value of the series.

(5) How many terms are needed to find a partial sum of the series $\sum_{n=1}^{\infty} (-1)^{n+1} \frac{3n}{n^2 + 1}$ that falls within 0.005 of the true value of the series?

Alternating Series — Practice Problems

Try these as you get the hang of the You Try It problems. Solutions to these problems are available in Sec. A.4.5.

(1) Investigate the convergence of these series:

$$(A) \sum_{n=1}^{\infty} (-1)^n \frac{\ln(n)}{\ln(2n)} \quad ; \quad (B) \sum_{n=0}^{\infty} (-1)^{n+1} \frac{n^{944} + \pi}{n^{946}}$$

(2) Investigate the convergence of $\sum_{n=1}^{\infty} (-1)^n \frac{\sqrt{n}}{1 + 2\sqrt{n}}$. If this series converges, find how many terms are necessary for the error to be less than 0.0001.

(3) Investigate the convergence of $\sum_{n=0}^{\infty} (-1)^{n+1} \frac{n^2 + 1}{2n^3 + 2}$. If this series converges, find the maximum error between the partial sum S_{40} and the true value of the series.

(4) Investigate the convergence of $\sum_{n=0}^{\infty}(-1)^n \dfrac{e^{2n}}{e^{3n}}$. If this series converges, find how many terms are necessary for the error to be less than 0.0001.

(5) The series $\sum_{n=0}^{\infty}(-1)^n$ itself is an alternating series, but not much of one. Your somewhat distracted friend claims that the series obviously converges to 0 since it expands as $1 + (-1) + 1 + (-1) + \ldots$. Do you agree? Why or why not?

Alternating Series — Challenge Problems

Try these problems to test your skills with the ideas in this section. Solutions to these problems are available in Sec. B.4.5.

(1) Investigate the convergence of $\sum_{n=0}^{\infty} \dfrac{(-1)^n}{4^n n!}$. If this series converges, find how many terms are necessary for the error to be less than 0.0001.

(2) Investigate the convergence of $\sum_{n=0}^{\infty}(-1)^n \cos\left((2n+1)\dfrac{\pi}{2}\right)$ and

$\sum_{n=0}^{\infty}(-1)^n \cos\left((2n+1)\dfrac{\pi}{2}\right)$.

(3) Investigate the convergence of $\sum_{n=0}^{\infty}(-1)^n \dfrac{|\sin(n+1)|}{n+1}$.

Alternating Series — You Try It — Solved

(1) One of these alternating series converges and one diverges. Which is which, and how do you know?

$$(A) \sum_{n=1}^{\infty} (-1)^n \frac{n^2 + 2n + 100}{n^3} \quad ; \quad (B) \sum_{n=0}^{\infty} (-1)^n \frac{e^n}{n^2}$$

□ Series (B) clearly fails the alternating series test; with $a_n = e^n/n^2$, we have $\lim_{n \to \infty} a_n = \infty$ (bust out L-Hopital's rule if you need to), which is as bad a fail of the second condition as we can get. So that series (A) as the convergent series, but we should be good and check the conditions to be sure. Certainly, with $a_n = (n^2 + 2n + 100)/n^3$, we have $\lim_{n \to \infty} a_n = 0$. But is $a_{n+1} < a_n$ for all n? Let's see, by creating a cascade of equivalent statements:

$$a_{n+1} < a_n$$
$$\frac{(n+1)^2 + 2(n+1) + 100}{(n+1)^3} < \frac{n^2 + 2n + 100}{n^3}$$
$$n^3(n^2 + 4n + 103) < (n^2 + 2n + 100)(n^3 + 3n^2 + 3n + 1)$$
$$n^5 + 4n^4 + 103n^3 < n^5 + 5n^4 + 109n^3 + 307n^2 + 302n + 100$$
$$0 \leq n^4 + 6n^3 + 307n^2 + 302n + 100$$

The last statement is true for all $n > 0$ so $a_{n+1} < a_n$ for $n > 0$, and both conditions of the alternating series test hold. ∎

(2) Investigate the convergence of $\displaystyle\sum_{n=0}^{\infty} \frac{(-1)^n}{n^2 + 1}$.

□ With $a_n = 1/(n^2 + 1)$, we have $\lim_{n \to \infty} a_n = 0$. It's pretty obvious we also have $a_{n+1} < a_n$, but maybe we should check to be sure. Since $a_{n+1} = 1/((n+1)^2 + 1)$, then the inequality looks like

$$a_{n+1} < a_n$$
$$\frac{1}{(n+1)^2 + 1} < \frac{1}{n^2 + 1}$$
$$(n+1)^2 + 1 > n^2 + 1$$
$$(n^2 + 2n + 1) + 1 > n^2 + 1$$
$$2n + 1 > 0$$

We have a string of equivalent statements from top to bottom and also bottom to top of that stack, so since $2n + 1 > 0$ for all $n \geq 0$, then $a_{n+1} < a_n$ for all $n \geq 0$. Both conditions of the alternating series test are met, and this series converges.   ∎

---

(3) Investigate the convergence of $\displaystyle\sum_{n=1}^{\infty}(-1)^{n+1}\frac{3n}{n^2+1}$.

☐ This series does not start at $n = 0$, but the first term in that case would be 0 anyway, so it doesn't matter. With $a_n = 3n/(n^2 + 1)$, we have $\lim_{n\to\infty} a_n = 0$. We can also check $a_{n+1} < a_n$ by making a cascade of equivalent statements:

$$a_{n+1} < a_n$$
$$\frac{3(n+1)}{(n+1)^2+1} < \frac{3n}{n^2+1}$$
$$(3n+3)(n^2+1) < 3n(n^2+2n+2)$$
$$3n^3 + 3n^2 + 3n + 3 < 3n^3 + 6n^2 + 6n$$
$$3 < 3n^2 + 3n$$

The bottom statement is true for all $n \geq 1$, so the top is also. Thus, $a_{n+1} < a_n$. Both conditions of the alternating series test are met, and this series converges.   ∎

---

(4) Find the 25th partial sum S_{25} of the series $\displaystyle\sum_{n=1}^{\infty}(-1)^{n+1}\frac{3n}{n^2+1}$ and determine the maximum possible error between S_{25} and the true value of the series.

☐ YTI 2 confirmed this series converges. The 25th partial sum is (by spreadsheet, not shown) $S_{25} \approx 0.8675$. The maximum possible error between S_{25} and the true sum is $|a_{26}| = 0.1152$, which is still a pretty significant portion of the current result!   ∎

---

(5) How many terms are needed to find a partial sum of the series $\sum_{n=1}^{\infty}(-1)^{n+1}\dfrac{3n}{n^2+1}$ that falls within 0.005 of the true value of the series?

☐ Based on the error estimate in YTI 3, this is probably going to take a lot of terms! We need to find which value of n will produce $a_n < 0.005$. This is:

$$\frac{3n}{n^2+1} < 0.005$$

$$3n < \frac{5}{1000}(n^2+1)$$

$$3000n < 5n^2 + 5$$

$$5n^2 - 3000n + 5 > 0$$

$$n^2 - 600n + 1 > 0$$

With a little electronic help, we can find that $n^2 - 600n + 1 = 0$ at about $n = 600$, so if we want $n^2 - 600n + 1 > 0$, we need n to be a bit bigger — since n needs to be an integer, we'll say that we must go out to $n = 601$ to find a partial sum within 0.005 of the true value of the series. ■

Chapter 11

The Fear of All Sums

11.1   The Integral Test and p-Series

Introduction

The big question about any infinite series is: will the series converge or diverge? There are many tests that can be applied to infinite series to check this, and you've seen a couple of them already (the divergence test, the geometric series test). With the exception of the geometric series test, such tests will rarely tell you what value a series converges *to*, but they do tell you whether or not you can expect a finite sum.

We'll see here a test that exploits the link between sequences and functions that we also saw when determining the limit of a sequence.

The Integral Test

This convergence test relies on the link between a sequence / series and a function. Consider the series

$$\sum_{n=1}^{\infty} \frac{1}{n^2}$$

The fate of this series is the same as the fate as the sequence of its partial sums. But remember that the fate (limit) of a sequence with terms a_n is very often the same as the fate of a function $f(x)$ that generates the sequence via $a_n = f(n)$. So you may imagine that the function $f(x) = 1/x^2$ might help us figure out what happens to the above series. And it does!

Let's define $x_1 = 1$, $x_2 = 2$, $x_3 = 3$, and so on. Then $a_1 = f(x_1)$, $a_2 = f(x_2)$, etc. Also note that the distance in between each x_n and x_{n+1}

is $\Delta x = 1$. We can update our series recipe with this information:

$$\sum_{n=1}^{\infty} \frac{1}{n^2} = \sum_{n=1}^{\infty} f(x_n) \cdot (1) = \sum_{n=1}^{\infty} f(n)\Delta x$$

Hopefully, you recognize the latter versio as a Riemann Sum. What we've learned is that our series is actually an estimate (a bad one, granted) to the area under $f(x) = 1/x^2$ on the interval $[1, \infty)$. So you may imagine that if that area under $f(x)$ is finite, our series should be finite (i.e. converge), too. And that's the idea behind a convergence test called the integral test:

Useful Fact 11.1. *Suppose $f(x)$ is a continuous, positive, decreasing function on $[a, \infty)$, and let $a_n = f(n)$. Then the convergence or divergence of $\sum_{n=a}^{\infty} a_n$ matches the convergence or divergence of $\int_a^{\infty} f(x)\,dx$. This is called the* **integral test***.*

BEWARE! Here are three things to realize about the integral test:

(1) The integral test doesn't work for just any old function / series, it requires the function $f(x)$ you associate with your series to be (ultimately) decreasing and continuous on an interval $[a, \infty)$. If the function $f(x)$ you end up with is not decreasing and continuous on the matching interval, then forget the integral test and go about your day.

(2) The integral test can determine if a series convergences, but it does NOT tell you what the series converges to. The integral and the series may both converge, but they don't necessarily converge to the same thing!

(3) The integral test requires that you do an improper integral; these must be done correctly, with limits, as usual.

So I guess we might as well finish up with the example series used above:

$\boxed{\text{EX 1}}$ Investigate the convergence of $\sum_{n=1}^{\infty} \dfrac{1}{n^2}$.

The function that generates the terms in this series is $f(x) = 1/x^2$, and $a_n = f(n)$. This function is continuous and decreasing on $[1, \infty)$, so we can apply the integral test. The integral test tells us the fate of our series and integral $\int_1^\infty f(x)\,dx$ are the same. The improper integral is evaluated[1] as:

$$\lim_{c\to\infty} \int_1^c \frac{1}{x^2}\,dx = \lim_{c\to\infty} \left.\frac{-1}{x}\right|_1^c = \lim_{c\to\infty} \left(-\frac{1}{c}+1\right)$$

The integral converges (we don't care what value it converges to), and therefore the improper integral AND series both converge. ∎

You Try It

(1) Investigate the convergence of $\displaystyle\sum_{n=1}^{\infty} \frac{1}{3n+1}$.

(2) Investigate the convergence of $\displaystyle\sum_{n=1}^{\infty} \frac{n}{n^2+1}$.

The p-Series Test

We'll get even more mileage out of the series in EX 1, because it is one instance of a larger class of series. You might imagine that since the series with terms $1/n^2$ converges, then so should a series with terms $1/n^3$ or $1/n^4$, etc. But what about a series with terms $1/n$? Well, that's the harmonic series — and we know it diverges. These are all instances of a category called *p*-**series**, which are of the form

$$\sum_{n=1}^{\infty} \frac{1}{n^p}$$

A value of $p = 1$ leads to divergence, but as seen in the above example, a series with $p = 2$ converges. Where is the cutoff? Let's see:

EX 2 Investigate the convergence of $\displaystyle\sum_{n=1}^{\infty} \frac{1}{n^p}$.

The function that generates the terms in this series is $f(x) = 1/x^p$, with $a_n = f(n)$. As long as $p > 0$, this function is continuous and decreasing on $[1, \infty)$, so we can apply the integral test. The integral test tells us the fate of our series and the integral $\int_1^\infty \frac{1}{x^p}\,dx$ are the same. This improper

---

[1]Properly!

integral is evaluated (properly!) as:

$$\lim_{c \to \infty} \int_1^c \frac{1}{x^p}\, dx = \lim_{c \to \infty} \frac{-1}{(p-1)x^{p-1}}\Big|_1^c = \lim_{c \to \infty} \left(-\frac{1}{(p-1)c^{p-1}} + \frac{1}{p-1} \right)$$

Since we can rewrite the first term as

$$\lim_{c \to \infty} \frac{-1}{(p-1)c^{p-1}} = \lim_{c \to \infty} \frac{-1}{p-1}c^{-(p-1)} = \lim_{c \to \infty} \frac{-1}{p-1}c^{1-p}$$

we can see that this limit will converge if $p > 1$ and will diverge if p is less than 1. (Be sure you know why.) Did you also notice that in the process of doing that calculation, we eliminated from consideration the case in which $p = 1$? As soon as we developed a $p - 1$ in the denominator, we ruled out use of $p = 1$. So we have to check that case separately. But if $p = 1$, we have

$$\lim_{c \to \infty} \int_1^c \frac{1}{x^1}\, dx = \lim_{c \to \infty} \ln|x|\Big|_1^c = \lim_{c \to \infty} (\ln c)$$

which diverges. Not only does this final calculation fill in a gap in the overall analysis, it's also our first direct demonstration that the harmonic series diverges. ∎

In all, what we learned falls under the broad category of the integral test, but is specific enough we can give it its own name:

Useful Fact 11.2. *A series of the form* $\displaystyle\sum_{n=1}^{\infty} \frac{1}{n^p}$ *is called a p-series. Such a series converges converges if $p > 1$ and diverges if $p \le 1$. This is called the* **p-series test**.

$\boxed{\text{EX 3}}$ Investigate the convergence of $\displaystyle\sum_{n=1}^{\infty} \frac{1}{\sqrt{n}}$.

Since this is a p-series with $p = 1/2$, it diverges according to the p-series test (we need $p > 1$ for convergence). ∎

You Try It

(3) Investigate the convergence of $\displaystyle\sum_{n=1}^{\infty} \frac{1}{5n^{1.2}}$.

(4) Investigate the convergence of $1 + \dfrac{1}{8} + \dfrac{1}{27} + \dfrac{1}{64} + \cdots$.

The Remainder Formula for the Integral Test

In Sec. 10.5, we saw the remainder formula for alternating series: $R_N < a_{N+1}$. This remainder formula allowed us to address two different questions: (1) Given that we sum only the first N terms of a convergent series, what is the bound on our error? (2) If we want our error to be less than some specific value, how many terms in the series must we add?

Series which are determined to converge with the integral test can also be analyzed with a corresponding remainder formula.

Useful Fact 11.3. *Suppose we have a convergent series for which the integral test is appropriate (i.e. for which the surrogate function $f(n) = a_n$ is continuous, positive, and decreasing on $[1, \infty)$). Then*

$$R_N \leq \int_N^{\infty} f(x)\, dx$$

Remember that when the integral test is in play, there is an association between the value of an infinite series $\displaystyle\sum_{n=1}^{\infty} a_n$ and the area under the function $f(n) = a_n$. This remainder formula, then, relates the error incurred by truncating the series after N terms to the corresponding area that we are discarding.

EX 4 Consider the series $\displaystyle\sum_{n=1}^{\infty} \frac{1}{n\sqrt{n}}$.

(a) What is the remainder between the partial sum S_{50} and the true value of the series?

(b) How many terms do we need to sum in order to achieve accuracy of 0.001?

(a) The function $f(x) = 1/(x\sqrt{x^2 + 1})$ is continuous, positive and decreasing for $x > 1$, so it meets the conditions of the integral test, and so the integral test remainder formula applies:

$$R_{50} \le \int_{50}^{\infty} \frac{1}{x\sqrt{x^2+1}}\, dx \approx 0.02$$

(This integral was estimated using Wolfram Alpha.) Therefore, an upper bound on our error after 50 terms is roughly 0.02.

For part (b), a value of R_N is given, and N itself is the unknown. So we need to solve

$$0.001 = \int_{N}^{\infty} \frac{1}{x\sqrt{x^2+1}}\, dx$$

Using Wolfram Alpha, we can find the integral and update our equation to

$$0.001 = \sinh^{-1} \frac{1}{N}$$

And sure, we just got an awful equation back, but we're going to use the same computing tools to solve that equation as we did to solve the integral in the first place, so who cares! For example, Wolfram Alpha reports the solution to this equation as $N \approx 1000$. And so we need $1,000$ terms to achieve a partial sum within 10^{-3} of the true value of the series. ∎

Don't forget that N must be an integer, so if we solve a problem like this get back a decimal approximation for N, we must round N up to the next integer. For example, if we get $N = 98.6$, then 98 terms would not be enough to get the desired accuracy, so we'd report that we need $N = 99$.

You Try It

(5) Consider the series $\displaystyle\sum_{n=1}^{\infty} \frac{4}{n^5}$.

    (a) Estimate the 100th partial sum of this series.

    (b) Find an upper bound on the error between this result and the true value of the series.

    (c) Determine how many terms would be needed for an accuracy of 10^{-6}.

Comparison of Series

At this point, you have seen and should be able to recognize / analyze the following special types of series (in addition to series in general):

- geometric series
- harmonic series
- p-series

Many infinite series can be recognized as "close" to one of these types, while not exactly fitting the mold. For example, the series $\sum_{n=1}^{\infty} \frac{1}{n^3 + n}$ does not match the definitive form of a p-series, but we can still determine its convergence by noting one specific trait. Let a_n be the term recipe in the p-series $\sum_{n=1}^{\infty} \frac{1}{n^3}$, that is, $a_n = 1/n^3$. Let b_n be the term recipe in the almost-but-not-quite p-series $\sum_{n=1}^{\infty} \frac{1}{n^3 + n}$, that is, $b_n = 1/(n^3 + n)$. For every n $(n \geq 1)$, we know that $b_n < a_n$. So, the sum of all the a_n's should be larger than the sum of all the b_n's. And if the sum of all the a_n's converges to a finite value, it stands to reason that the sum of all the b_n's (which will be smaller) does as well. This can be described as the "comparison test"; it's not a test with a specific formula or calculation to do (such as with the geometric series or integral test), but rather, it's just formalizing a rule of thumb that's fairly obvious:

Useful Fact 11.4. *Suppose we have two infinite series,* $S_a = \sum_{n=c}^{\infty} \frac{1}{a_n}$ *and* $S_b \sum_{n=c}^{\infty} \frac{1}{b_n}$.

- *If series S_b converges, and $a_n \leq b_n$ for every n, then the series S_a converges, too.*
- *If series S_b diverges, and $a_n \geq b_n$ for every n, then the series S_a diverges, too.*

Let's call this the **direct comparison test**.

EX 5 Investigate the convergence of $\sum_{n=0}^{\infty} \frac{3}{n^2 + 3}$.

The $+3$ in the denominator keeps this from being a p-series by strict definition, but gosh, it's awfully close. Really, since $\sum_{n=0}^{\infty} \frac{3}{n^2}$ converges, then surely this new series should converge, too! The direct comparison test formalizes this. Let's assign $b_n = 3/n^2$ from the convergent series, and $a_n = 3/(n^2 + 3)$ from the mystery series. If $a_n \leq b_n$, then the mystery series converges, too. Let's set up a cascade of equivalent statements:

$$a_n \leq b_n$$
$$\frac{3}{n^2 + 3} \leq \frac{3}{n^2}$$
$$3(n^2) \leq (3)(n^2 + 3)$$
$$3n^2 \leq 3n^2 + 9$$

If any one of these statements is true for all n, then all are true. The final statement is definitively true, so $a_n \leq b_n$ for all n. Therefore, since $\sum_{n=0}^{\infty} \frac{3}{n^2}$ converges, so does $\sum_{n=0}^{\infty} \frac{3}{n^2 + 3}$. ∎

You Try It

(6) Investigate the convergence of $\sum_{n=1}^{\infty} \frac{1}{n^4 + \ln(n)}$.

BEWARE! Like other mathematical statements, the direct comparison test works only the way it is written, and you can't try to draw broader conclusions or reverse conclusions. If $a_n \leq b_n$ for every n, but S_b diverges, then we still know nothing about S_a. If $a_n \geq b_n$ for every n, and S_b converges, we still know nothing about S_a. For example, while EX 5 showed that $\sum_{n=1}^{\infty} \frac{3}{n^3 + 3}$ converges, we still know nothing about $\sum_{n=1}^{\infty} \frac{3}{n^3 - 3}$.

Now, you might be thinking, "Come on now, $\sum_{n=1}^{\infty} \frac{3}{n^3 - 3}$ still looks a lot like $\sum_{n=1}^{\infty} \frac{3}{n^3}$, so it should converge, too!" And you're right, but the direct comparison test in Useful Fact 11.4 doesn't do the job.

When the direct comparison test cannot be used, but we still have a hunch that our given series is enough "like" a known series to determine its convergence or divergence, we need a way to quantify that "like"-ness. This is a job for the *limit* comparison test!

Useful Fact 11.5. *Suppose we have two infinite series,* $S_a = \sum\limits_{n=c}^{\infty} \dfrac{1}{a_n}$ *and* $S_b \sum\limits_{n=c}^{\infty} \dfrac{1}{b_n}$ *for which* a_n *and* b_n *are eventually all positive. Then if* $\lim\limits_{n\to\infty} \dfrac{a_n}{b_n} = c$, *where* $0 < c < \infty$, *then either both series converge or both series diverge.*

The limit comparison test encapsulates the idea that if two series look enough "alike", then they'll both have the same eventual behavior, either convergence or divergence. The likeness is determined by their relative behavior "way out there" as n increases. The finite limit of the ratio of terms is like a mathematical rope that ties the two series together; it doesn't matter which terms are bigger than which, but ultimately, the two series will have the same behavior.

Note that the limit comparison test supercedes the direct comparison test ... meaning if the direct comparison test worked for a pair of series, then the limit comparison test would have worked as well, but not vice versa. However, the direct comparison test can sometimes be simpler to carry out, so let's not forget about it.

EX 6 Investigate the convergence of $\sum_{n=0}^{\infty} \frac{3}{n^2-3}$.

The direct comparison test does not apply, since with $a_n = 3/(n^2 - 3)$ and $b_n = 3/n^2$, a_n is not less than b_n, so convergence of the latter series does not force convergence of the former. But still, we think they're enough alike that they should do the same thing, and the limit comparison test is how we demonstrate that:

$$\lim_{n\to\infty} \frac{a_n}{b_n} = \lim_{n\to\infty} \frac{3/(n^2 - 3)}{3/n^2} = \lim_{n\to\infty} \frac{n^2}{n^2 - 3} = 1$$

Since the limit of the ratio of terms is a finite positive constant, then the two series will exhibit the same behavior. Because S_b converges according to the p-series test, then S_a converges as well. ∎

You Try It

(7) Investigate the convergence of $\displaystyle\sum_{n=1}^{\infty} \frac{n}{n^2 + \cos^2(n)}$.

The Integral Test and p-series — Problem List

The Integral Test and p-series — You Try It

These appeared above; solutions begin on the next page.

(1) Investigate the convergence of $\displaystyle\sum_{n=1}^{\infty} \frac{1}{3n+1}$.

(2) Investigate the convergence of $\displaystyle\sum_{n=1}^{\infty} \frac{n}{n^2+1}$.

(3) Investigate the convergence of $\displaystyle\sum_{n=1}^{\infty} \frac{1}{5n^{1.2}}$.

(4) Investigate the convergence of $1 + \dfrac{1}{8} + \dfrac{1}{27} + \dfrac{1}{64} + \cdots$.

(5) Consider the series $\displaystyle\sum_{n=1}^{\infty} \frac{4}{n^5}$.

   (a) Estimate the 100th partial sum of this series.
   (b) Find an upper bound on the error between this result and the true value of the series.
   (c) Determine how many terms would be needed for an accuracy of 10^{-6}.

(6) Investigate the convergence of $\displaystyle\sum_{n=1}^{\infty} \frac{1}{n^4 + \ln(n)}$.

(7) Investigate the convergence of $\displaystyle\sum_{n=1}^{\infty} \frac{n}{n^2 + \cos^2(n)}$.

The Integral Test and p-series — Practice Problems

Try these as you get the hang of the You Try It problems. Solutions to these problems are available in Sec. A.5.1.

(1) Investigate the convergence of $\displaystyle\sum_{n=1}^{\infty} e^{-n}$.

(2) Investigate the convergence of $1 + \dfrac{1}{2\sqrt{2}} + \dfrac{1}{3\sqrt{3}} + \dfrac{1}{4\sqrt{4}} + \cdots$.

(3) Investigate the convergence of $\displaystyle\sum_{n=1}^{\infty} \frac{1}{n^2 - 4n + 5}$. (Hint: Complete the square in the denominator.)

(4) Consider the series $\displaystyle\sum_{n=1}^{\infty} \frac{1}{n^2 + 1}$.

    (a) Estimate the 100th partial sum of this series.

    (b) Find an upper bound on the error between this result and the true value of the series.

    (c) Determine how many terms would be needed for an accuracy of 10^{-4}.

(5) Investigate the convergence of $\displaystyle\sum_{n=0}^{\infty} \frac{1}{4} \cdot \left(\frac{2}{3} - \frac{1}{\pi}\right)^n$.

(6) Investigate the convergence of $\displaystyle\sum_{n=1}^{\infty} \frac{1}{2^n - n}$.

(7) Investigate the convergence of $\displaystyle\sum_{n=3}^{\infty} \frac{5}{(n-2)^2}$.

The Integral Test and p-series — Challenge Problems

Try these problems to test your skills with the ideas in this section. Solutions to these problems are available in Sec. B.5.1.

(1) Investigate the convergence of $\displaystyle\sum_{n=1}^{\infty} n e^{-n}$.

(2) Investigate the convergence of $\displaystyle\sum_{n=1}^{\infty} \frac{3n^2 + 4n}{\sqrt{n^5 + 2}}$.

(3) Investigate the convergence of $\displaystyle\sum_{n=1}^{\infty} \frac{\ln n}{n^2 + 1}$.

The Integral Test and p-Series — You Try It — Solved

---

(1) Investigate the convergence of $\displaystyle\sum_{n=1}^{\infty} \frac{1}{3n+1}$.

☐ Applying the integral test, this series will behave the same as the corresponding integral:

$$\int_1^{\infty} \frac{1}{3x+1}\, dx = \lim_{c\to\infty} \frac{1}{3}\ln(3x+1)\Big|_1^c = \lim_{c\to\infty} \frac{1}{3}(\ln(3c+1) - \ln 4)$$

Since the limit diverges, so do the integral AND series. ∎

---

(2) Investigate the convergence of $\displaystyle\sum_{n=1}^{\infty} \frac{n}{n^2+1}$.

☐ Applying the integral test, this series will behave the same as the corresponding integral:

$$\int_1^{\infty} \frac{x}{x^2+1}\, dx = \lim_{c\to\infty} \frac{1}{2}\ln(x^2+1)\Big|_1^c = \lim_{c\to\infty} \frac{1}{2}(\ln(c^2+1) - \ln 2)$$

Since the limit diverges, so do the integral AND series. ∎

---

(3) Investigate the convergence of $\displaystyle\sum_{n=1}^{\infty} \frac{1}{5n^{1.2}}$.

☐ The constant 5 can be factored out and doesn't affect convergence. This series is a p-series with $p = 1.2$; by the p-series test, it converges since $p > 1$. ∎

---

(4) Investigate the convergence of $1 + \dfrac{1}{8} + \dfrac{1}{27} + \dfrac{1}{64} + \cdots$.

☐ We can write this in series notation as $\displaystyle\sum_{n=1}^{\infty} \frac{1}{n^3}$. So it's recognizable as a p-series with $p = 3$; since $p > 1$, the series converges. ∎

---

(5) Consider the series $\displaystyle\sum_{n=1}^{\infty} \frac{4}{n^5}$.

    (a) Estimate the 100th partial sum of this series.

    (b) Find an upper bound on the error between this result and the true value of the series.

(c) Determine how many terms would be needed for an accuracy of 10^{-6}.

□ (a) Using a spreadsheet (not shown), we can find $S_{100} \approx 4.147711$.

(b) The function $f(x) = \dfrac{4}{x^5}$ is continuous, positive, and decreasing or $x > 1$, so the integral test would apply, and so the integral test remainder formula is appropriate. By Useful Fact 11.3, the remainder at S_{100} is bounded as follows:

$$R_{100} \leq \int_{100}^{\infty} \frac{4}{x^5}\, dx \approx 10^{-8}$$

(c) To achieve an accuracy of 10^{-6}, we must find the value N for which

$$\int_{N}^{\infty} \frac{4}{x^5}\, dx = 10^{-6}$$

or

$$\frac{1}{N^4} = 10^{-6}$$

The genuine numerical solution to this is $N = 10\sqrt{10} \approx 31.623$. But since N must be an integer, we now know that $N = 31$ is not sufficient and we need to use 32 terms or more to have an accuracy of 10^{-6}. ■

---

(6) Investigate the convergence of $\displaystyle\sum_{n=1}^{\infty} \frac{1}{n^4 + \ln(n)}$.

□ We know that $\dfrac{1}{n^4 + \ln(n)} < \dfrac{1}{n^4}$ for every $n \geq 1$. And we know $\displaystyle\sum_{n=1}^{\infty} \frac{1}{n^4}$ converges, because it is a p-series with $p = 4$. So by the comparison test, $\displaystyle\sum_{n=1}^{\infty} \frac{1}{n^4 + \ln(n)}$ converges, too. ■

11.2 The Ratio Test

Introduction

If a series has a chance to converge, then the sequence of magnitudes of its terms *must* be approaching 0. If the terms do not approach zero, the series will diverge (that's what the Divergence Test tells us). But even when the terms in the series do approach 0, there is no guarantee of convergence. We've discovered in Sec. 10.5 that if a series has terms which alternate in sign, then we need one other condition for convergence: not only does the sequence of individual terms need to be approaching zero, but it must be doing so *monotonically* — that is, the magnitude of any one term must be smaller than that of the preceding term. But what if the series is not alternating? If having the terms themselves heading for zero in magnitude isn't enough to force convergence, what else is needed? Why does the harmonic series $\sum_{n=1}^{\infty} \frac{1}{n}$ diverge while the almost-identical series $\sum_{n=1}^{\infty} \frac{1}{n^2}$ converge? The answer is, very loosely, that the terms in the latter series are heading for zero "faster" than those in the harmonic series. A series will converge if the terms getting small enough "fast enough". The *ratio test* is a diagnosis of the terms in a series that helps us decide if those terms are getting small enough fast enough to generate convergence.

The Ratio Test

We need to quantify a measure of "how fast" terms in a series are shrinking in magnitude. Let's rewrite $a_{n+1} < a_n$ as $a_{n+1}/a_n < 1$; we can even throw some absolute value bars in there to get $|a_{n+1}/a_n|$, so that we don't have to make a distinction between series that alternate in sign and those that don't.

Is the condition $|a_{n+1}/a_n| < 1$ any better than what we had before? No! That dang harmonic series *still* passes this test (make sure you see why). But this was only a cosmetic change to the condition; it still tests only two neighboring terms, like terms 100 and 101, or terms 245 and 246. What we *really* need to do is look waaaay out in the series and ask: How does the ratio $|a_{n+1}/a_n|$ behave, not for just any two neighboring terms, but as we go farther and farther out in the series? Well, we've known since Chapter 2 how to ask, "what's going on way out there for larger and larger values of n?" This is the job of a limit! Now we're on to something good:

Useful Fact 11.6. *Given the series* $\displaystyle\sum_{n=1}^{\infty} a_n$, *let* $L = \displaystyle\lim_{n\to\infty}\left|\dfrac{a_{n+1}}{a_n}\right|$. *Then,*

- *If* $L < 1$, *the series is convergent.*
- *If* $L > 1$, *the series is divergent.*
- *If* $L = 1$, *there is no conclusion, and you have to try something else.*

This is called the **Ratio Test.**

The routine for performing the Ratio Test on a series is to

(1) Use the series recipe a_n to build the ratio $|a_{n+1}/a_n|$.
(2) Find the limit of this expression as $n \to \infty$ using regular limit procedures.
(3) Decide if the limit is greater than, less than, or equal to 1 and draw the appropriate conclusion.

The ratio test does not work for ALL series, but it is one of the most common tests. It works really well for series recipes that contain factorials or powers of n. It can work for some alternating series as well as "regular" series. When in doubt, try the ratio test! Here are three examples in a row:

$\boxed{\textbf{EX 1}}$ Investigate the convergence of $\displaystyle\sum_{n=1}^{\infty}\dfrac{(-1)^n 2^n}{n^2}$.

In this series, we have $a_n = (-1)^n 2^n / n^2$. Since absolute value bars will remove the $(-1)^n$, we have

$$|a_n| = \frac{2^n}{n^2} \text{ and } |a_{n+1}| = \frac{2^{n+1}}{(n+1)^2}$$

so that

$$\left|\frac{a_{n+1}}{a_n}\right| = \left|\frac{2^{n+1}/(n+1)^2}{2^n/n^2}\right| = \left|\frac{2^{n+1}}{(n+1)^2}\cdot\frac{n^2}{2^n}\right| = 2\left(\frac{n}{n+1}\right)^2$$

(since $n > 1$, the absolute value bars are not needed at the latter end of that calculation). The limit of this expression is

$$L = \lim_{n\to\infty} 2\left(\frac{n}{n+1}\right)^2 = 2\left(\lim_{n\to\infty}\frac{n}{n+1}\right)^2 = 2\cdot 1 = 2$$

Since $L > 1$, the ratio test tells us that this series diverges. ■

EX 2 Investigate the convergence of $\displaystyle\sum_{n=1}^{\infty} \frac{3^n}{n!}$.

In this series, we have

$$|a_n| = \frac{3^n}{n!} \text{ and } |a_{n+1}| = \frac{3^{n+1}}{(n+1)!}$$

so that

$$\left|\frac{a_{n+1}}{a_n}\right| = \left|\frac{3^{n+1}/(n+1)!}{3^n/n!}\right| = \left|\frac{3^{n+1}}{(n+1)!} \cdot \frac{n!}{3^n}\right| = \frac{3}{n+1}$$

(again, since $n > 1$, the absolute value bars are not needed). The limit of this expression is

$$L = \lim_{n\to\infty} \left(\frac{3}{n+1}\right) = 0$$

Since $L < 1$, the ratio test tells us that this series converges. ∎

And this final example is where you find out that I probably shouldn't have been using the harmonic series in any of the above motivation for the ratio test:

EX 3 Investigate the convergence of $\displaystyle\sum_{n=1}^{\infty} \frac{1}{n}$.

In this series, we have

$$|a_n| = \frac{1}{n} \text{ and } |a_{n+1}| = \frac{1}{n+1}$$

so that

$$\left|\frac{a_{n+1}}{a_n}\right| = \left|\frac{1/n+1}{1/n}\right| = \left|\frac{1}{n+1} \cdot \frac{n}{1}\right| = \frac{n}{n+1}$$

The limit of this expression is

$$L = \lim_{n\to\infty} \left(\frac{n}{n+1}\right) = 1$$

Since $L = 1$, the ratio test is inconclusive, and we've learned nothing about the convergence or divergence of the harmonic series by using the ratio test. ∎

You Try It

(1) Investigate the convergence of $\displaystyle\sum_{n=0}^{\infty} \frac{(-10)^n}{n!}$.

(2) Investigate the convergence of $\displaystyle\sum_{n=1}^{\infty} (-1)^{n+1} \frac{3^n}{n^2}$.

(3) Investigate the convergence of $\displaystyle\sum_{n=1}^{\infty} \left(\frac{2}{3}\right)^n$.

The Ratio Test — Problem List

The Ratio Test — You Try It

These appeared above; solutions begin on the next page.

(1) Investigate the convergence of $\displaystyle\sum_{n=0}^{\infty} \frac{(-10)^n}{n!}$.

(2) Investigate the convergence of $\displaystyle\sum_{n=1}^{\infty} (-1)^{n+1} \frac{3^n}{n^2}$.

(3) Investigate the convergence of $\displaystyle\sum_{n=1}^{\infty} \left(\frac{2}{3}\right)^n$.

The Ratio Test — Practice Problems

Try these as you get the hang of the You Try It problems. Solutions to these problems are available in Sec. A.5.2.

(1) Investigate the convergence of $\displaystyle\sum_{n=1}^{\infty} (-1)^{n-1} \frac{2^n}{n^4}$.

(2) Investigate the convergence of $\displaystyle\sum_{n=1}^{\infty} e^{-n} n! = \sum_{n=1}^{\infty} \frac{n!}{e^n}$.

(3) Investigate the convergence of $\displaystyle\sum_{n=1}^{\infty} \frac{1}{(2n)!}$.

The Ratio Test — Challenge Problems

Try these problems to test your skills with the ideas in this section. Solutions to these problems are available in Sec. B.5.2.

(1) Investigate the convergence of $\displaystyle\sum_{n=1}^{\infty} (-1)^{n+1} \frac{n^2 2^n}{n!}$.

(2) Investigate the convergence of $\displaystyle\sum_{n=1}^{\infty} \frac{n!}{n^n}$. (Hint: You will need to know

$$\lim_{n\to\infty} \left(\frac{n}{n+1}\right)^n = \frac{1}{e}$$)

(3) Investigate the convergence of $\displaystyle\sum_{n=1}^{\infty} \frac{3^n n!}{(2n)!}$.

The Ratio Test — You Try It — Solved

---

(1) Investigate the convergence of $\displaystyle\sum_{n=0}^{\infty} \frac{(-10)^n}{n!}$.

☐ The series contains factorials and powers of n, so the ratio test should work well.

$$L = \lim_{n\to\infty} \left| \frac{a_{n+1}}{a_n} \right| = \lim_{n\to\infty} \left| \frac{\left(\frac{(-10)^{n+1}}{(n+1)!}\right)}{\left(\frac{(-10)^n}{n!}\right)} \right|$$

$$= \lim_{n\to\infty} \left| \frac{(-10)^{n+1}}{(n+1)!} \frac{n!}{(-10)^n} \right| = \lim_{n\to\infty} \left| \frac{-10}{n+1} \right| = 0$$

Since $L < 1$ the series converges. ∎

---

(2) Investigate the convergence of $\displaystyle\sum_{n=1}^{\infty} (-1)^{n+1} \frac{3^n}{n^2}$.

☐ The series contains factorials and powers of n, so the ratio test should work well.

$$L = \lim_{n\to\infty} \left| \frac{a_{n+1}}{a_n} \right| = \lim_{n\to\infty} \left| \frac{\left(\frac{3^{n+1}}{(n+1)^2}\right)}{\left(\frac{3^n}{n^2}\right)} \right|$$

$$= \lim_{n\to\infty} \left| \frac{3^{n+1}}{(n+1)^2} \frac{n^2}{3^n} \right| = 3 \lim_{n\to\infty} \left(\frac{n}{n+1} \right)^2 = 3$$

Since $L > 1$ the series diverges. ∎

---

(3) Investigate the convergence of $\displaystyle\sum_{n=1}^{\infty} \left(\frac{2}{3} \right)^n$.

☐ We can already recognize that this is a convergent geometric series (make sure you can!), but let's run the ratio test anyway:

$$L = \lim_{n\to\infty} \left| \frac{a_{n+1}}{a_n} \right| = \lim_{n\to\infty} \left| \frac{(2/3)^{n+1}}{(2/3)^n} \right| = \lim_{n\to\infty} \left| \frac{2}{3} \right| = \frac{2}{3}$$

Since $L < 1$ the series converges. ∎

11.3 Power Series

Introduction

Compare these two expressions:

$$\sum_{n=0}^{\infty} 2\left(\frac{3}{4}\right)^n \quad \text{vs} \quad \sum_{n=0}^{\infty} 2\left(\frac{x}{4}\right)^n$$

We've seen series similar to both before. The series on the left is a geometric series, and it converges. The second series is an actual *function*; if we plug in different values for x, different things can happen — convergence or divergence. So, given a series which represents a function, we have a nice mystery: how do we decide which values of x will lead to convergence? In this example, you should recognize that the second series, i.e. the *function of x*, will converge for $|x| < 4$.

Let's explore this relationship between series and functions a bit (or a lot!) more.

Power Series

A series whose terms contain powers of the variable x, usually structured as

$$\sum_{n=0}^{\infty} c_n x^n \quad \text{or} \quad \sum_{n=0}^{\infty} c_n (x - x_0)^n$$

(where the c_ns are constants) is called a **power series**. In other words, *a power series is a polynomial with an infinite number of terms*. Sometimes a power series is an approximating polynomial (a Taylor polynomial) developed to its full extent, but a power series is a perfectly good function in its own right — whether we know it "agrees" with a more conventionally written function or not.

A power series may or may not converge. If it does converge, it may not converge for all values of x. The collection of values of x for which a power series does converge are called its *interval of convergence*. The distance from the center of the interval of convergence to either endpoint of the interval is called its **radius of convergence**.[2] Questions about any given power series can include:

---

[2]This terminology comes from the study of functions of a complex variable, where regions of convergence are circular. It's good stuff, you should be sure to take a course in Complex Analysis!

- What is the interval and/or of convergence of the power series?
- Does the power series represent a function also known in a different (algebraic) form?

We'll learn to answer the first question here, and the second question later.

Finding the interval of convergence for a power series usually involves an application of the ratio test via the following steps:

(1) Apply the ratio test to the series and find an expression for L. Note that L will be written in terms of x, rather than as a single value as in previous cases. This is helpful though, because...

(2) The ratio test says that in order for a series to converge, the value of L must be less than 1. So, find the values of x which will make $L < 1$. This gives you the interval of convergence, although you don't know yet whether to include the endpoints or not because...

(3) Remember that the ratio test is inconclusive if $L = 1$. The values of x that make $L = 1$ are the endpoints of the interval of convergence found in the previous step. So, you have to test the endpoints individually to see if either makes the power series converge.

Here are three examples in a row.

EX 1 Find the interval and radius of convergence of the power series
$$\sum_{n=0}^{\infty} \frac{(x-1)^n}{n!}.$$

Let's set up the ratio test:

$$\left|\frac{a_{n+1}}{a_n}\right| = \left|\frac{(x-1)^{n+1}/(n+1)!}{(x-1)^n/n!}\right| = \left|\frac{(x-1)^{n+1}}{(n+1)!}\frac{n!}{(x-1)^n}\right| = |x-1|\frac{1}{n+1}$$

In order to have convergence, we need the limit of this expression to be less than 1. But when we compute this limit, we get:

$$L = \lim_{n\to\infty} |x-1|\frac{1}{n+1} = |x-1| \lim_{n\to\infty} \frac{1}{n+1} = |x-1| \cdot (0) = 0$$

The limit L is always 0, regardless of what x is, and so the power series converges for all values of x. The interval of convergence is $(-\infty, \infty)$, and the radius of convergence is ∞. ∎

EX 2 Find the interval and radius of convergence of the power series
$$\sum_{n=0}^{\infty} \frac{(-1)^n x^n}{n \cdot 2^n}.$$

Let's set up the ratio test:
$$\left| \frac{a_{n+1}}{a_n} \right| = \left| \frac{x^{n+1}/(n+1)2^{n+1}}{x^n/n2^n} \right| = \left| \frac{x^{n+1}}{(n+1)2^{n+1}} \frac{n2^n}{x^n} \right| = \left| \frac{x}{2} \right| \frac{n}{n+1}$$

In order to have convergence, we need the limit of this expression to be less than 1. The limit is:
$$L = \lim_{n \to \infty} \left| \frac{x}{2} \right| \frac{n}{n+1} = \left| \frac{x}{2} \right| \lim_{n \to \infty} \frac{n}{n+1} = \left| \frac{x}{2} \right| \cdot (1)$$

To have $L < 1$, we need $|x/2| < 1$, i.e. $|x| < 2$. This is the interval $-2 < x < 2$. Now we still need to test the endpoints of that interval, since the endpoints are where $L = 1$ and the ratio test is inconclusive. At the endpoint $x = -2$, the power series becomes
$$\sum_{n=0}^{\infty} \frac{(-1)^n (-2)^n}{n \cdot 2^n} = \sum_{n=0}^{\infty} \frac{(-1)^n (-1)^n 2^n}{n2^n} = \sum_{n=0}^{\infty} \frac{1}{n}$$

But this is the harmonic series, which diverges. Since $x = -2$ produces a divergent series, that point is NOT in the interval of convergence of the power series. How about the other endpoint, $x = 2$? Well, at that point, the series becomes
$$\sum_{n=0}^{\infty} \frac{(-1)^n 2^n}{n \cdot 2^n} = \sum_{n=0}^{\infty} \frac{(-1)^n}{n}$$

This is the alternating harmonic series, which converges. Since $x = 2$ produces a convergent series, that point is in the interval of convergence of the power series. The whole interval of convergence of the power series is then $-2 < x \le 2$, or $(-2, 2]$. The radius of convergence is 2. ∎

EX 3 Find the interval and radius of convergence of the power series
$$\sum_{n=0}^{\infty} \left(\frac{2}{3} \right)^n (x+3)^n.$$

Let's set up the ratio test:
$$\left| \frac{a_{n+1}}{a_n} \right| = \left| \frac{(2/3)^{n+1}(x+3)^{n+1}}{(2/3)^n (x+3)^n} \right| = \left| \frac{2}{3}(x+3) \right| = \frac{2}{3}|x+3|$$

In order to have convergence, we need the limit of this expression to be less than 1. Since n no longer appears in the expression, we have

$$L = \lim_{n\to\infty} \frac{2}{3}|x+3| = \frac{2}{3}|x+3|$$

To have $L < 1$, we need $2|x+3|/3 < 1$, or $|x+3| < 3/2$. This is equation is resolved by opening it up without the absolute value bars:

$$-\frac{3}{2} < x+3 < \frac{3}{2}$$

$$-\frac{3}{2} - 3 < x < \frac{3}{2} - 3$$

$$-\frac{9}{2} < x < -\frac{3}{2}$$

We still need to test the endpoints of that interval, since the endpoints are where $L = 1$ and the ratio test is inconclusive. At the endpoint $x = -9/2$, the power series becomes

$$\sum_{n=0}^{\infty} \left(\frac{2}{3}\right)^n \left(-\frac{9}{2}+3\right)^n = \sum_{n=0}^{\infty} \left(\frac{2}{3}\right)^n \left(-\frac{3}{2}\right)^n = \sum_{n=0}^{\infty}(-1)^n$$

which is a divergent series. At the other endpoint, $x = -3/2$, we get

$$\sum_{n=0}^{\infty} \left(\frac{2}{3}\right)^n \left(-\frac{3}{2}+3\right)^n = \sum_{n=0}^{\infty} \left(\frac{2}{3}\right)^n \left(\frac{3}{2}\right)^n = \sum_{n=0}^{\infty}(1)$$

which is also a divergent series. Both endpoints produce a divergent series, so neither is in the interval of convergence, which is $-9/2 < x < -3/2$. The radius of convergence is $3/2$. ∎

Note that this last example could have been done another way. We could actually write the power series as a geometric series,

$$\sum_{n=0}^{\infty} \left(\frac{2}{3}\right)^n (x+3)^n = \sum_{n=0}^{\infty} \left(\frac{2}{3} \cdot (x+3)\right)^n$$

in which $a = 1$ and $r = 2(x+3)/3$. For a geometric series to converge, we need $|r| < 1$, or here,

$$\left|\frac{2}{3}(x+3)\right| < 1$$

and that puts us right in the middle of the above computation.

You Try It

(1) Find the interval and radius of convergence of the power series
$$\sum_{n=1}^{\infty} \frac{(-1)^{n-1}x^n}{n^3}.$$

(2) Find the interval and radius of convergence of the power series
$$\sum_{n=1}^{\infty} \sqrt{n}(x-1)^n.$$

(3) Find the interval and radius of convergence of the power series
$$\sum_{n=1}^{\infty} \frac{(x-2)^n}{n^n}.$$

Power Series — Problem List

Power Series — You Try It

These appeared above; solutions begin on the next page.

Find the interval and radius of convergence of the following power series:

(1) $\displaystyle\sum_{n=1}^{\infty} \frac{(-1)^{n-1}x^n}{n^3}$.　　　(2) $\displaystyle\sum_{n=1}^{\infty} \sqrt{n}(x-1)^n$.　　　(3) $\displaystyle\sum_{n=1}^{\infty} \frac{(x-2)^n}{n^n}$.

Power Series — Practice Problems

Try these as you get the hang of the You Try It problems. Solutions to these problems are available in Sec. A.5.3.

Find the interval and radius of convergence of the following power series:

(1) $\displaystyle\sum_{n=0}^{\infty} \frac{x^n}{n!}$.　　　(2) $\displaystyle\sum_{n=0}^{\infty} n^3(x-5)^n$.　　　(3) $\displaystyle\sum_{n=1}^{\infty} \frac{(3x-2)^n}{n3^n}$.

Power Series — Challenge Problems

Try these problems to test your skills with the ideas in this section. Solutions to these problems are available in Sec. B.5.3.

(1) Find the interval and radius of convergence of the power series $\displaystyle\sum_{n=1}^{\infty} \frac{x^n}{n3^n}$.

(2) Find the interval and radius of convergence of the power series $\displaystyle\sum_{n=1}^{\infty} \frac{(-2)^n}{\sqrt{n}}(x+3)^n$.

(3) The following series defines a *Bessel Function*.[3] What is its interval of convergence?

$$\sum_{n=0}^{\infty} \frac{(-1)^n x^{2n+1}}{n!(n+1)!2^{2n+1}}$$

---

[3] These are useful in solving Partial Differential Equations; a good overview can be found in any edition of *Fourier Series and Boundary Value Problems*, by Brown & Churchill.

Power Series — You Try It — Solved

---

(1) Find the interval and radius of convergence of the power series
$$\sum_{n=1}^{\infty} \frac{(-1)^{n-1}x^n}{n^3}.$$

☐ Starting up the ratio test,

$$\left|\frac{a_{n+1}}{a_n}\right| = \left|\frac{\left(\frac{x^{n+1}}{(n+1)^3}\right)}{\left(\frac{x^n}{n^3}\right)}\right| = \left|\frac{x^{n+1}}{(n+1)^3} \cdot \frac{n^3}{x^n}\right| = \left(\frac{n}{n+1}\right)^3 |x|$$

To have convergence, we need the limit of this expression to be less than 1; that limit is

$$L = \lim_{n\to\infty} \left(\frac{n}{n+1}\right)^3 |x| = |x| \lim_{n\to\infty} \left(\frac{n}{n+1}\right)^3 = |x| \cdot (1) = |x|$$

We have $L < 1$ when when $|x| < 1$, giving the interval $-1 < x < 1$. We also have to test the endpoints of this interval. When $x = -1$, we get

$$\sum_{n=1}^{\infty} \frac{(-1)^{n-1}(-1)^n}{n^3} = -\sum_{n=1}^{\infty} \frac{1}{n^3}$$

which is a p-series with $p = 3$, and so converges. For $x = +1$, we get

$$\sum_{n=1}^{\infty} \frac{(-1)^{n-1}(1)^n}{n^3} = \sum_{n=1}^{\infty} \frac{(-1)^{n-1}}{n^3}$$

which converges by the alternating series test. The interval of convergence is $-1 \le x \le 1$, or $[-1, 1]$. The radius of convergence is 1. ∎

---

(2) Find the interval and radius of convergence of the power series
$$\sum_{n=1}^{\infty} \sqrt{n}(x - 1)^n.$$

☐ Starting up the ratio test,

$$\left|\frac{a_{n+1}}{a_n}\right| = \left|\frac{\sqrt{n+1}(x-1)^{n+1}}{\sqrt{n}(x-1)^n}\right| = |x - 1|\sqrt{\frac{n+1}{n}}$$

To have convergence, we need the limit of this expression to be less than 1. That limit is

$$L = \lim_{n\to\infty} |x - 1|\sqrt{\frac{n+1}{n}} = |x - 1| \lim_{n\to\infty} \sqrt{\frac{n+1}{n}} = |x - 1|(1) = |x - 1|$$

To have $L < 1$, we need $|x - 1| < 1$, or $0 < x < 2$. We also have to test the endpoints of this interval. When $x = 0$, we get

$$\sum_{n=1}^{\infty} (-1)^n \sqrt{n}$$

which diverges (the divergence test). For $x = 2$, we get

$$\sum_{n=1}^{\infty} \sqrt{n}$$

which also diverges. The interval of convergence is $0 < x < 2$; the radius of convergence is 1. ∎

---

(3) Find the interval and radius of convergence of the power series $\sum_{n=1}^{\infty} \dfrac{(x-2)^n}{\sqrt{n}}$.

☐ Starting up the ratio test,

$$\left| \frac{a_{n+1}}{a_n} \right| = \left| \frac{(x-2)^{n+1}/\sqrt{n+1}}{(x-2)^n/\sqrt{n}} \right| = \left| \frac{(x-2)^{n+1}}{\sqrt{n+1}} \frac{\sqrt{n}}{(x-2)^n} \right| = |x - 2| \sqrt{\frac{n}{n+1}}$$

To have convergence, we need the limit of this expression to be less than 1. That limit is

$$L = \lim_{n \to \infty} |x - 2| \sqrt{\frac{n}{n+1}} = |x - 2| \lim_{n \to \infty} \sqrt{\frac{n}{n+1}} = |x - 2|(1) = |x - 2|$$

So $L < 1$ when $|x - 2| < 1$, i.e. $1 < x < 3$. We also have to test the endpoints of this interval. When $x = 1$, the power series becomes

$$\sum_{n=1}^{\infty} \frac{(-1)^n}{\sqrt{n}}$$

which converges according to the alternating series test. For $x = 3$, we get

$$\sum_{n=1}^{\infty} \frac{1}{\sqrt{n}}$$

which is a p-series with $p = \dfrac{1}{2}$ and therefore diverges. The interval of convergence of our series is thus $1 \leq x < 3$, or $[1, 3)$. The radius of convergence is 1. ∎

11.4 Functions in Disguise

Introduction

The "big deal" about a power series is that within its interval of convergence, a power series is itself a perfectly good function. Sometimes this function is also known in a familiar algebraic form,[4] and sometimes it's not. We can approach this link between functions and series in two ways:

(1) Given a power series which converges in some interval around a point x_0, can we determine the algebraic form of the function it represents there?
(2) Given a function, can we determine its power series form and interval of convergence centered around a given point x_0?

We will resolve the general answer to these questions soon, by returning, finally, to the topic of Taylor and Maclaurin Series. But first, let's play around with some intermediate examples, in which we can take the easy way out and use what we know about geometric series and their convergence.

Determining the Series Representation of a Function — The Easy Way!

Suppose I gave you any old function, say, $f(x) = x^2 + \sin x$ and asked you for its power series (i.e. Taylor series) around $x_0 = 1$. You know how to do that, and it involves computing several derivatives of $f(x)$, plugging in $x = 1$, and using those values in the expansion

$$T(x) = \sum_{n=0}^{\infty} \frac{f^{(n)}(x_0)}{n!}(x - x_0)^n$$

But that's no fun, so we'll save it for the next, final, section on series. In the meantime, we can use our knowledge of the geometric series and its convergence formula to build a power series more quickly for *some* (not all!) functions.

Recall that when a geometric series $\sum_{n=0}^{\infty} ar^n$ converges ($|r| < 1$), it converges to $a/(1 - r)$. Conversely, if we can recognize a function as having the form of $a/(1 - r)$ in disguise, then we can reconstruct the series it goes

---

[4]See EX 6 and YTI 6 in Sec. 10.4.

with. All of the following functions can be related to an $a/(1-r)$, in one way or another. Here is a set of examples of this situation.

EX 1 Find the power series representation and interval of convergence of
$$f(x) = \frac{1}{1+x}.$$

The expression $1/(1+x)$ matches the form $a/(1-r)$ directly with $a=1$ and $r=-x$. Therefore, this function is also known as the following geometric series for values of x in its interval of convergence:

$$\sum_{n=0}^{\infty} ar^n = \sum_{n=0}^{\infty} (1)(-x)^n = \sum_{n=0}^{\infty} (-1)^n x^n$$

The interval of convergence occurs where $|r| < 1$, i.e. where $|-x| < 1$, so it is the interval $-1 < x < 1$. ∎

You Try It

   (1) Find the power series representation and interval of convergence of
$$f(x) = \frac{1}{x-5}.$$

EX 2 Find the power series representation and interval of convergence of
$$f(x) = \frac{5}{1-(x-1)^2}.$$

This match between this function and the form $a/(1-r)$ is fairly straightforward, with $a=5$ and $r=(x-1)^2$. Therefore, within the interval of convergence, this function is also known as the geometric series

$$\sum_{n=0}^{\infty} ar^n = \sum_{n=0}^{\infty} (5)((x-1)^2)^n = \sum_{n=0}^{\infty} 5(x-1)^{2n}$$

The interval of convergence occurs where $|r| < 1$, i.e. where $|(x-1)^2| < 1$, so it is the interval $0 < x < 2$. ∎

You Try It

   (2) Find the power series representation and interval of convergence of
$$f(x) = \frac{1}{1-x^3}.$$

EX 3 Find the power series representation and interval of convergence of
$$f(x) = \frac{x}{4 - x^2}.$$

This function needs a slight makeover before we see a match to the form $a/(1 - r)$ is not so obvious. Specifically, the function has a 4 rather than a 1 leading the denominator. That's pretty easy to fix:

$$\frac{x}{4 - x^2} = \frac{x}{4(1 - x^2/4)} = \frac{x/4}{1 - (x/2)^2}$$

Now the with the identification $a = x/4$ and $r = (x/2)^2$, we can identify

$$\frac{x}{4 - x^2} = \sum_{n=0}^{\infty} ar^n = \sum_{n=0}^{\infty} \frac{x}{4} \left[\left(\frac{x}{2}\right)^2 \right]^n = \sum_{n=0}^{\infty} \frac{x^{2n+1}}{2^{2n+2}}$$

(Make sure you understand how the exponents have worked out.) The interval of convergence is found where $|r| < 1$, i.e. where $|x/2|^2 < 1$, i.e. $-2 < x < 2$. ∎

You Try It

(3) Find the power series representation and interval of convergence of
$$f(x) = \frac{x^2}{x - 3}.$$

Determining New Series From Known Series

As we develop our library of known pairings between traditional functions and their alter ego power series, we can use them to create more. Here's a quick example of using a known series to create a very similar second series:

EX 4 Find the power series representation and interval of convergence of
$$g(x) = \frac{1}{1 + 2x}.$$

Note that $g(x) = f(2x)$, where $f(x) = 1/(1 + x)$. We already know the series for $f(x)$ from EX 1, so

$$g(x) = f(2x) = \sum_{n=0}^{\infty} (-1)^n (2x)^n$$

The interval of convergence is now where $|2x| < 1$, i.e. $|x| < 1/2$. ∎

You Try It

(4) Find the power series representation and interval of convergence of
$f(x) = \dfrac{1}{3x - 5}$. (HINT: See YTI 1.)

The next trick we have is to not relate a function itself to the form $a/(1 - r)$, but its derivative or antiderivative.

EX 5 Find the power series representation and interval of convergence of $f(x) = \tan^{-1}(x)$.

Recall that

$$\frac{d}{dx}\tan^{-1}(x) = \frac{1}{1 + x^2} \quad \text{so that} \quad \tan^{-1}(x) = \int \frac{1}{1 + x^2}\, dx$$

We don't know much about the series for $\tan^{-1}(x)$ yet, but we can easily put together the series for $1/(1 + x^2)$, which matches the form $a/(1 - r)$ with $a = 1$ and $r = -x^2$. Its series representation is

$$\frac{1}{1 + x^2} = \sum_{n=0}^{\infty}(1)(-x^2)^n = \sum_{n=0}^{\infty}(-1)^n x^{2n}$$

Therefore,

$$\tan^{-1}(x) = \int \frac{1}{1 + x^2}\, dx = \int \sum_{n=0}^{\infty}(-1)^n x^{2n}\, dx = \sum_{n=0}^{\infty}(-1)^n \frac{x^{2n+1}}{2n + 1}$$

At this point, I'll pause for you to say, "Hey, what about the $+C$ that's supposed to be part of the antiderivative?" Well, that's an excellent question, and I'm glad you asked! What we *really* get is

$$\tan^{-1}(x) = \sum_{n=0}^{\infty}(-1)^n \frac{x^{2n+1}}{2n + 1} + C$$

We can determine the value of C needed by balancing the two expressions at any one point, such as $x = 0$. When $x = 0$, the left side of this expression is $\tan^{-1}(0) = 0$; the right side is $0 + C$. This only works when $C = 0$. See YTI 5 for a different outcome.

Anyway, the interval of convergence for our series occurs where $|r| < 1$, i.e. where $|-x^2| < 1$, i.e. $-1 < x < 1$. ∎

> **You Try It**
>
> (5) Find the power series representation and interval of convergence of $f(x) = \ln(5 + x)$.

In this next example, we'll see a case where the *antiderivative* of a given function looks very much $a/(1 - r)$:

EX 6 Find a power series representation and interval of convergence of $f(x) = \sinh(x)$.

In Challenge Problem 1 in Sec. 10.2, we learned that a series representation of $\cosh(x)$ is

$$\cosh(x) = \sum_{n=0}^{\infty} \frac{x^{2n}}{(2n)!}$$

Now $\sinh(x)$ is the derivative of $\cosh(x)$, so it's very tempting to just proceed like this:

$$\sinh(x) = \frac{d}{dx} \sum_{n=0}^{\infty} \frac{x^{2n}}{(2n)!} = \sum_{n=0}^{\infty} (2n) \frac{x^{2n-1}}{(2n)!} = \sum_{n=0}^{\infty} \frac{x^{2n-1}}{(2n-1)!}$$

This isn't really a good answer, since now the leading term (for $n = 0$) produces a term with x^{-1}, as well as a $(-1)!$. We don't want either of those. This shows the danger of plowing ahead with a derivative operation on a series without keeping an eye on the terms themselves. What if we open the series up and watch the individual terms?

$$\begin{aligned}
\frac{d}{dx} \cosh(x) &= \frac{d}{dx} \sum_{n=0}^{\infty} \frac{x^{2n}}{(2n)!} \\
&= \frac{d}{dx} \left(1 + \frac{x^2}{2!} + \frac{x^4}{4!} + \frac{x^6}{6!} + \cdots \right) \\
&= \frac{2x}{2!} + \frac{4x^3}{4!} + \frac{6x^5}{6!} + \cdots \\
&= \frac{x}{1!} + \frac{x^3}{3!} + \frac{x^5}{5!} + \cdots = \sum_{n=1}^{\infty} \frac{x^{2n-1}}{(2n-1)!}
\end{aligned}$$

Here we see explicitly that the first term of the original series (which was just the constant 1) has vanished under the derivative operation — so that the $n = 0$ term is gone. This is a better result. If needed, we can re-index the new series so that it starts with $n = 0$, so that altogther,

$$\sinh(x) = \sum_{n=0}^{\infty} \frac{x^{2n+1}}{(2n+1)!}$$

Now, notice that back in Sec. 10.2, we did not say anything about the interval of convergence of the series representation of $\cosh(x)$, and the series here for $\sinh(x)$ is certainly not a geometric series. So to find its interval of convergence, we should probably apply the ratio test — because the term recipe is heavily dependent on powers and factorials. With $a_n = x^{2n+1}/(2n+1)!$, we have

$$a_{n+1} = \frac{x^{2(n+1)+1}}{(2(n+1)+1)!} = \frac{x^{2n+3}}{(2n+3)!}$$

so that

$$\left| \frac{a_{n+1}}{a_n} \right| = \left| \frac{x^{2n+3}}{(2n+3)!} \cdot \frac{(2n+1)!}{x^{2n+1}} \right| = \left| \frac{x^2}{(2n+3)(2n+2)} \right|$$

and

$$L = \lim_{n \to \infty} \left| \frac{a_{n+1}}{a_n} \right| = 0$$

Since $L < 0$ for all x, the interval of convergence here is all real numbers. ∎

You Try It

(6) Find the power series representation and interval of convergence of
$$f(x) = \frac{1}{(1+x)^2}.$$

Functions in Disguise — Problem List

Functions in Disguise — You Try It

These appeared above; solutions begin on the next page.

Find the power series representation and interval of convergence of the following functions:

(1) $f(x) = \dfrac{1}{x - 5}$.

(2) $f(x) = \dfrac{1}{1 - x^3}$.

(3) $f(x) = \dfrac{x^2}{x - 3}$.

(4) $f(x) = \dfrac{1}{3x - 5}$.

(5) $f(x) = \ln(5 + x)$.

(6) $f(x) = \dfrac{1}{(1 + x)^2}$.

Functions in Disguise — Practice Problems

Try these as you get the hang of the You Try It problems. Solutions to these problems are available in Sec. A.5.4.

Find the power series representation and interval of convergence of the following functions:

(1) $f(x) = \dfrac{1}{1 + 9x^2}$.

(2) $f(x) = \dfrac{x}{9 + x^2}$.

(3) $f(x) = \dfrac{1}{(1 + x)^3}$. (Hint: Can you use YTI 6?)

(4) $f(x) = \dfrac{x^3}{(x - 2)^2}$.

Functions in Disguise — Challenge Problems

Try these problems to test your skills with the ideas in this section. Solutions to these problems are available in Sec. B.5.4.

Find the power series representation and interval of convergence of the following functions:

(1) $f(x) = \dfrac{x}{4x + 1}$.

(2) $f(x) = \dfrac{x^2}{(1 + x)^3}$.

(3) $f(x) = \tan^{-1}\left(\dfrac{x}{3}\right)$.

Functions in Disguise — You Try It — Solved

---

(1) Find the power series representation and interval of convergence of
$f(x) = \dfrac{1}{x-5}$.

☐ With a little work, we can get this to look like the $a/(1-r)$ form of a geometric series:

$$\frac{1}{x-5} = \frac{1}{1-(6-x)}$$

This matches with $a = 1$ and $r = 6 - x$. Thus a series form would be

$$\sum_{n=0}^{\infty} a(r)^n = \sum_{n=0}^{\infty} (x-6)^n$$

This will converge when $|r| = |x - 6| < 1$, i.e. when $5 < x < 7$. (It will not converge for $x = 5$ or $x = 7$.)

Alternately, we can solve it like this:

$$\frac{1}{x-5} = \frac{1}{5(x/5-1)} = \frac{-1/5}{1-x/5}$$

This matches the geometric series form with $a = -1/5$ and $r = x/5$. Thus a series form would be

$$\sum_{n=0}^{\infty} a(r)^n = \sum_{n=0}^{\infty} -\frac{1}{5}\left(\frac{x}{5}\right)^n = \sum_{n=0}^{\infty} -\frac{x^n}{5^{n+1}}$$

This will converge when $|r| = |x/5| < 1$, i.e. when $|x| < 5$ or $-5 < x < 5$.

We now have two different series representations of the same function, with different radii of convergence. That's pretty cool!   ∎

---

(2) Find the power series representation and interval of convergence of
$f(x) = \dfrac{1}{1-x^3}$.

☐ That form looks like the $a/(1-r)$ form of a geometric series with $a = 1$ and $r = x^3$. Thus a series form would be

$$\sum_{n=0}^{\infty} a(r)^n = \sum_{n=0}^{\infty} (1)(x^3)^n = \sum_{n=0}^{\infty} x^{3n}$$

This will converge when $|r| = |x^3| < 1$, i.e. when $-1 < x < 1$. (It will not converge for $x = -1$ or $x = 1$.) ∎

---

(3) Find the power series representation and interval of convergence of $f(x) = \dfrac{x^2}{x-3}$.

☐ We can rearrange things a bit to match to the form $a/(1-r)$:

$$\frac{x^2}{x-3} = \frac{1}{3}\frac{x^2}{x/3-1} = \frac{-x^2/3}{1-x/3}$$

Now with $a = -x^2/3$ and $r = x/3$ we have a match to

$$\sum_{n=0}^{\infty} a(r)^n = \sum_{n=0}^{\infty}\left(-\frac{x^2}{3}\right)\left(\frac{x}{3}\right)^n = \sum_{n=0}^{\infty}\left(-\frac{x^{n+2}}{3^{n+1}}\right)$$

and convergence for $|r| < 1$, i.e. $|x/3| < 1$, or $-3 < x < 3$. ∎

---

(4) Find the power series representation and interval of convergence of $g(x) = \dfrac{1}{3x-5}$.

☐ Note that $g(x) = f(3x)$, where $f(x) = 1/(x-5)$ is in YTI 1. So we can take either series from YTI 1 and plug $3x$ in for x. Choosing the first version,

$$\frac{1}{3x-5} = \sum_{n=0}^{\infty}(3x-6)^n$$

We also need to re-scale the interval of convergence accordingly: now, $5 < 3x < 7$, or $5/3 < x < 7/3$. ∎

---

(5) Find the power series representation and interval of convergence of $f(x) = \ln(5+x)$.

☐ That form does not look like the $a/(1-r)$ form of a geometric series. However, note that

$$\ln(5+x) = \int \frac{1}{5+x}dx$$

and the integrand $1/(5+x)$ can be put in geometric series form:

$$\frac{1}{5+x} = \frac{1}{5(1+x/5)}$$

which is in geometric form with $a = 1/5$ and $r = -x/5$. So,

$$\frac{1}{5+x} = \sum_{n=0}^{\infty} \frac{1}{5}\left(\frac{-x}{5}\right)^n = \sum_{n=0}^{\infty} \frac{(-1)^n x^n}{5^{n+1}}$$

Then,

$$\ln(5+x) = \int \sum_{n=0}^{\infty} \frac{(-1)^n x^n}{5^{n+1}}\, dx = \sum_{n=0}^{\infty} \frac{(-1)^n x^{n+1}}{(n+1)5^{n+1}} + C$$

Let's use $x = 0$ as a test point to resolve C. From the function itself, $f(0) = \ln(5)$. So we need to ensure that we get $\ln 5$ when we plug $x = 0$ into the series, too. That means we need $C = \ln 5$. Altogether,

$$\ln(5+x) = \ln 5 + \sum_{n=0}^{\infty} \frac{(-1)^n x^{n+1}}{(n+1)5^{n+1}}$$

The interval of convergence is $|r| < 1$, i.e. $|x/5| < 1$, or $-5 < x < 5$. ∎

---

(6) Find the power series representation and interval of convergence of $f(x) = \dfrac{1}{(1+x)^2}$.

☐ That form does not look like $a/(1-r)$. However, note that

$$f(x) = \frac{1}{(1+x)^2} = \frac{d}{dx}\frac{-1}{1+x}$$

and the latter function is in geometric series form, with $a = -1$ and $r = -x$. Convergence occurs when $|x| < 1$. So

$$f(x) = \frac{d}{dx}\frac{-1}{1+x} = \frac{d}{dx}\sum_{n=0}^{\infty}(-1)(-x)^n$$

$$= \frac{d}{dx}\sum_{n=0}^{\infty}(-1)^{n+1}x^n = \sum_{n=1}^{\infty}(-1)^{n+1}nx^{n-1}$$

$$= \sum_{n=0}^{\infty}(-1)^{n+2}(n+1)x^n$$

Together,

$$\frac{1}{(1+x)^2} = \sum_{n=0}^{\infty}(-1)^n(n+1)x^n$$

for $|x| < 1$. (Be sure you locate and understand the shifting of the starting value of the index during the manipulation of the series.) ∎

11.5   Taylor Series Part 2

Our Story So Far....

It's time to close the loop on the topics that began in Chap. 10: Interlude — Introduction to Approximation. Our overall story started when we looked at how we can develop a simple polynomial approximation to a function around a chosen point. We started with linear (first order) approximations, then explored quadratic (second order) and cubic (third order) approximations. We noted that there was no need to stop at these, we could develop any n-th order approximation we wanted. This led to the idea of Maclaurin and Taylor polynomials. And the recognition that n can actually approach infinity led to Maclaurin and Taylor series. The open questions in this process were

- How good is a polynomial approximation to a function, and how can you measure that?
- If the order of the approximation increases, does the agreement improve?
- When does the approximating polynomial become completely indistinguishable from the function?
- As the agreement between the approximating polynomial and the function improves, do they become indistinguishable everywhere, or just on a certain interval?

In order to investigate these questions, we took a detour into intermediate topics regarding infinite series in general. We learned to identify several basic types of series: p-series, geometric series, and alternating series. Each of those specific types of series comes with its own convergence test, and beyond those, we also saw three more generic convergence tests that can be applied to miscellaneous series that don't fall into one of the above categories; these more general tests are the divergence test, the integral test, and the ratio test. Finally, as much fun as infinite series are, we recognized we are limited in practice to computing the nth partial sum of a series rather than the full series, but we saw that it is possible to measure an upper bound on the error between the nth partial sum of a series and the true value of the series — so we know how bad our approximations can be when we are forced to limit them to a finite number of terms. The alternating series test and integral test come with such remainder formulas.

After all that, we learned that if we take a regular infinite series and place a variable in it, say x, we can create a "power series" that becomes an actual function. Power series will often converge for only a limited band of x values, and not for others. The interval of convergence describes the set of x values that produce a convergent series.

We now have the information we need to close the loop. Back in Sec. 10.2, we learned how to compute the Taylor series for any $f(x)$ function around an anchor point x_0. We now know that a Taylor series is a power series, and therefore anything we know about power series applies to Taylor series. Specifically (1) we now know how to compute the interval of convergence of many Taylor series, and (2) we should expect that there is a remainder formula for Taylor series, so that we know just how awful our approximations to a given function happen to be when we truncate a Taylor series into an approximating Taylor polynomial. We've already seen that remainder formula: it's Taylor's Inequality, as seen in Useful Fact 10.1 of Chap. 10.

In this final section related to infinite series,[5] we'll concentrate on forming Taylor series and determining their intervals of convergence; it's a short section, because we've already built up the necessary ideas.

Picking Up Where We Left Off

If you need to, go back and review Sec. 10.2: Taylor Series Part 1. We will use many of the Taylor and Maclaurin series created in Part 1 without redeveloping them.

EX 1 Recover the Taylor series for $f(x) = \dfrac{1}{1-x}$ around $x_0 = 3$ and find its interval of convergence.

This Taylor series was already developed in EX 1 of Sec. 10.2: Taylor Series Part 1:

$$\frac{1}{1-x} = \sum_{n=0}^{\infty} \frac{(-1)^{n+1}}{2^{n+1}} (x-3)^n$$

Now we know how to compute the interval of convergence of this Taylor series with the ratio test:

$$\left| \frac{a_{n+1}}{a_n} \right| = \left| \frac{(x-3)^{n+1}/2^{n+2}}{(x-3)^n/2^{n+1}} \right| = \left| \frac{(x-3)^{n+1}}{2^{n+2}} \cdot \frac{2^{n+1}}{(x-3)^n} \right| = \frac{1}{2}|x-3|$$

---

[5]Don't you dare cheer!

In order to have convergence, we need the limit of this expression to be less than 1. But when we compute this limit, we get:

$$L = \lim_{n \to \infty} \frac{1}{2}|x - 3| = \frac{1}{2}|x - 3|$$

This is less than 1 when $|x - 3|/2 < 1$, or $|x - 3| < 2$. This translates to the interval $1 < x < 5$.

Now we still need to test the endpoints of that interval, since the endpoints are where the ratio test is inconclusive because $L = 1$. At the endpoint $x = 1$, the Taylor series becomes

$$\sum_{n=0}^{\infty} \frac{(-1)^{n+1}}{2^{n+1}}(x - 3)^n = \sum_{n=0}^{\infty} \frac{(-1)^{n+1}}{2^{n+1}}(-2)^n = \sum_{n=0}^{\infty} \frac{(-1)}{2}$$

This is no longer an alternating series since every term is negative, and the divergence test tells us it will diverge. At the endpoint $x = 5$, the Taylor series becomes

$$\sum_{n=0}^{\infty} \frac{(-1)^{n+1}}{2^{n+1}}(x - 3)^n = \sum_{n=0}^{\infty} \frac{(-1)^{n+1}}{2^{n+1}}(2)^n = \sum_{n=0}^{\infty} \frac{(-1)^{n+1}}{2}$$

This is an alternating series but the terms do not decrease to 0, and therefore the series diverges. Neither endpoint is within the interval of convergence. The full interval of convergence is therefore $1 < x < 5$. Flip back and look at Fig. 10.9 of Sec. 10.2, which shows the Taylor polynomial out to x^6 for this function; then see Fig. 11.1 here, which extends this out to the term $(x - 3)^8$. In this new figure, you can really see the interval of convergence in action: the truncated Taylor series fits $f(x)$ quite well around $x = 3$, but goes off on its merry way at $x = 5$, and also towards $x = 1$. ∎

$\boxed{\text{EX 2}}$ Recover the Maclaurin series for $f(x) = \tan^{-1}(x)$ and find its interval of convergence.

This Maclaurin series was already developed in EX 2 of Sec. 10.2. We also found it another way in EX 5 of Sec. 11.4. It is:

$$\tan^{-1}(x) = \sum_{n=0}^{\infty} \frac{(-1)^n}{2n + 1}x^{2n+1}$$

Now we know how to compute the interval of convergence of this series with the ratio test:

$$\left| \frac{a_{n+1}}{a_n} \right| = \left| \frac{x^{2n+3}/(2n+3)}{x^{2n+1}/(2n+1)} \right| = \left| \frac{x^{2n+3}}{2n+3} \cdot \frac{2n+1}{x^{2n+1}} \right| = \left(\frac{2n+1}{2n+3} \right)|x^2|$$

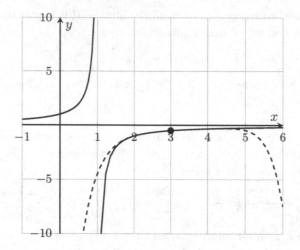

Fig. 11.1 $f(x) = 1/(1-x)$ (solid) and its Taylor polynomial around $x_0 = 3$ for $n = 8$.

In order to have convergence, we need the limit of this expression to be less than 1. But when we compute this limit, we get:

$$L = \lim_{n \to \infty} \left(\frac{2n+1}{2n+3} \right) |x^2| = x^2$$

This is less than 1 when $x^2 < 1$, or $-1 < x < 1$. Now we still need to test the endpoints of that interval, since the endpoints are where the ratio test is inconclusive because $L = 1$. At the endpoint $x = -1$, the Maclaurin series becomes

$$\sum_{n=0}^{\infty} \frac{(-1)^n}{2n+1}(-1)^{2n+1} = \sum_{n=0}^{\infty} \frac{(-1)^{n+1}}{2n+1}$$

The simplification of the powers of -1 happens like this:

$$(-1)^n(-1)^{2n+1} = (-1)^n(-1)^{2n}(-1) = (-1)^n(1)(-1) = (-1)^{n+1}$$

This is an alternating series whose terms decrease to 0; the series converges and $x = -1$ is within the interval of convergence. At the endpoint $x = 1$, the Maclaurin series becomes

$$\sum_{n=0}^{\infty} \frac{(-1)^n}{2n+1}(1)^{2n+1} = \sum_{n=0}^{\infty} \frac{(-1)^n}{2n+1}$$

This is also an alternating series whose terms decrease to 0; the series converges and $x = 1$ is within the interval of convergence. The full interval of convergence includes both endpoints, and is $-1 \le x \le 1$. Figure 11.2

shows the function $f(x) = \tan^{-1}(x)$ and the Maclaurin polynomial out to x^{11}. The agreement is perfect on this visual scale until we approach the ends of the interval of convergence, where the approximating polynomial (dashed) goes away from $f(x)$. ∎

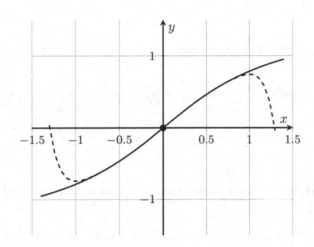

Fig. 11.2 $f(x) = \tan^{-1}(x)$ (solid) and its Maclaurin polynomial out to x^{11}.

You Try It

(1) Recover the Taylor series for $f(x) = e^x$ around $x_0 = 3$ and find the interval of convergence.
(2) Recover the Maclaurin series for $f(x) = \cos x$ and find the interval of convergence.
(3) Recover the Maclaurin series for $f(x) = \ln(1 + x)$ and find the interval of convergence.

Note: The series in each of these problems have already been determined in the You Try It problems of Sec. 10.2.

Building New Series From Old

Given that we've been assembling quite a collection of Taylor and Maclaurin series, it sometimes possible to create new series from known series rather than from scratch, as seen in Sec. 11.4. We'll build more on that here. Since Maclaurin series are easier to summarize, we'll stick with those. Here are a few Maclaurin series, most of which are already known or will be known after completion of the problems for this section:

$$\frac{1}{1-x} = \sum_{n=0}^{\infty} x^n \quad \text{for} \quad -1 < x < 1 \tag{11.1}$$

$$\ln(1+x) = \sum_{n=1}^{\infty} (-1)^{n+1} \frac{x^n}{n} \quad \text{for} \quad -1 < x < 1 \tag{11.2}$$

$$e^x = \sum_{n=0}^{\infty} \frac{x^n}{n!} \quad \text{for} \quad -\infty < x < \infty \tag{11.3}$$

$$\sin x = \sum_{n=0}^{\infty} (-1)^n \frac{x^{2n+1}}{(2n+1)!} \quad \text{for} \quad -\infty < x < \infty \tag{11.4}$$

$$\cos x = \sum_{n=0}^{\infty} (-1)^n \frac{x^{2n}}{(2n)!} \quad \text{for} \quad -\infty < x < \infty \tag{11.5}$$

EX 3 Find the Maclaurin series and its interval of convergence for $f(x) = e^{-x}$.

We can substitute $-x$ for x in the known series for e^x in (11.3):

$$e^{-x} = \sum_{n=0}^{\infty} \frac{(-x)^n}{n!} = \sum_{n=0}^{\infty} (-1)^n \frac{x^n}{n!}$$

The interval of convergence of the original series is $-\infty < x < \infty$, and remains so for this new series. ∎

EX 4 Find the Maclaurin series and its interval of convergence for $f(x) = x\cos(x)$.

We can multiply the known series for $\cos x$, shown in (11.5) by x:

$$x\cos x = x \cdot \sum_{n=0}^{\infty} (-1)^n \frac{x^{2n}}{(2n)!} = \sum_{n=0}^{\infty} (-1)^n \frac{x^{2n+1}}{(2n)!}$$

The interval of convergence of the original series is $-\infty < x < \infty$, and remains so for this new series. ∎

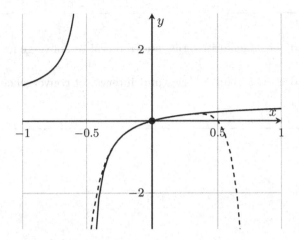

Fig. 11.3 $f(x) = x/(1+2x)$ (solid) and 8th order Maclaurin polynomial (dashed).

EX 5 Find the Maclaurin series and its interval of convergence for $f(x) = \dfrac{x}{1+2x}$.

Let's do this in two steps. First, note that

$$\frac{1}{1+2x} = \frac{1}{1-(-2x)}$$

so we can plug $-2x$ into the known Maclaurin series for $\dfrac{1}{1-x}$ in (11.1) to get

$$\frac{1}{1+2x} = \sum_{n=0}^{\infty}(-2x)^n = \sum_{n=0}^{\infty}(-1)^n 2^n x^n$$

and now multiplying by x we get

$$\frac{x}{1+2x} = x \cdot \frac{1}{1+2x} = x \cdot \sum_{n=0}^{\infty}(-1)^n 2^n x^n = \sum_{n=0}^{\infty}(-1)^n 2^n x^{n+1}$$

Since the original interval of convergence required $|x| < 1$, we now require $|-2x| < 1$, or $-1/2 < x < 1/2$. And just in case you don't believe me, Fig. 11.3 showing $f(x)$ and its Maclaurin polynomial out to x^8. The function has an asymptote at $x = -1/2$; you can see the agreement between the approximating polynomial and $f(x)$ breaking down at the ends of the interval of convergence. ∎

You Try It

(4) Find the Maclaurin series and interval of convergence for $f(x) = xe^{2x}$.

(5) Find the Maclaurin series and interval of convergence for $f(x) = \ln\left(1 - \dfrac{x}{2}\right)$.

Taylor and Maclaurin Series Part 2 — Problem List

For these problems, you should be efficient and refer to any series already created in any previous problem. Otherwise, you must create the series yourself ... and who needs that!

Taylor and Maclaurin Series Part 2 — You Try It

These appeared above; solutions begin on the next page.

(1) Recover the Taylor series for $f(x) = e^x$ around $x_0 = 3$ and find the interval of convergence.
(2) Recover the Maclaurin series for $f(x) = \cos x$ and find the interval of convergence.
(3) Recover the Maclaurin series for $f(x) = \ln(1 + x)$ and find the interval of convergence.
(4) Find the Maclaurin series and interval of convergence for $f(x) = xe^{2x}$.
(5) Find the Maclaurin series and interval of convergence for $f(x) = \ln\left(1 - \dfrac{x}{2}\right)$.

Taylor and Maclaurin Series Part 2 — Practice Problems

Try these as you get the hang of the You Try It problems. Solutions to these problems are available in Sec. A.5.5.

(1) Find the Maclaurin series and interval of convergence for $f(x) = \cos \pi x$.
(2) Find the Taylor series and interval of convergence for $f(x) = \ln x$ around $x_0 = 2$.
(3) Find the Maclaurin series and interval of convergence for $f(x) = \dfrac{1}{(1 + x)^2}$.
(4) Find the Taylor series and interval of convergence for $f(x) = \dfrac{1}{x}$ around $x = -2$.

Taylor and Maclaurin Series Part 2 — Challenge Problems

Try these problems to test your skills with the ideas in this section. Solutions to these problems are available in Sec. B.5.5.

(1) Find the Taylor series and interval of convergence for $f(x) = \cos x$ around $x_0 = \pi/2$.

(2) Find the Maclaurin series and interval of convergence for $f(x) = x^2 \tan^{-1}(x^2)$.

(3) Find the (fully simplified!) Maclaurin series and its interval of convergence for $f(x) = e^x + 1/e^x$. (Hint: There's the hard way, and then there's the easy way ... use your instincts!)

Taylor Series Part 2 — You Try It — Solved

(1) Find the Taylor series and interval of convergence for $f(x) = e^x$ around $x_0 = 3$.

☐ We found this Taylor Series in You Try It 1 of Sec. 10.2:

$$e^x = \sum_{n=0}^{\infty} \frac{e^3}{n!}(x-3)^n$$

The interval of convergence is determined by the ratio test:

$$\left|\frac{a_{n+1}}{a_n}\right| = \left|\frac{(x-3)^{n+1}/(n+1)!}{(x-3)^n/n!}\right| = \left|\frac{(x-3)^{n+1}}{(n+1)!} \cdot \frac{n!}{(x-3)^n}\right| = \frac{1}{n+1}|x-3|$$

In order to have convergence, we need the limit of this expression to be less than 1. But when we compute this limit, we get:

$$L = \lim_{n \to \infty} \frac{1}{n+1}|x-3| = 0$$

Since $L < 1$ for all values of x, the interval of convergence is $-\infty < x < \infty$. ∎

(2) Find the Maclaurin series and interval of convergence for $f(x) = \cos x$.

☐ We found this Maclaurin Series in You Try It 2 of Sec. 10.2:

$$\cos(x) = \sum_{n=0}^{\infty} \frac{(-1)^n}{(2n)!}x^{2n}$$

The interval of convergence is determined by the ratio test:

$$\left|\frac{a_{n+1}}{a_n}\right| = \left|\frac{x^{2n+2}/(2n+2)!}{x^{2n}/(2n)!}\right| = \left|\frac{x^{2n+2}}{(2n+2)!} \cdot \frac{(2n)!}{x^{2n}}\right| = \frac{1}{(2n+2)(2n+1)}|x^2|$$

In order to have convergence, we need the limit of this expression to be less than 1. But when we compute this limit, we get:

$$L = \lim_{n \to \infty} \frac{1}{(2n+2)(2n+1)}|x^2| = 0$$

Since $L < 1$ for all values of x, the interval of convergence is $-\infty < x < \infty$. ∎

(3) Find the Maclaurin series and interval of convergence for $f(x) = \ln(1 + x)$.

We found this Maclaurin Series in You Try It 3 of Sec. 10.2:

$$\ln(1 + x) = \sum_{n=1}^{\infty} (-1)^{n+1} \frac{x^n}{n}$$

The interval of convergence is determined by the ratio test:

$$\left| \frac{a_{n+1}}{a_n} \right| = \left| \frac{x^{n+1}/(n+1)}{x^n/n} \right| = \left| \frac{x^{n+1}}{(n+1)} \cdot \frac{n}{x^n} \right| = \frac{n}{n+1}|x|$$

In order to have convergence, we need the limit of this expression to be less than 1. But when we compute this limit, we get:

$$L = \lim_{n \to \infty} \frac{n}{n+1}|x| = |x|$$

To have $L < 1$ we need $-1 < x < 1$. Since the function itself is not defined for $x = -1$, the series cannot converge there. At the other endpoint $x = 1$, the series becomes a convergent alternating series — so the overall interval of convergence is $-1 < x \leq 1$, or $(-1, 1]$. ■

(4) Find the Maclaurin series and interval of convergence for $f(x) = xe^{2x}$.

☐ We can build on the Maclaurin series for e^x shown in (11.3):

$$xe^{2x} = x \cdot \sum_{n=0}^{\infty} \frac{(2x)^n}{n!} = \sum_{n=0}^{\infty} \frac{2^n x^{n+1}}{n!}$$

The original interval of convergence was $-\infty < x < \infty$, and remains so for this new series. ■

(5) Find the Maclaurin series and interval of convergence for $f(x) = \ln\left(1 - \frac{x}{2}\right)$.

☐ We can build on the Maclaurin series for $\ln(1 + x)$ shown in (11.2) by replacing x with $-x/2$:

$$\ln\left(1 - \frac{x}{2}\right) = \sum_{n=1}^{\infty} (-1)^{n+1} \frac{(-x/2)^n}{n} = \sum_{n=1}^{\infty} (-1)^{n+1}(-1)^n \frac{x^n}{n2^n} = -\sum_{n=1}^{\infty} \frac{x^n}{n2^n}$$

The original interval of convergence required $|x| < 1$, so we now require $|-x/2| < 1$ or $-2 < x < 2$. ■

Chapter 12

A Change in Graph Paper

12.1 Introduction to Parametric Equations

Introduction

Once students complete the delight that is Single Variable Calculus (as Calculus I and II, perhaps), the next math course in line is usually either (Multivariable Calculus (often Calculus III) or Differential Equations (or both, for the lucky ones!). While One of the main themes of Multivariable Calculus is the re-envisioning of space and construction of functions. Up until now, you have been restricted to considering just two-dimensional space in Cartesian (rectangular) coordinates, and your functions have been scalar functions — scalar goes in, scalar comes out. In Multivariable Calculus, you learn about other ways to describe 2D space, but also step up into 3D space, and think about other ways to assemble functions and curves in those spaces — such as, "scalar goes in, vector comes out". Some of these functions are based on parametric equations. And so ... here we go.

Parametric Equations: Sounds Scary, But Isn't!

Back in the days just before you hit puberty, when you still drank apple juice in the middle school cafeteria out of of a squishy plastic bag with a straw poked into it, someone will have asked you to graph one of those $y = f(x)$ things. In those days, you did it this way: you (a) selected a bunch of x values, (b) found the y-values that went with them, (c) plotted the resulting (x, y) pairs, and (d) played connect the dots — and thus, a graph of $y = f(x)$ was born.

A *parametric curve* is one in which BOTH the x and y values are generated by a third variable, or *parameter*, very often named t. We plot parametric curves in the xy-plane (for now), but neither x nor y are independent variables. Instead of a curve being a collection of points $(x, f(x))$, we now have a collection of points of the form $(x(t), y(t))$. The functions $x(t)$ and $y(t)$ that determine the points on the curve are *parametric equations*. They often come with minimum and maximum values of t to use, say $a \leq t \leq b$; in the absence of those limits, we assume we want to allow $-\infty < t < \infty$. Altogether, a parametric curve is fully described in this notation:

$$\{(x, y) : x = x(t), y = y(t), a \leq t \leq b\}$$

This says, "Hey, we're collecting a bunch of points (x, y) in which x and y are generated as functions of some parameter t, and t is allowed to take on any value between a and b."

Some of the benefits of using parametric equations to draw curves are:

- We can create much more complicated and interesting curves
- It is easy to describe only a certain segment of a curve
- Parametric curves have *direction* associated with them (we follow the curve in the direction of the points created as t increases). Curves of the $y = f(x)$ variety have "direction", too, but it's strictly left-to-right; parametric curves can go all over the place.

EX 1 Graph the parametric curve $\{(x, y) : x = t^2, y = t - 1, 0 \leq t \leq 4\}$

We can get an idea of what the curve looks like by plotting individual points. We select several t values between 0 and 4, then compute the associated x and y values:

t	0	1	2	3	4
x	0	1	4	9	16
y	-1	0	1	2	3

and therefore we can draw, and connect in order, the points $(x, y) = (0, -1), (1, 0), (4, 1), (9, 2), (16, 3)$. The final curve is shown in Fig. 12.1. ∎

You Try It

(1) Plot the curve given by the parametric equations $x = 3t - 5, y = 2t + 1$.

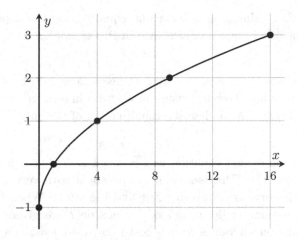

Fig. 12.1 The parametric curve $x = t^2, y = t - 1$ for $0 \leq t \leq 4$.

NOTE: The parametric description of a curve is not unique. For example, the curve given in Example 1 above can also be written $\{(x, y) : x = (t+1)^2, y = t, -1 \leq t \leq 3\}$. Make sure you see why!

Eliminating the Parameter

Very often, a curve can be given in both "regular" $y = f(x)$ form as well as parametric form. If we need to take a "regular" curve $y = f(x)$ and put it in parametric form, we can just declare $x = t$, and then write $y = f(t)$; we can also pick any bounds for t that we need (or specify no bounds and so let t take on all real values).

$\boxed{\textbf{EX 2}}$ Write the function $y = \sin(x^2 + \pi)$ between $(0, 0)$ and $(\sqrt{\pi}, 0)$ in parametric form.

If we declare $x = t$, then the function becomes $y = \sin(x^2 + \pi)$. The first endpoint $(0, 0)$ corresponds to $t = 0$, and the second endpoint $(\sqrt{\pi}, 0)$ corresponds to $t = \sqrt{\pi}$. Altogether, we can write the parametric version of the curve as:

$$\begin{cases} x = t \\ y = \sin(t^2 + 1) \end{cases} \quad -\infty < t < \infty \quad \blacksquare$$

When we need to find the "regular" form $y = f(x)$ or $x = g(y)$ of a parametric curve, we need to find a way to connect the two parametric equations through the parameter t. There is no one magic process for this.

Sometimes it's as simple as solving one equation for t and plugging the result into the other equation, as shown in this example:

EX 3 Eliminate the parameter from the equations $x = t^2, y = t - 1$ to see the what the relationship in its "also known as" form $y = f(x)$ or $x = g(y)$. Also, describe the direction of the curve.

If we rewrite the second equation as $t = y + 1$ and plug it into the first, we get $x = (y + 1)^2$. This is a sideways parabola with vertex $(0, -1)$. A *portion* of this curve is shown in EX 1 and Fig. 12.1. When the curve is presented in parametric form and no bounds on t are given, we assume that t can take on all values from $-\infty$ to ∞ (or at least can take on all values in the more restrictive of $x(t)$ or $y(t)$ ... in this case, we do have $-\infty < t < \infty$. When t is negative, we get positive x coordinates and negative y coordinates. For positive t, we have positive x and y. The larger t is in magnitude, the farther from the vertex we are. So we traverse the parabola from the lower right, clockwise to the vertex, and then out into the upper right. This direction is shown in Fig. 12.2. Please also use this Figure to consider why we are not using the phrase "parametric functions" to describe these curves.

■

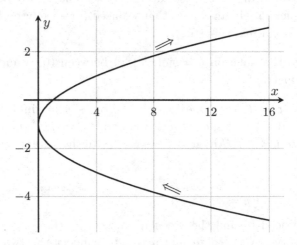

Fig. 12.2   The directed parametric curve $x = t^2, y = t - 1$ for $-\infty \le t \le \infty$.

You Try It

(2) Eliminate the parameter from the parametric equations $x = 3t - 5, y = 2t + 1$. Describe the direction of the curve.

Here are two new terms to consider before getting to EX 4:

- EX 3 showed how we can **eliminate the parameter** by solving either $x(t)$ or $y(t)$ for the parameter t and plugging that result into the other. In other cases, it may be more efficient to rely on a known identity.
- When we assert a direction based on how increasing t "drives" us along the curve. we say we have a *directed curve*. When a curve bends in precisely or approximately a circular way, we can describe the directionality, or **orientation**, as clockwise or counterclockwise.

EX 4 Identify the curve $x = \cos t, y = \sin t$ for $\pi/2 \le t \le 3\pi/2$ and state the orientation of the curve.

I don't think we want to solve *either* of the given equations for t! But look at what happens when we square both equations and add them together: we get $x^2 + y^2 = \cos^2 t + \sin^2 t$, so by invoking our favorite trigonometric identity, we now identify this curve as $x^2 + y^2 = 1$, which is the unit circle. Or rather, the curve only part of the unit circle because of the restrictions on t. We start at $t = \pi/2$, which gives the point $(x, y) = (0, 1)$, and we end at $t = 3\pi/2$, which gives the point $(x, y) = (0, -1)$. Altogether, these parametric equations result in the left half of the unit circle. Since we start at the top of the circle (12:00 on a clock) and then move leftwards around to the bottom (6:00 on a clock), the curve is oriented counterclockwise. ∎

You Try It

(3) Identify the curve $x = \sin\theta, y = \cos\theta$; $0 \le \theta \le \pi$ and state its orientation.

Here's one last question to illuminate a difference between parametric curves and the usual kind:

EX 5 Are the following two parametric curves the same curve?

- Curve A: $x = \cos t, y = \sin t$ from $t = 0$ to $t = 2\pi$

- Curve B: $x = \cos t, y = \sin t$ from $t = 2\pi$ to $t = 0$

These are not the same curve, as long as we consider the curve as distinct from the drawing on a piece of paper. The collection of points in both comprises the entire unit circle, so their "images" are the same. However, Curve A traverses that circle in the counterclockwise direction, while Curve B traverses the circle in the clockwise direction. Therefore, they are not the same curve. ∎

   The last thing to do here is (finally!) reveal the secret behind the naming of the hyperbolic trigonometric functions that have been showing up here and there throughout this text. What we've seen about them so far is that they are a set of functions which behave a whole lot like trigonometric functions in terms of algebraic properties (such as being even or odd functions), identities they obey (such as $\cosh^2(x) - \sinh^2(x) = 1$, and derivative relationships among themselves (such as $d/dx(\sinh(x)) = \cosh(x)$). Why in the world is the word "hyperbolic" tied these functions? Well, suppose we set two parametric equations as $x(t) = \cosh(t)$ and $y(t) = \sinh(t)$ for $-\infty < t < \infty$. What does the graph of this parametric function look like? The key is in the identity that I just reminded you about: we can eliminate the parameter by constructing $x^2(t) - y^2(t) = \cosh^2(t) - \sinh^2(t)$ which simplifies to (drumroll) $x^2 - y^2 = 1$ ... a hyperbola! Just like the pair $x(t) = \cos(t), y(t) = \sin(t)$ forms a circle, the pair $x(t) = \cosh(t), y(t) = \sinh(t)$ forms a hyperbola. There, you waited through two semesters' worth of calculus content to see that, I hope it was worth it.

Introduction to Parametric Equations — Problem List

Introduction to Parametric Equations — You Try It

Try these problems, appearing above, for which solutions are available at the end of this section.

(1) Plot the curve given by the parametric equations $x = 3t - 5, y = 2t + 1$ for $-2 \le t \le 2$.
(2) Eliminate the parameter from the parametric equations $x = 3t - 5, y = 2t + 1$. Describe the direction of the curve.
(3) Identify the curve $x = \sin\theta, y = \cos\theta$; $0 \le \theta \le \pi$ and state its orientation.

Introduction to Parametric Equations — Practice Problems

Try these as you get the hang of the You Try It problems. Solutions to these problems are available in Sec. A.6.1.

(1) Graph the parametric curve $x = \sqrt{t}, y = 1 - t$ for $0 \le t \le 4$. Eliminate the parameter to give the curve's $y = f(x)$ form.
(2) Eliminate the parameter to give the curve's $y = f(x)$ form for:
$$\begin{cases} x = -2t^3 + 4 \\ y = \sqrt{t^9 + 1} \end{cases} \quad -\infty < t < \infty$$
(3) Identify the curve $x = \sec\theta, y = \tan\theta$ for $-\pi/2 < \theta < \pi/2$ and state its orientation.
(4) Eliminate the parameter to identity the curve $x = 2\sinh(t)$, $y = 2\cosh(t)$ for $-\infty < t < \infty$.

Introduction to Parametric Equations — Challenge Problems

Try these problems to test your skills with the ideas in this section. Solutions to these problems are available in Sec. B.6.1.

(1) Eliminate the parameter and identify the curve $x = 1 + 3t, y = 2 - t^2$.
(2) Eliminate the parameter and identify the curve $x = e^t + 1, y = e^{2t}$.
(3) Identify the curve $x = 4\cos\theta, y = 5\sin\theta$ for $-\pi/2 < \theta < \pi/2$ and state its orientation.

Introduction to Parametric Equations — You Try It — Solved

(1) Plot the curve given by the parametric equations $x = 3t - 5, y = 2t + 1$ for $-2 \le t \le 2$.

□ Let's make a table of representative points,

t	-2	-1	0	1	2
x	-11	-8	-5	-2	1
y	-3	-1	1	3	5

When we plot the resulting (x, y) pairs, we'll see a straight line segment from $(-11, 3)$ to $(1, 5)$.   ∎

----

(2) Eliminate the parameter from the parametric equations $x = 3t - 5, y = 2t + 1$. Describe the direction of the curve.

□ Solving the first equation for t gives $t = \dfrac{x + 5}{3}$. Plugging this into the second equation:

$$y = 2t + 1 = 2\left(\frac{x + 5}{3}\right) + 1 = \frac{2}{3}x + \frac{5}{3} + 1$$

and so the curve is also known as the straight line $y = 2x/ + 13/3$. We follow the line from the lower left (Quadrant 3) to the upper right (Quadrant 1).   ∎

----

(3) Identify the curve $x = \sin\theta, y = \cos\theta$; $0 \le \theta \le \pi$ and state its orientation.

□ If we square both parametric equations, we have $x^2 = \sin^2\theta$ and $y^2 = \cos^2\theta$. Adding these together gives

$$x^2 + y^2 = \sin^2\theta + \cos^2\theta = 1$$

and so we have (at least part of) the unit circle. At $t = 0$ we're at the point $(0, 1)$; at $t = \pi$, we're at the point $(0, -1)$. So, we have the left half of the unit circle, oriented counterclockwise.   ∎

12.2   Calculus of Parametric Equations

Introduction

The early portion of Calc II is concerned with applications of partitioning and Riemann sums, which allow us to make measurements related to functions and graphs — such as slopes of tangent lines, arc length, area between curves, and so on. This will continue in Calc III. These measurements rely on the ability to find derivatives and integrals of functions. Before now, all we could work with were functions of one of the coordinate variables x or y. Now that we have curves being described parametrically, perhaps we can expect modifications to our graphical measurements.

Tangent Lines

Even though parametric curves are created in a strange fashion, they are still plotted in the xy-plane. So, in order to find the slope of a tangent line at a point on a parametric curve, we still need to know how fast the y coordinate changes with respect to changes in the x-coordinate. That is, we still need to know dy/dx. For parametric curves, though, we only know the equations $x = f(t)$ and $y = g(t)$, and thus only know the derivatives $x'(t) = dx/dt$ and $y'(t) = dy/dt$. So how can we create dy/dx? The chain rule to the rescue! You may recall that one version of the chain rule allows us to lay out a set of derivatives like this, from Eq. (4.37):

$$\frac{dy}{dx} \cdot \frac{dx}{dt} = \frac{dy}{dt}$$

A rearrangement of that will produce what we need:

Useful Fact 12.1. *Given parametric equations $x(t)$ and $y(t)$, the slope of a line tangent to the associated parametric curve comes from*

$$\frac{dy}{dx} = \frac{dy/dt}{dx/dt}$$

There is a possible complication to this. The expressions dy/dt and dx/dt are going to be written in terms of the parameter t, and so we'll have an expression for dy/dx written in terms of t. That's a bit awkward, since we'd usually expect dy/dx to be written in terms of x (or maybe y, too, if its an implicit derivative).

In some cases, that complication might not be an issue. We can use the above formula to compute dy/dt at a specific point using *values* of dy/dt and dx/dt rather than full expressions, as seen here:

EX 1 Find the slope of the line tangent to the parametric curve $x = t^2, y = (t-1)^2$ at the point $(1,0)$.

First, note that since $x = t^2$, we have $dx/dt = 2t$, and since $y = (t-1)^2$, we have $dy/dt = 2t - 2$. Next, recognize that the point $(1,0)$ corresponds to $t = 1$. So at that point, we have $dx/dt = 2(1) = 2$ and $dy/dt = 2(1)-2 = 0$. So then by Useful Fact 12.1, the slope of the tangent line at that specific point is:

$$\frac{dy}{dx} = \frac{dy/dt}{dx/dt} = \frac{0}{2} = 0 \quad \blacksquare$$

Now, if we're interested in an *expression* for dy/dx rather than an individual value, we can try to employ the parametric equations $x = f(t)$ and $y = g(t)$ to help eliminate t in the derivative expression — but that's often inconvenient. We can also just ignore the awkwardness of having dy/dx in terms of t, and just live with it, as in this next example:

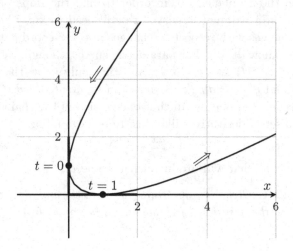

Fig. 12.3   The (directed) parametric curve $x = t^2, y = (t-1)^2$.

EX 2 Find an expression for the slope of the line tangent to the parametric curve $x = t^2, y = (t-1)^2$ at any point, and learn some things from it.

We have $dx/dt = 2t$ and $dy/dt = 2t - 2$, so

$$\frac{dy}{dx} = \frac{dy/dt}{dx/dt} = \frac{2t-2}{2t} = \frac{t-1}{t} = 1 - \frac{1}{t}$$

While this expression for dy/dx is in terms of t, we can still learn quite a bit from it. For example, the slope of the tangent line is undefined at $t = 0$, i.e. the point $(0,1)$. As t increases to ∞, i.e. as x and y approach ∞, the slopes approaches 1. And we can get the answer to EX 1 by plugging in $t = 1$ (for the point $(1,0)$) to get a tangent slope of 0 there. All of this should be evident in the graph of this parametric curve, shown in Fig. 12.3.

■

Another complication that can arise when computing of slopes of tangent lines of parametric curves is that parametric curves can have loops or crossings, and they can even redraw the same shape over and over. These features can lead to the existence of more than one tangent line at a single point (like in a criss-cross), or multiple values of t giving the same point in space.

EX 3 The parametric curve $x = 1 - t^2, y = t^3 - t$ is shown in Fig. 12.4. Find the slope of the tangent lines at the origin.

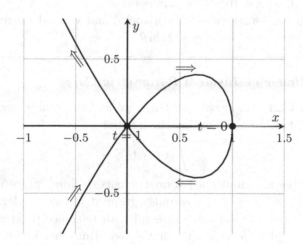

Fig. 12.4   The (directed) parametric curve $x = 1 - t^2, y = t^3 - t$.

Useful Fact 12.1 tell us that

$$\frac{dy}{dx} = \frac{dy/dt}{dx/dt} = \frac{3t^2 - 1}{-2t} = -\frac{3}{2}t + \frac{1}{2t}$$

Figure 12.4 shows that this curve crosses the origin twice, so there must be *two* t values that give the point $(0,0)$. From $x = 1 - t^2$, we see that x is 0 whenever $t = -1$ or $t = 1$, and both of these values also yield $y = 0$. So at $t = -1$ we have

$$\frac{dy}{dx}\bigg|_{t=-1} = -\frac{3}{2}(-1) + \frac{1}{2(-1)} = 1$$

and at $t = 1$ we have

$$\frac{dy}{dx}\bigg|_{t=1} = -\frac{3}{2}(1) + \frac{1}{2(1)} = -1$$

So the two tangent lines at the origin have slope -1 and 1.   ∎

Given the slope of a tangent line at a point, it should be no trouble to obtain the equation of the tangent lines.

You Try It

(1) Find the equation of the line tangent to $x = t^4 + 1; y = t^3 + t$ at $t = -1$.

(2) Find the locations of all horizontal and vertical tangent lines on the curve $x = 10 - t^2; y = t^3 - 12t$

(3) Find the locations of all horizontal and vertical tangent lines on the curve $x = \cos 3\theta; y = 2\sin \theta$.

The Area Under or Inside a Parametric Curve

For a regular Cartesian curve $y = f(x)$, we find the area under the curve on the interval $[a, b]$ by computing the integral

$$\int_a^b y(x)\, dx$$

Can we find the area under a parametric curve? Sure! Since parametric equations $x(t), y(t), a \leq t \leq b$ are also graphed in the xy-plane, we just need to (re)construct that same integral with reference to $x(t)$ and $y(t)$. First, we can replace y with $y(t)$, that's easy. But what about dx? Well, using $x = x(t)$, we can denote $dx/dt = x'(t)dt$; rewritten, that tells us that $dx = x'(t)\, dt$. We can put this all together as follows:

Useful Fact 12.2. *Given a parametric curve with equations $x = x(t), y = t(t)$, the area A under the arc traced out for $a \leq t \leq b$ is given by*

$$A = \int_a^b y(t) \cdot x'(t)\, dt$$

NOTE: The integrals for area under a parametric curve can often become a bit unpleasant, so it is fair to get numerical estimates when exact answers cannot reasonably be found.

EX 4 Find the area under the parametric curve $x = e^t$; $y = 4 - t^2$ and above the x-axis.

This curve intersects the x axis when $y = 0$, i.e. at $t = \pm 2$. From the equation $x = e^t$, we get $dx = e^t\, dt$, and the integral for area under this curve is:

$$\int_a^b y(t)x'(t)\, dt = \int_{-2}^2 (4 - t^2)e^t dt = 2e^2 + \frac{6}{e^2} \approx 15.59 \quad \blacksquare$$

You Try It

(4) Find the area inside $x = a\cos\theta$; $y = b\sin\theta$; $0 \le \theta \le 2\pi$. (The answer will be in terms of the constants a and b.)

Here's a heads up on the limits of integration: in a regular definite integral over x in $[a, b]$, we examine areas from left to right. But often in parametric curves, as t increases we move from right to left (counterclockwise). If you ever compute an area that should be positive but comes out negative, it could be that your limits need to be reversed.

Arc Length

In Sec. 8.2, we learned that the arc length of a Cartesian curve $y(x)$ over the interval $x \in [c, d]$ is given by

$$L = \int_c^d \sqrt{1 + [y'(x)]^2}\, dx$$

The arc length of a parametric curve found in a similar fashion; the derivation is below (in the Pit!), but the formulation and an example is right here:

Useful Fact 12.3. *Given a parametric curve with equations $x = x(t), y = t(t)$, the length L of the arc traced out for $a \le t \le b$ is given by*

$$L = \int_a^b \sqrt{[x'(t)]^2 + [y'(t)]^2}\, dt$$

Note again that the integrals for arc length can get gruesome, so it is fair to get numerical estimates when exact answers cannot reasonably be found.

EX 5 Find the arc length of the portion of the parametric curve $x(t) = e^t$; $y(t) = 4 - t^2$ that is above the x-axis.

From EX 4, we know that this curve intersects the x axis when $t = \pm 2$. From the equation $x(t) = e^t$, we get $x'(t) = e^t$; from the equation $y(t) = 4 - t^2$, we get $y'(t) = -2t$. The integral for arc length of this curve is, by Useful Fact 12.3:

$$L = \int_a^b \sqrt{[x'(t)]^2 + [y'(t)]^2} \, dt = \int_{-2}^{2} \sqrt{(e^t)^2 + (-2t)^2} \, dt = \int_{-2}^{2} \sqrt{e^{2t} + 4t^2} \, dt$$

This integral cannot be solved by hand,[1] so we can use a CAS to approximate the result as $L \approx 11.86$. ∎

You Try It

(5) Find the arc length of the curve $x = t - t^2$; $y = \dfrac{4}{3}t^{3/2}$ for $1 \le t \le 2$

Into the Pit!!

Derivation of the Arc Length Formula

Are you still here? Good! I'm glad you're interested in not just using the new arc length formula, but seeing how it's brought to life.

To find the arc length of a parametric curve $x(t), y(t)$ for $a \le t \le b$, we need to adapt the arc length integral. In the discussion above on area, we noted that

$$dx = x'(t) \, dt = \frac{dx}{dt} \, dt$$

From Useful Fact 12.1, we know that

$$\frac{dy}{dx} = \frac{dy/dt}{dx/dt}$$

So let's load these into the arc length integral L and shake it out. Note that we must exchange the limits of integration to the t values that start and end the arc; we associate $t = a$ with $x = c$ and $t = b$ with $x = d$. To

---

[1] This is not "we don't feel like it", but rather: we just can't!

get things started, let's arrange this:

$$L = \int_c^d \sqrt{1 + y'(x)}\, dx = \int_a^b \sqrt{1 + \left(\frac{dy/dt}{dx/dt}\right)^2} \frac{dx}{dt}\, dt$$

$$= \int_a^b \sqrt{1 + \left(\frac{dy}{dt}\right)^2 \cdot \left(\frac{dt}{dx}\right)^2} \frac{dx}{dt}\, dt$$

Now we can get ready to absorb the trailing dx/dt into the square root so that the derivative terms can clobber each other:

$$L = \int_a^b \sqrt{1 + \left(\frac{dy}{dt}\right)^2 \cdot \left(\frac{dt}{dx}\right)^2} \cdot \sqrt{\left(\frac{dx}{dt}\right)^2}\, dt$$

$$= \int_a^b \sqrt{\left(\frac{dx}{dt}\right)^2 + \left(\frac{dy}{dt}\right)^2 \cdot \left(\frac{dt}{dx}\right)^2 \cdot \left(\frac{dx}{dt}\right)^2}\, dt$$

$$= \int_a^b \sqrt{\left(\frac{dx}{dt}\right)^2 + \left(\frac{dy}{dt}\right)^2}\, dt$$

If you prefer the final version of that integral, then go with it, but we can shrink it by returning to prime notation as shown in Useful Fact 12.3.

Calculus of Parametric Equations — Problem List

Calculus of Parametric Equations — You Try It

Try these problems, appearing above, for which solutions are available at the end of this section.

(1) Find the equation of the line tangent to $x = t^4 + 1; y = t^3 + t$ at $t = -1$.
(2) Find all locations of all horizontal and vertical tangent lines on the curve $x = 10 - t^2; y = t^3 - 12t$.
(3) Find the locations of all horizontal and vertical tangent lines on the curve $x = \cos 3\theta; y = 2 \sin \theta$.
(4) Find the area inside $x = a \cos \theta; y = b \sin \theta; 0 \leq \theta \leq 2\pi$. (The answer will be in terms of the constants a and b.)
(5) Find the arc length of the curve $x = t - t^2; y = 4t^{3/2}/3$ for $1 \leq t \leq 2$.

Calculus of Parametric Equations — Practice Problems

Try these as you get the hang of the You Try It problems. Solutions to these problems are available in Sec. A.6.2.

(1) Find the equation of the line tangent to $x = e^{\sqrt{t}}; y = t - \ln t^2$ at $t = 1$.
(2) Find the locations of all horizontal and vertical tangent lines on the curve $x = 2t^3 + 3t^2 - 12t; y = 2t^3 + 3t^2 + 1$.
(3) Find the point where the curve $x = \cos t; y = \sin t \cos t$ has more than one tangent line, and find the slopes of those tangent lines.
(4) Find the area bounded by the curve $x = t - 1/t; y = t + 1/t$ and the line $y = 2.5$.
(5) Find the arc length of the curve $x = \cos t; y = \sin t$ for $0 \leq t \leq \pi$.

Calculus of Parametric Equations — Challenge Problems

Try these problems to test your skills with the ideas in this section. Solutions to these problems are available in Sec. B.6.2.

(1) Find the locations on the parametric curve $x = a \cos^3 \theta, y = a \sin^3 \theta$ where (a) tangent lines are only horizontal, (b) tangent lines are only vertical, and (c) where the curve forms a cusp, and therefore has no unique tangent line.
(2) Find the area inside the curve $x = a \cos^3 \theta; y = a \sin^3 \theta$.
(3) Find the arc length of $x = 1 + 3t^2; y = 4 + 2t^3$ for $0 \leq t \leq 1$.

Calculus of Parametric Equations — You Try It — Solved

(1) Find the equation of the line tangent to $x = t^4 + 1; y = t^3 + t$ at $t = -1$.

☐ Using Useful Fact 12.1, we have

$$\frac{dy}{dx} = \frac{dy/dt}{dx/dt} = \frac{3t^2 + 1}{4t^3}$$

and so

$$m_{tan} = \frac{3(-1)^2 + 1}{4(-1)^3} = \frac{4}{-4} = -1$$

At $t = -1$, we get $x = (-1)^4 + 1 = 2$ and $y = (-1)^3 + (-1) = -2$, so we hit the point $(2, -2)$. With a point and a slope, we get the equation of the tangent line:

$$y - (-2) = (-1)(x - 2)$$
$$y + 2 = -x + 2$$
$$y = -x \quad \blacksquare$$

(2) Find the locations of all horizontal and vertical tangent lines on the curve $x = 10 - t^2; y = t^3 - 12t$.

☐ Slopes of the tangent lines are given by

$$\frac{dy}{dx} = \frac{dy/dt}{dx/dt} = \frac{3t^2 - 12}{-2t}$$

Tangent lines are horizontal when $dy/dx = 0$, i.e. when $t = \pm 2$. When $t = 2$, the point is $(6, -16)$; when $t = -2$, the point is $(6, 16)$.

Tangent lines are vertical when dy/dx is undefined, i.e. at $t = 0$, which is the point $(10, 0)$. $\quad \blacksquare$

(3) Find the locations of all horizontal and vertical tangent lines on the curve $x = \cos 3\theta; y = 2 \sin \theta$.

☐ Slopes of tangent lines are given by

$$\frac{dy}{dx} = \frac{dy/d\theta}{dx/d\theta} = -\frac{2 \cos \theta}{3 \sin 3\theta}$$

Tangent lines are horizontal when $dy/dx = 0$, i.e. when θ is any odd multiple of $\pi/2$, that is $\pm \pi/2, \pm 3\pi/2, \pm 5\pi/2, \ldots$. All of these θ values give $x = 0$ since $\cos \theta = 0$ for each. On the other hand, for these values we can get $\cos\theta = 1$ OR $\cos\theta = -1$, and so we'll have y coordinates of either -2 or 2. We are getting multiple instances of two points, $(0, -2)$ and $(0, 2)$. (A more precise description is that when $\theta = \pi/2 + 2n\pi$, the point is $(0, 2)$; when $\theta = 3\pi/2 + 2n\pi$, the point is $(0, -2)$.)

Tangent lines are vertical when dy/dx is undefined, i.e. at $3\theta = n\pi$ or $\theta = n\pi/3$. The x-coordinates of these points will either be 1 or -1. Distinct y-coordinates will match $2\sin(\pi/3), 2\sin(2\pi/3)$, and $2\sin(\pi)$, which are $\sqrt{3}, -\sqrt{3}$ and 0. In all, points with vertical tangent lines are $(\pm 1, \pm\sqrt{3})$ and $(\pm 1, 0)$. ∎

---

(4) Find the area inside $x = a\cos\theta; y = b\sin\theta; 0 \le \theta \le 2\pi$.

☐ This is a full ellipse. From the equation for x, we have $dx = -a\sin\theta\, d\theta$. So the area enclosed by this curve is, using Useful Fact 12.2,

$$A = \int_a^b y(t)x'(t)dt = \int_0^{2\pi} (b\sin\theta)(-a\sin\theta\, d\theta) = \int_0^{2\pi} ab\sin^2\theta\, d\theta$$

so that the area is $A = \pi ab$. ∎

---

(5) Find the arc length of the curve $x = t - t^2; y = 4t^{3/2}/3$ for $1 \le t \le 2$.

☐ Note that

$$x'(t) = 1 - 2t \qquad \text{and} \qquad y'(t) = \frac{3/2}{3} \cdot 4t^{1/2} = 2\sqrt{t}$$

and so by Useful Fact 12.3,

$$L = \int_a^b \sqrt{[x'(t)]^2 + [y'(t)]^2}\, dt = \int_1^2 \sqrt{(1-2t)^2 + \left(2\sqrt{t}\right)^2}\, dt$$

$$= \int_1^2 \sqrt{1 - 4t + 4t^2 + 4t}\, dt = \int_1^2 \sqrt{1 + 4t^2}\, dt$$

This would be solvable by a trigonometric substitution, but the exact answer is hideous. So let's approximate it and get $L \approx 3.17$. ∎

12.3   Introduction to Polar Coordinates

Introduction

Cartesian (rectangular) coordinates give you a roadmap of how to get from the origin to another point. The rectangular coordinates (3,4) mean you start at the origin, go three units right and 4 units up, and there you are! You know, we could probably think of lots of ways to describe how to get from the origin to some second point. Suppose I gave you directions in this form: from the origin, face down the positive x-axis. Then turn left (counterclockwise) by a certain angle and then move forwards a certain distance. Those directions could get you anywhere! If I wanted you end up at the Cartesian point (0,2), I would tell you look along the positive x-axis, turn left (counterclockwise) by an angle of $\pi/2$, and then move forward by a distance of 2 units. If I wanted you to end up at the Cartesian point (1,1), I would tell you to turn counterclockwise by an angle of $\pi/4$, then move forward by a distance of $1/\sqrt{2}$. If I wanted you to end up at the Cartesian point $(-5,5)$, I would tell you to turn counterclockwise by an angle of $3\pi/4$, then move forward by a distance of $1/\sqrt{2}$.

If this makes sense, then you already understand polar coordinates.

Points and Regions in Polar Coordinates

Any point in the plane can be located through its rectangular coordinates (x, y) and its polar coordinates (r, θ). In polar coordinates, θ is the counterclockwise angle you turn away from the positive x-axis, and r is the distance you move out in the direction specified by θ. Using the examples above, the Cartesian point (0,2) has polar coordinates $(2, \pi/2)$. The Cartesian point (1,1) has polar coordinates $(\sqrt{2}, \pi/4)$ (see Fig. 12.5). The Cartesian point $(-5,5)$ has polar coordinates $(5\sqrt{2}, 3\pi/4)$. If I had been in charge when polar coordinates were invented, I would have insisted they be given as (θ, r) rather than (r, θ), because it is efficient to consider the angular rotation by θ as the first "move" when going from the origin to the point. But, sadly, no one asked me.

One fundamental difference between polar and rectangular coordinates is that the rectangular coordinates of a point are unique, whereas a point's polar coordinates are not. If I describe a point in rectangular coordinates as $(1, -4)$, there are no other rectangular coordinates that put us in the same

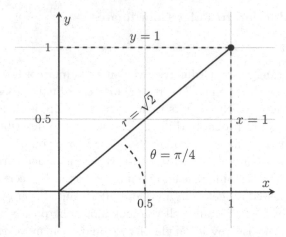

Fig. 12.5   Rectangular coords $(1, 1)$ vs polar coords $\left(\dfrac{1}{\sqrt{2}}, \dfrac{\pi}{4}\right)$.

place. But polar coordinates are NOT unique. Here are three *other* polar representations of the point known by rectangular coordinates of $(1, 1)$:

$$\left(\sqrt{2}, \frac{9\pi}{4}\right) \quad ; \quad \left(-\sqrt{2}, \frac{5\pi}{4}\right) \quad ; \quad \left(\sqrt{2}, -\frac{7\pi}{4}\right)$$

If a point is located at a direction specified by an angle θ, then any other angle $\theta \pm 2n\pi$ will point in that same direction. Further, we can reach the same point by going in the opposite angular direction (such as $\theta + \pi$) but then moving backwards in the negative r direction. As an example, both $(3, \pi/2)$ and $(-3, 3\pi/2)$ lead to the same point. (Do you know what that point is in rectangular form?)

You Try It

(1) Plot the polar point $(r, \theta) = (2, 2\pi/3)$ and provide two other sets of polar coordinates that describe the same point.

In rectangular coordinates, we can describe regions of the plane using bounding values of the variables x and y. For example, the rectangular region $-1 < x < 3; 0 \le y \le 4$ describes a square whose left and right edges are not included in the region, but whose top and bottom edges are included. Regions described by polar coordinate bounds tend to be pie shaped. (As you get into Calc III, be prepared for a lot of diagrams that look like food.)

You Try It

(2) Describe the region given by the polar bounds $0 \leq r < 4$, $-\pi/2 \leq \theta < \pi/6$.

Converting Between Rectangular and Polar Coordinates

Having alternate coordinate systems is handy because sometimes equations of curves and/or calculations are simpler in different coordinate systems. For example, the equation of the unit circle in Cartesian coordinates is $x^2 + y^2 = 1$. But in polar coordinates, it's simply $r = 1$. Why? Because the equation of a curve tells you what all the points on the curve have in common. The unit circle is the collection of points one unit from the origin, and the polar coordinate r describes distance from the origin. Therefore collecting all points for which $r = 1$ uniquely describes the unit circle.

It is important to be able to move back and forth between Cartesian coordinates and polar coordinates, both for individual points and for equations of curves. In order to do this, we need conversion equations. But don't panic, all you need to know about converting between polar coordinates and Cartesian coordinates is the central triangle of Fig. 12.5 and your memory of most basic trigonometry. If we toss out the specific values of x, y, r, θ in Fig. 12.5, we still know the following relationships:

$$\cos \theta = \frac{x}{r} \qquad \text{and} \qquad \sin \theta = \frac{y}{r}$$

which, with a bit of rearrangement, are the same as

$$x = r \cos \theta \qquad \text{and} \qquad y = r \sin \theta$$

These equations allow us to convert (r, θ) values into (x, y) values. Also from the figure, and your memory of the Pythagorean Theorem, we have

$$r^2 = x^2 + y^2 \qquad \text{and} \qquad \tan \theta = \frac{y}{x}$$

These equations allow us to convert (x, y) values into (r, θ) values. Put together, we have...

Useful Fact 12.4. *Given rectangular coordinates (x, y) and polar coordinates $(r, theta)$, we can convert back and forth using the relationships*

$$x = r \cos \theta \quad ; \quad y = r \sin \theta \quad ; \quad r^2 = x^2 + y^2 \quad ; \quad \tan \theta = \frac{y}{x}$$

EX 1 Convert these points from rectangular to polar coordinates: (a) $(1, -\sqrt{3})$, (b) (-2,2).

For point (a), we have $r^2 = x^2 + y^2 = (1)^2 + (-\sqrt{3})^2 = 4$, so that $r = 2$. Also, we have

$$\tan\theta = \frac{y}{x} = \frac{-\sqrt{3}}{1} = -\sqrt{3}$$

so that $\theta = \tan^{-1}(-\sqrt{3}) = -\pi/3$. Together, the polar coordinates of point (a) are $(r,\theta) = (2, -\pi/3)$. For point (b), we have $r^2 = x^2 + y^2 = (-2)^2 + (2)^2 = 8$, so that $r = 2\sqrt{2}$. Also, we have

$$\tan\theta = \frac{y}{x} = \frac{2}{-2} = -1$$

so that $\theta = \tan^{-1}(-1) = 3\pi/4$. Together, the polar coordinates of point (b) are $(2\sqrt{2}, 3\pi/4)$. But wait! In point (b), we used $\theta = \tan^{-1}(-1)$, and isn't $\tan^{-1}(-1)$ also equal to $-\pi/4$? Yes, it is! But from the original rectangular coordinates of the point, we know that point (b) is in Quadrant II, whereas an angle of $\theta = -\pi/4$ would put us in Quadrant IV. We chose the appropriate value of $\tan^{-1}(-1)$ to land us in the correct quadrant. Reference angles rule! ∎

You Try It

(3) Convert these points from polar to rectangular coordinates:
(a) $(3, \pi/2)$, (b) $(2\sqrt{2}, 3\pi/4)$, (c) $(-1, \pi/3)$.

Converting Between Rectangular and Polar Equations

The equations given in Useful Fact 12.4 for converting coordinates of individual points can also be used to convert entire functions.

EX 2 Convert the rectangular equation $x^2 + y^2 = 2xy$ to polar form.

We can recognize the left side immediately as r^2. In the right side, we substitute the individual conversion expressions for x and y to get

$$r^2 = 2(r\cos\theta)(r\sin\theta)$$
$$1 = 2\cos\theta\sin\theta$$

We'd probably want to present this polar equation as $\cos\theta\sin\theta = 1/2$. ∎

You Try It

(4) Convert $r = 3\sin\theta$ to Cartesian form and identify it.

Common Polar Curves

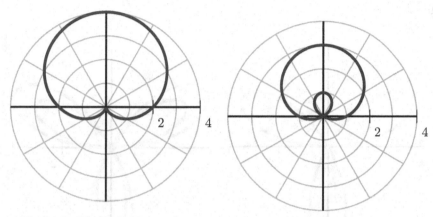

Fig. 12.6 The cardioid $r = 2(1 + \sin\theta)$.

Fig. 12.7 A curve $r = a + b\sin\theta$.

Circles are the primary shape of concern when considering polar coordinates, just as rectangles (any right triangles) are when considering rectangular (Cartesian) coordinates. So, our need for polar coordinates comes mostly in the contexts of circles and regions, as seen above. But there are a few other graphs that are the "usual suspects" that come up in discussion of polar curves. A few of them are shown in Figs. 12.6, 12.7, 12.8, and 12.9. They can pop up in applications in curious ways.

A *cardioid* is shown in Fig. 12.6, which shows $r = 2(1 + \sin\theta)$. An example of their use is in live audio engineering, specifically the use of *cardioid* microphones. This style of microphone is designed for the live setting because it is more sensitive to sound from the forwards and side directions than the rear. Imagine a microphone at the origin, facing a stage in the direction of the positive y-axis, with the audience towards the negative y-axis. The cardioid then represents the locations of equal sensitivity of the microphone. 〔O〕 FFT: Given that this figure shows, can you imagine what $r = a(1 + \sin\theta)$ shows in general? How about $r = a(1 + \cos\theta)$ 〔O〕

In Fig. 12.6, there is just a dimple at the origin due to the value of the function at $\theta = 3\pi/2$. If we generalize $r = a(1 + \sin\theta)$ (which has only one constant, a, to help shape the curve) to $r = a + b\sin\theta$ (which now has two constants, a and b, to help shape the curve), we can generate loops inside loops. Such a curve is shown in Fig. 12.7. [IOI] FFT: Can you use points on the curve to decide what values of a and b are used to generate this plot? [IOI]

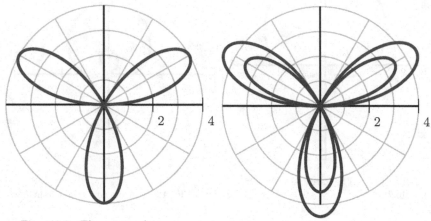

Fig. 12.8   The rose plot $r = 4\sin 3\theta$.

Fig. 12.9   A plot of $r = a + b\sin 3\theta$.

Rose plots can come in different varieties, $r = a\sin(n\theta)$ and $r = b\cos(m\theta)$. Figure 12.8 shows $r = 2\sin 3\theta$. A graph of $r = 2\cos 3\theta$ would be similar, but rotated. [IOI] In Sec. 12.4, we will see shapes like these, and will have a need to discover the values of θ at which any one loop closes out. Do you know how to determine that value? [IOI]

If we combine the equations of the last two sample plots into $r = a + b\sin(n\theta)$ or $r = a + b\cos(m\theta)$, things start to get a little crazy — such as multiple loops inside of other loops! Figure 12.9 shows a curve $r = a + b\sin 3\theta$. [IOI] FFT: Can you determine the values of a and b? [IOI] I encourage you to experiment with a plotting utility to discover the effects of the constants a, b, n, and m. For example, you can call up polar plots in Wolfram Alpha with a command like "`polar plot r = 1 + 2 * cos(3*theta)`".

Introduction to Polar Coordinates — Problem List

Introduction to Polar Coordinates — You Try It

Try these problems, appearing above, for which solutions are available at the end of this section.

(1) Plot the polar point $(r, \theta) = (2, 2\pi/3)$ and provide two other sets of polar coordinates that describe the same point.

(2) Describe the region given by the polar bounds $0 \leq r < 4$, $-\pi/2 \leq \theta < \pi/6$.

(3) Convert these points from polar to rectangular coordinates: (a) $(3, \pi/2)$, (b) $(2\sqrt{2}, 3\pi/4)$, (c) $(-1, \pi/3)$.

(4) Convert $r = 3\sin\theta$ to Cartesian form and identify it.

Introduction to Polar Coordinates — Practice Problems

Try these as you get the hang of the You Try It problems. Solutions to these problems are available in Sec. A.6.3.

(1) Describe the region given by the polar bounds $2 < r \leq 5$, $3\pi/4 < \theta < 5\pi/4$.

(2) Convert these points from polar to rectangular coordinates: (a) $(2, 2\pi/3)$, (b) $(4, 3\pi)$, (c) $(-2, -5\pi/6)$.

(3) Convert $r = 2\sin\theta + 2\cos\theta$ to Cartesian form and identify it.

(4) Convert $x^2 + y^2 = 9$ to polar form.

Introduction to Polar Coordinates — Challenge Problems

Try these problems to test your skills with the ideas in this section. Solutions to these problems are available in Sec. B.6.3.

(1) Describe the region given by the polar bounds $-1 \leq r \leq 1$, $\pi/4 < \theta < 3\pi/4$.

(2) Convert $r = \tan\theta\sec\theta$ to Cartesian form and identify it.

(3) Convert $x + y = 9$ to polar form.

Introduction to Polar Coordinates — You Try It — Solved

---

(1) Plot the polar point $(r, \theta) = (2, 2\pi/3)$ and provide two other sets of polar coordinates that describe the same point.

☐ Since $\theta = 2\pi/3$, we can expect this point in Quadrant II. We go counterclockwise by an angle of $2\pi/3$ from the positive x-axis, then move along that direction a distance of 2 units. We actually end up at the rectangular point $(-1, \sqrt{3})$. Two other polar coordinates that would put us at the same location are $(2, -4\pi/3)$ or $(-2, -\pi/3)$. ■

---

(2) Describe the region given by the polar bounds $0 \le r < 4$, $-\pi/2 \le \theta < \pi/6$.

☐ This is the region inside a circle of radius 4 starting at the negative y-axis and sweeping counterclockwise to an angle of $\pi/6$ above the x-axis. The outer edge of the circle is not included and neither is the side of the region in Quadrant I. (See Fig. 12.10.) ■

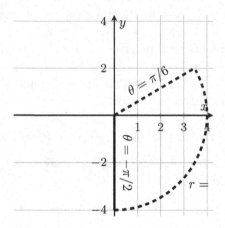

Fig. 12.10   Polar wedge $0 \le r < 4$, $-\dfrac{\pi}{2} \le \theta < \dfrac{\pi}{6}$.

---

(3) Convert these points from polar to rectangular coordinates: (a) $(3, \pi/2)$, (b) $(2\sqrt{2}, 3\pi/4)$, (c) $(-1, \pi/3)$.

☐ (a) The Cartesian coordinates of $(3, \pi/2)$ are $(x, y) = (0, 3)$. You should be able to tell that without using conversion equations.

(b) The Cartesian coordinates of $(2\sqrt{2}, 3\pi/4)$ are $(x, y) = (-2, 2)$ because:

$$x = r\cos\theta = 2\sqrt{2}\cos\frac{3\pi}{4} = -2$$

$$y = r\sin\theta = 2\sqrt{2}\sin\frac{3\pi}{4} = 2$$

(c) The Cartesian coordinates of $(-1, \pi/3)$ are $(x, y) = (-1/2, -\sqrt{3}/2)$ because

$$x = r\cos\theta = -\cos\frac{\pi}{3} = -\frac{1}{2}$$

$$y = r\sin\theta = -\sin\frac{\pi}{3} = -\frac{\sqrt{3}}{2} \quad \blacksquare$$

---

(4) Convert $r = 3\sin\theta$ to Cartesian form and identify it.

☐ Useful Fact 12.4 tells us that $y = r\sin\theta$, or $\sin\theta = y/r$, so $r = 3\sin\theta$ becomes $r = 3(y/r)$, or $r^2 = 3y$. We can then turn r^2 into $x^2 + y^2$ and do some completing the square:

$$x^2 + y^2 = 3y$$
$$x^2 + y^2 - 3y = 0$$
$$x^2 + y^2 - 3y + \frac{9}{4} = \frac{9}{4}$$
$$x^2 + \left(y - \frac{3}{2}\right)^2 = \frac{9}{4}$$

Our polar function is a circle with center $(0, 3/2)$ and radius $3/2$. $\quad \blacksquare$

12.4   Calculus of Polar Coordinates

Introduction

Graphs of polar functions have tangent lines, they can surround areas, and have arc length. As we did with parametric equations, let's see how to adapt the rectangular versions of these quantities to polar coordinates.

Tangent Lines

The two-dimensional plane in which we plot graphs is the same regardless of whether our "road map" around the plane is given in rectangular or polar coordinates. We still consider the primary directions to be left / right and up / down; the slope of a horizontal line is zero, the slope of a vertical line is undefined. Ultimately, the slope of a line, or more specifically a line tangent to a curve, should be related back to the quantity dy/dx. In polar coordinates, we don't have x or y directly available, but we can use the same strategy as for parametric equations. Since the chain rule tells us that

$$\frac{dy}{dx} \cdot \frac{dx}{d\theta} = \frac{dy}{d\theta}$$

we can arrange this to say:

$$\frac{dy}{dx} = \frac{dy/d\theta}{dx/d\theta} \tag{12.1}$$

So to find dy/dx, all we have to do is find $dy/d\theta$ and $dx/d\theta$, right? And since $y = r\sin\theta$, then $dy/d\theta = r\cos\theta$, right? And finding $dx/d\theta$ would be just as easy, right? Nope!

Remember that polar curves are written as $r = f(\theta)$. So r itself is considered a function of θ, and $\sin\theta$ is clearly a function of θ. So the expression $r\sin\theta$ is actually the *product* of two functions of θ. Therefore, if we want the derivative of $y = r\sin\theta$ with respect to θ, the product rule is required!

$$\frac{dy}{d\theta} = \left(\frac{dr}{d\theta}\right)\sin\theta + r \cdot \frac{d}{d\theta}(\sin\theta) = \frac{dr}{d\theta}\sin\theta + r\cos\theta$$

Similarly, since $x = r\cos\theta$,

$$\frac{dx}{d\theta} = \left(\frac{dr}{d\theta}\right)\cos\theta + r \cdot \frac{d}{d\theta}(\cos\theta) = \frac{dr}{d\theta}\cos\theta - r\sin\theta$$

Now we're ready to put these together to form dy/dx using Eq. (12.1):

Useful Fact 12.5. *Given a polar curve* $r = f(\theta)$, *the slope of a line tangent to the the curve comes from*

$$\frac{dy}{dx} = \frac{dy/d\theta}{dx/d\theta} = \frac{\frac{dr}{d\theta}\sin\theta + r\cos\theta}{\frac{dr}{d\theta}\cos\theta - r\sin\theta}$$

This formula is messy, but not difficult — it just requires book-keeping. Note that just as with parametric equations, polar curves can have more than one tangent line at a single point!

| EX 1 | Find the slope of the line tangent to $r = \theta$ at $\theta = \pi/4$:

In order to use Useful Fact 12.5, we have the choice of (a) designing the numerator and denominator specifically for the given point, or (b) creating a general expression in r and θ, and then plugging in the values specific to the given point. In this case, we anticipate the expressions required for the numerator and denominator will be a mess, so we'll compute specific values first. We'll need the quantities $dr/d\theta$, r, $\sin\theta$, and $\cos\theta$. With the function $r = \theta$ and the point given by $\theta = \pi/4$, we'll have

$$\frac{dr}{d\theta} = 1 \quad ; \quad r = \theta = \frac{\pi}{4} \quad ; \quad \sin\theta = \sin\frac{\pi}{4} = \frac{1}{\sqrt{2}} \quad ; \quad \cos\theta = \cos\frac{\pi}{4} = \frac{1}{\sqrt{2}}$$

So, the numerator of the expression that builds dy/dx is:

$$\frac{dr}{d\theta}\sin\theta + r\cos\theta = (1)\frac{1}{\sqrt{2}} + \frac{\pi}{4}\cdot\frac{1}{\sqrt{2}}$$

and the denominator is

$$\frac{dr}{d\theta}\cos\theta - r\sin\theta = (1)\frac{1}{\sqrt{2}} - \frac{\pi}{4}\cdot\frac{1}{\sqrt{2}}$$

Putting them together,

$$\frac{dy}{dx} = \frac{\frac{dr}{d\theta}\sin\theta + r\cos\theta}{\frac{dr}{d\theta}\cos\theta - r\sin\theta} = \frac{\frac{1}{\sqrt{2}} + \frac{\pi}{4}\cdot\frac{1}{\sqrt{2}}}{\frac{1}{\sqrt{2}} - \frac{\pi}{4}\cdot\frac{1}{\sqrt{2}}} = \frac{4+\pi}{4-\pi} \quad \blacksquare$$

You Try It

(1) Find the slope of the line tangent to $r = 2\sin\theta$ at $\theta = \pi/6$.

The Area Under or Inside a Polar Curve

Useful Fact 12.6. *Given a polar curve $r = f(\theta)$, the area A under the arc traced out for $a \le \theta \le b$ is given by*

$$A = \int_a^b \frac{1}{2}[r(\theta)]^2 \, d\theta$$

Remember that in polar coordinates, a curve of the form $\theta = c$ is a ray through the origin so the limits of integration $\theta = a$ and $\theta = b$ are the rays that bound the region of interest. A quick tour through partitioning in polar coordinates is shown at the end of this section; perhaps you can use that to derive this equation yourself!

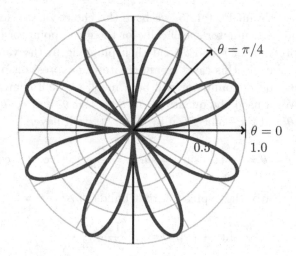

Fig. 12.11   The polar curve $r = \sin 4\theta$, with bounds for one loop.

<hr>

EX 2   Find the area inside one loop of $r = \sin 4\theta$.

This curve is shown in Fig. 12.11. The limits of integration will be the θ values that start and close a loop. But the start / close of a loop is characterized by $r = 0$. Therefore, we need two consecutive θ values which will give $r = 0$. We know $\sin 0 = 0$ and $\sin \pi = 0$, so we need values of θ such that $4\theta = 0$ and $4\theta = \pi$. These are $\theta = 0$ and $\theta = \pi/4$, so the area inside this region is

$$A = \int_a^b \frac{1}{2}r^2 \, d\theta = \int_0^{\pi/4} \frac{1}{2}(\sin 4\theta)^2 \, d\theta = \frac{\pi}{16} \quad \blacksquare$$

> **You Try It**
>
> (2) Find the area inside one loop of $r = \sin 2\theta$.
> (3) Find the area inside $r = 4\sin\theta$ and outside $r = 2$. (Hint: At which values of θ do these curves intersect?)

Arc Length

Useful Fact 12.7. *Given a polar curve $r = f(\theta)$, the length L of the arc traced out for $a \le \theta \le b$ is given by*

$$L = \int_a^b \sqrt{[r(\theta)]^2 + \left(\frac{dr}{d\theta}\right)^2}\, d\theta$$

If you are in a traditional three-term calculus sequence, you are probably very close to the end of your semester in Calc II. Perhaps you're ready to try to derive this for yourself? I wonder if it can be converted from the formula for the arc length of a curve in rectangular coordinates, (8.7), using our standard conversion equations between rectangular and polar coordinates?

EX 3 Find the arc length of the spiral $r = \theta$ from $\theta = 0$ to $\theta = 2\pi$.

Since $r = \theta$ we have $\dfrac{dr}{d\theta} = 1$. By Useful Fact 12.7),

$$L = \int_a^b \sqrt{[r(\theta)]^2 + \left(\frac{dr}{d\theta}\right)^2}\, d\theta = \int_0^{2\pi} \sqrt{\theta^2 + ((1))^2}\, d\theta = \int_0^{2\pi} \sqrt{\theta^2 + 1}\, d\theta$$

which is approximately $L \approx 21.26$. ∎

> **You Try It**
>
> (4) Find the arc length of $r = 3\sin\theta$ for $0 \le \theta \le \pi/3$.

Partitioning in Polar Coordinates

Into the Pit!!

Here is the foundation for several equations we just saw and used. I encourage you to pick up the threads and weave together any complete derivation that looks interesting to you.

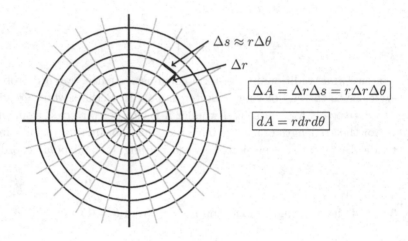

$$\Delta s \approx r\Delta\theta$$

$$\Delta r$$

$$\boxed{\Delta A = \Delta r \Delta s = r\Delta r\Delta\theta}$$

$$\boxed{dA = r\,dr\,d\theta}$$

Fig. 12.12   Partitioning in polar coordinates.

The formula for area inside a region bounded by a polar curve, shown in Useful Fact 12.6, can best be derived from scratch; Fig. 12.12 shows how partitioning works in the polar coordinate system. Rather than partitioning in a rectangular sense, we partition by r and θ, so that each individual ΔA is a small wedge, with inner and outer radii separated by Δr, and left & right edges separated by $\Delta\theta$. The edges of the wedge separated by $\Delta\theta$ are straight lines, but the edges separated by Δr are small circular arcs — do you remember that the length of any such arc would be $r\Delta\theta$? Altogether, the area of any tiny partitioned wedge, which is *almost* a rectangle, could then be approximated by $\Delta A = \Delta r \cdot (r^*\Delta\theta)$, where r^* is a representative radius value for the wedge — either the inner or outer radius, or perhaps the radius at the center? Can you use all this information and the diagram in Fig. 12.12 to derive Useful Fact 12.6?

(Note: Fig. 12.12 also visualizes why the area differential dA changes from $dA = dx\,dy$ in rectangular coordinates to $dA = r\,dr d\theta$ in polar coordinates.)

Calculus of Polar Coordinates — Problem List

Calculus of Polar Coordinates — You Try It

Try these problems, appearing above, for which solutions are available at the end of this section.

(1) Find the slope of the line tangent to $r = 2\sin\theta$ at $\theta = \pi/6$.
(2) Find the area inside one loop of $r = \sin 2\theta$.
(3) Find the area inside $r = 4\sin\theta$ and outside $r = 2$. (Hint: At which values of θ do these curves intersect?)
(4) Find the arc length of $r = 3\sin\theta$ for $0 \le \theta \le \pi/3$.

Calculus of Polar Coordinates — Practice Problems

Try these as you get the hang of the You Try It problems. Solutions to these problems are available in Sec. A.6.4.

(1) Find the slope of the line tangent to $r = 2 - \sin\theta$ at $\theta = \pi/3$.
(2) Find the area inside one loop of $r = 4\sin 3\theta$.
(3) Find the area inside $r = 1 - \sin\theta$ and outside $r = 1$.
(4) Find the arc length of both $r = e^\theta$ and $r = e^{\theta+1}$ for $0 \le \theta \le 2\pi$. How many times longer is the latter arc length?

Calculus of Polar Coordinates — Challenge Problems

Try these problems to test your skills with the ideas in this section. Solutions to these problems are available in Sec. B.6.4.

(1) Find the slope of $r = \ln\theta$ at $\theta = e$.
(2) Find the area inside one loop of $r = 2\cos 4\theta$.
(3) Find the arc length of $r = \theta^2$ for $0 \le \theta \le 2\pi$.

Calculus of Polar Coordinates — You Try It — Solved

---

(1) Find the slope of the line tangent to $r = 2\sin\theta$ at $\theta = \pi/6$.

☐ This curve is shown in Fig. 12.13. In EX 1, we used the given value of θ to compute the specific quantities needed for dy/dx ahead of time; in this case, let's build the full expression for dy/dx first, and then plug in $\theta = \pi/6$. Starting with Useful Fact 12.5,

$$\frac{dy}{dx} = \frac{\frac{dr}{d\theta}\sin\theta + r\cos\theta}{\frac{dr}{d\theta}\cos\theta - r\sin\theta} = \frac{2\cos\theta\sin\theta + 2\sin\theta\cos\theta}{2\cos\theta\cos\theta - 2\sin\theta\sin\theta} = \frac{2\cos\theta\sin\theta}{\cos^2\theta - \sin^2\theta}$$

So when $\theta = \pi/6$,

$$\frac{dy}{dx} = \frac{2\cos\frac{\pi}{6}\sin\frac{\pi}{6}}{\cos^2\frac{\pi}{6} - \sin^2\frac{\pi}{6}} = \sqrt{3} \quad \blacksquare$$

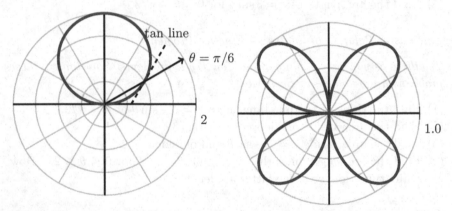

Fig. 12.13 $r = 2\sin\theta$ and tangent line at $\theta = \pi/6$.

Fig. 12.14   The polar curve $r = \sin 2\theta$.

---

(2) Find the area inside one loop of $r = \sin 2\theta$.

☐ This curve is shown in Fig. 12.14. One loop of the curve is delineated by two consecutive θ values which produce $r = 0$. Here, we get $r = 0$ when $2\theta = 0$ and when $2\theta = \pi$, so our bounds are $\theta = 0$ and $\theta = \pi/2$. The area is, then,

$$A = \int_a^b \frac{1}{2}r^2\,d\theta = \int_0^{\pi/2} \frac{1}{2}(\sin 2\theta)^2\,d\theta = \frac{\pi}{8} \quad \blacksquare$$

(3) Find the area inside $r = 4\sin\theta$ and outside $r = 2$. (Hint: At which values of θ do these curves intersect?)

☐ These curves are shown in Fig. 12.15. Like with rectangular coordinates, we can consider finding the total area inside the outer curve, then subtracting away the area inside the inner curve. That is,

$$Atot = \int_a^b \frac{1}{2}(r_{out})^2 \, d\theta - \int_a^b \frac{1}{2}(r_{in}^2 \, d\theta = \int_a^b \frac{1}{2}[(r_{out})^2 - (r_{in}^2] \, d\theta$$

We must find the θ values which mark the intersections of these two curves. We find $4\sin\theta = 2$ when $\sin\theta = 1/2$, so we have intersections when $\theta = \pi/6, 5\pi/6$. So,

$$A = \int_{\pi/6}^{5\pi/6} \frac{1}{2}\left[(4\sin\theta)^2 - (2)^2\right] \, d\theta = \frac{4\pi}{3} + 2\sqrt{3} \quad \blacksquare$$

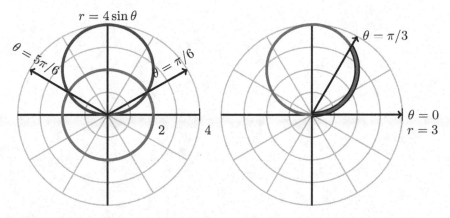

Fig. 12.15   The polar curves $r = 4\sin\theta$ and $r = 2$.

Fig. 12.16   The polar curve $r = 3\sin\theta$ for $\theta = 0$ to $\theta = \pi/3$.

(4) Find the arc length of $r = 3\sin\theta$ for $0 \le \theta \le \pi/3$.

☐ Using Useful Fact 12.7,

$$L = \int_a^b \sqrt{r^2 + \left(\frac{dr}{d\theta}\right)^2}\, d\theta = \int_0^{\pi/3} \sqrt{(3\sin\theta)^2 + (3\cos\theta)^2}\, d\theta$$

$$= \int_0^{\pi/3} 3\sqrt{\sin^2\theta + \cos^2\theta}\, d\theta = 3\int_0^{\pi/3} d\theta$$

$$= 3\theta\Big|_0^{\pi/3} = \pi \quad \blacksquare$$

Appendix A

Solutions to All Practice Problems

A.1 Chapter 7: Practice Problem Solutions

A.1.1 *Int. by Substitution — Practice — Solved*

---

(1) Find $\displaystyle\int \sqrt{2-x}\,dx$ and evaluate $\displaystyle\int_0^2 \sqrt{2-x}\,dx$.

□ With $2-x=u$ and so also $dx=-du$, we get

$$\int \sqrt{2-x}\,dx = -\int \sqrt{u}\,du = -\frac{2}{3}u^{3/2}+C = -\frac{2}{3}(2-x)^{3/2}+C$$

For the definite integral, the same substitution provides new endpoints:

$$x=0 \rightarrow u=2$$
$$x=2 \rightarrow u=0$$

and so borrowing a step from the indefinite integral,

$$\int \sqrt{2-x}\,dx = -\frac{2}{3}u^{3/2}\Big|_2^0 = \frac{2}{3}u^{3/2}\Big|_0^2 = \frac{2}{3}2^{3/2} = \frac{4}{3}\sqrt{2} \quad \blacksquare$$

---

(2) Find $\displaystyle\int \frac{x}{1+x^4}\,dx$ and evaluate $\displaystyle\int_0^{\sqrt[4]{3}} \frac{x}{1+x^4}\,dx$.

□ A first possible choice for substitution is $1+x^4=u$, but this would come with $4x^3\,dx=du$ — and there is no x^3 in the integral to make this work. Rather, we need a different strategy: the x in the numerator likely needs to be part of du, which means u might be based on an x^2. How can we make that work? Examine the denominator and rewrite it:

$$\int \frac{x}{1+x^4}\,dx = \int \frac{x}{1+(x^2)^2}\,dx$$

Then with $x^2 = u$, and also $x\,dx = \dfrac{1}{2}\,du$,

$$\int \frac{x}{1+(x^2)^2}\,dx = \frac{1}{2}\int\frac{1}{1+u^2}\,du = \frac{1}{2}\tan^{-1}(u) + C = \frac{1}{2}\tan^{-1}(x^2) + C$$

For the definite integral, the same substitution affects the endpoints as follows:

$$x = 0 \rightarrow u = 0$$
$$x = \sqrt[4]{3} \rightarrow u = \sqrt{3}$$

so that:

$$\int_0^{\sqrt[4]{3}} \frac{x}{1+(x^2)^2}\,dx = \frac{1}{2}\int_0^{\sqrt{3}}\frac{1}{1+u^2}\,du = \frac{1}{2}\tan^{-1}(u)\Big|_0^{\sqrt{3}}$$
$$= \frac{1}{2}\left(\tan^{-1}(\sqrt{3}) - \tan^{-1}(0)\right) = \frac{1}{2}\left(\frac{\pi}{3} - 0\right) = \frac{\pi}{6}\quad\blacksquare$$

---

(3) Find $\displaystyle\int \cos\theta \sin^3\theta\,d\theta$ and evaluate $\displaystyle\int_{-\pi/2}^{\pi/2} \cos\theta\sin^3\theta\,d\theta$.

□ Let's try $\sin\theta = u$, so that $\cos\theta d\theta = du$:

$$\int \cos\theta\sin^3\theta\,d\theta = \int u^3\,du = \frac{1}{4}u^4 + C = \frac{1}{4}\sin^4\theta + C$$

Then with endpoints,

$$x = -\frac{\pi}{2} \rightarrow u = \sin\left(-\frac{\pi}{2}\right) = -1$$
$$x = \frac{\pi}{2} \rightarrow u = \sin\left(\frac{\pi}{2}\right) = 1$$

$$\int_{-\pi/2}^{\pi/2} \cos\theta\sin^3\theta\,d\theta = \int_{-1}^{1} u^3\,du = \frac{1}{4}u^4\Big|_{-1}^{1} = 0 \quad\blacksquare$$

---

(4) Evaluate $\displaystyle\int_0^{\pi/3} \frac{\sin t}{\cos^2 t}\,dt$.

□ Let's try $\cos t = u$, so that $\sin t\,dt = -du$. For the endpoints,

$$x = 0 \rightarrow u = 1$$
$$x = \pi/3 \rightarrow u = 1/2$$

and so

$$\int_0^{\pi/3} \frac{\sin t}{\cos^2 t}\,dt = -\int_1^{1/2}\frac{du}{u^2} = \int_{1/2}^{1}\frac{du}{u^2} = -\frac{1}{u}\Big|_{1/2}^{1}$$
$$= -\left(\frac{1}{1} - \frac{1}{1/2}\right) = -(1-2) = 1 \quad\blacksquare$$

(5) Find $\displaystyle\int \frac{3x}{(x^2+1)^4}\,dx$.

☐ We can choose $x^2 + 1 = u$, so that $x\,dx = du/2$, and

$$\int \frac{3x}{(x^2+1)^4}\,dx = 3 \cdot \frac{1}{2}\int u^{-4}\,du = \frac{3}{2}\cdot\left(-\frac{1}{3}u^{-3}\right)+C$$

$$= -\frac{1}{2(x^2+1)^3}+C \quad\blacksquare$$

(6) Evaluate $\displaystyle\int_0^{10} \cot\theta\tan\theta\,d\theta$.

☐ Remember, don't make *every* integration problem into a substitution problem. Are there other things you can do?

$$\int_0^{10} \cot\theta\tan\theta\,d\theta = \int_0^{10} \frac{\cos\theta}{\sin\theta}\cdot\frac{\sin\theta}{\cos\theta}\,d\theta$$

$$= \int_0^{10} (1)\,d\theta = \theta\Big|_0^{10} = 10 \quad\blacksquare$$

(7) Find $\displaystyle\int \frac{\sec^2\theta}{1+\tan\theta}\,d\theta$ and $\displaystyle\int \frac{\sec^2\theta}{(1+\tan\theta)^2}\,d\theta$.

☐ In both antiderivatives, we can use $1 + \tan\theta = u$, which gives $\sec^2\theta\,d\theta = du$. So,

$$\int \frac{\sec^2\theta}{1+\tan\theta}\,d\theta = \int \frac{1}{u}\,du = \ln|u| + C = \ln|1 + \tan\theta| + C$$

For the second, do NOT go on autopilot and say that

$$\int \frac{\sec^2\theta}{(1+\tan\theta)^2}\,d\theta = \ln|(1+\tan\theta)^2| + C$$

That would be incorrect. Rather, the same substitution leads to the power rule in the second:

$$\int \frac{\sec^2\theta}{(1+\tan\theta)^2}\,d\theta = \int \frac{1}{u^2}\,du = -\frac{1}{u} + C$$

$$= -\frac{1}{1+\tan\theta} + C \quad\blacksquare$$

(8) Evaluate $\displaystyle\int_0^\pi e^{3\cos x}\sin x\,dx$.

□ With $3\cos x = u$, we get $\sin x\,dx = -\dfrac{1}{3}du$. The endpoints are updated as:

$$x = 0 \to u = 3$$
$$x = \pi \to u = -3$$

And we get

$$\int_0^\pi e^{3\cos x}\sin x\,dx = \int_3^{-3} e^u\left(-\frac{1}{3}du\right)$$
$$= \frac{1}{3}\int_{-3}^3 e^u\,du = \frac{1}{3}(e^3 - e^{-3}) \quad\blacksquare$$

(9) Evaluate $\displaystyle\int_0^1 xe^{-x^2}\,dx$.

□ With $-x^2 = u$, so that $x\,dx = -du/2$, and endpoints changed as:

$$x = 0 \to u = 0$$
$$x = 1 \to u = -1$$

we get

$$\int_0^1 xe^{-x^2}\,dx = -\frac{1}{2}\int_0^{-1} e^u\,du = \frac{1}{2}\int_{-1}^0 e^u\,du = \frac{1}{2}e^u\Big|_{-1}^0$$
$$= \frac{1}{2}(e^0 - e^{-1}) = \frac{1}{2}\left(1 - \frac{1}{e}\right) \quad\blacksquare$$

(10) Evaluate $\displaystyle\int_0^{\pi/2} \cos x \sin(\sin x)\,dx$.

□ With $\sin x = u$ so that $\cos x\,dx = du$ and

$$x = 0 \to u = 0$$
$$x = \pi/2 \to u = 1$$

we get

$$\int_0^{\pi/2} \cos x \sin(\sin x)\,dx = \int_0^1 \sin u\,du = -\cos u\Big|_0^1$$
$$= -(\cos(1) - \cos 0) = 1 - \cos(1) \quad\blacksquare$$

(11) Find $\displaystyle\int \frac{1}{\sqrt{x}(1+\sqrt{x})}\,dx$.

□ Let's try $1 + \sqrt{x} = u$, so that $dx/\sqrt{x} = 2\,du$ to get

$$\int \frac{1}{\sqrt{x}(1+\sqrt{x})}\,dx = 2\int \frac{du}{u} = 2\ln|u| + C$$

$$= 2\ln|1 + \sqrt{x}| + C = 2\ln(1 + \sqrt{x}) + C$$

Do you know why we can remove the absolute value bars in the last step? ■

(12) Consider $\displaystyle\int_0^1 \frac{x+a}{x+b}\,dx$, where a can be any real number, but b is restricted. State the necessary restriction on b and evaluate the integral.

□ In order the use the Fundamental Theorem of Calculus, the integrand cannot have any discontinuities within the interval of integration. So, b cannot be in the interval between $x = -1$ and $x = 0$; if b is in that interval, then $\frac{x+a}{x+b}$ will have a discontinuity between the endpoints $x = 0$ and $x = 1$.

To solve the integral, we need to somehow break the function down a bit. Let's focus on the denominator with $x + b = u$ (so also $dx = du$). This isn't much of a substitution, but maybe it'll help. How do we change the numerator? Well, if $x + b = u$, then $x = u - b$, so $x + a = (u - b) + a = u + (a - b)$. Also for the endpoints,

$$x = 0 \rightarrow u = 0 + b = b$$

$$x = 1 \rightarrow u = 1 + b$$

For the rest of it, be ready with properties of the natural log:

$$\int_0^1 \frac{x+a}{x+b}\,dx = \int_b^{1+b} \frac{u + (a-b)}{u}\,du = \int_b^{1+b} \left(1 + \frac{a-b}{u}\right)\,du$$

$$= (u + (a-b)\ln|u|)\,\Big|_b^{1+b}$$

$$= (1 + b) + (a - b)\ln|1 + b| - (b + (a - b)\ln|b|)$$

$$= 1 + (a - b)\ln|1 + b| - (a - b)\ln|b| = 1 + (a - b)\ln\frac{|1 + b|}{|b|}$$

$$= 1 + (a - b)\ln\left|\frac{1 + b}{b}\right| \quad ■$$

(13) Find the area under the function $y = e^x/(1+e^{2x})$ from $x = 0$ to $x = 1$. Your answer must be in exact form.

☐ This area is encoded into the integral $\displaystyle\int_0^1 \frac{e^x}{1+e^{2x}}\, dx$.

Remembering that $e^{2x} = (e^x)^2$, we can use $e^x = u$ (with $e^x\, dx = du$) to get new endpoints

$$x = 0 \to u = 1$$
$$x = 1 \to u = e$$

and full definite integral,

$$\int_0^1 \frac{e^x}{1+e^{2x}}\, dx = \int_1^e \frac{du}{1+u^2} = \tan^{-1}(u)\Big|_1^e$$
$$= \tan^{-1}(e) - \tan^{-1}(1) = \tan^{-1}(e) - \frac{\pi}{4} \quad \blacksquare$$

A.1.2   *Int. by Parts — Practice — Solved*

---

(1) Find $\int x^5 \ln x \, dx$.

□ We get started with:

$$\text{choose} \to \quad f(x) = \ln x \quad\quad g'(x) = x^5$$
$$\text{compute} \to \quad f'(x) = \tfrac{1}{x} \quad\quad g(x) = \tfrac{1}{6}x^6$$

Then, using (7.2),

$$\int x^5 \ln x \, dx = \frac{1}{6}x^6 \ln x - \frac{1}{6}\int x^5 \, dx = \frac{1}{6}x^6 \ln x - \frac{1}{36}x^6 + C \quad \blacksquare$$

---

(2) Find $\int t \sin 2t \, dt$.

□ The makeover of this integral starts with:

$$\text{choose} \to \quad f(t) = t \quad\quad g'(t) = \sin 2t$$
$$\text{compute} \to \quad f'(t) = 1 \quad\quad g(t) = -\tfrac{1}{2}\cos 2t$$

Then, using (7.2),

$$\int t \sin 2t \, dt = -\frac{1}{2}t \cos 2t + \frac{1}{2}\int \cos 2t \, dt$$

$$= -\frac{1}{2}t \cos 2t + \frac{1}{2}\left(\frac{1}{2}\sin 2t\right) + C$$

$$= -\frac{1}{2}t \cos 2t + \frac{1}{4}\sin 2t + C \quad \blacksquare$$

---

(3) Find $\int x^2 \sin \pi x \, dx$.

□ We start with

$$\text{choose} \to \quad f(x) = x^2 \quad\quad g'(x) = \sin \pi x$$
$$\text{compute} \to \quad f'(x) = 2x \quad\quad g(x) = -\tfrac{1}{\pi}\cos \pi x$$

so that with (7.2),

$$\int x^2 \sin \pi x \, dx = -\frac{1}{\pi}x^2 \cos \pi x - \left(-\frac{2}{\pi}\int x \cos \pi x \, dx\right)$$

$$= -\frac{1}{\pi}x^2 \cos \pi x + \frac{2}{\pi}\int x \cos \pi x \, dx$$

Well, integration by parts once at least made it easier, but we're not done; the new integral needs to be evaluated. For the new one,

$$\text{choose} \to \quad f(x) = x \quad\quad g'(x) = \cos \pi x$$
$$\text{compute} \to \quad f'(x) = 1 \quad\quad g(x) = \tfrac{1}{\pi}\sin \pi x$$

so that

$$\int x \cos \pi x \, dx = \frac{1}{\pi} x \sin \pi x - \frac{1}{\pi} \int \sin \pi x \, dx$$

$$= \frac{1}{\pi} x \sin \pi x + \frac{1}{\pi^2} \cos \pi x + C$$

Putting it all together,

$$\int x^2 \sin \pi x \, dx = -\frac{1}{\pi} x^2 \cos \pi x + \frac{2}{\pi} \left(\frac{1}{\pi} x \sin \pi x + \frac{1}{\pi^2} \cos \pi x + C \right)$$

$$= -\frac{1}{\pi} x^2 \cos \pi x + \frac{2}{\pi^2} x \sin \pi x + \frac{2}{\pi^3} \cos \pi x + C \quad \blacksquare$$

---

(4) Find $\int \sin^{-1} x \, dx$.

☐ Here, there are not two parts to choose from, so we really have only one choice:

$$\text{choose} \rightarrow \quad f(x) = \sin^{-1}(x) \qquad g'(x) = 1$$
$$\text{compute} \rightarrow \quad f'(x) = \frac{1}{\sqrt{1-x^2}} \qquad g(x) = x$$

Then, using (7.2),

$$\int \sin^{-1} x \, dx = x \sin^{-1} x - \int \frac{x}{\sqrt{1 - x^2}} \, dx$$

For the new integral, consider the substitution $1 - x^2 = u$, which comes with $x \, dx = -du/2$. Then we have

$$\int \frac{x}{\sqrt{1 - x^2}} \, dx = -\frac{1}{2} \int \frac{1}{\sqrt{u}} \, du = -\frac{1}{2} \left(2\sqrt{u} \right) + C = -\sqrt{1 - x^2} + C$$

Finally,

$$\int \sin^{-1} x \, dx = x \sin^{-1} x + \sqrt{1 - x^2} + C \quad \blacksquare$$

(5) Find $\int x^2 \cosh(x)\, dx$.

☐ We start with

$$\text{choose} \rightarrow \quad f(x) = x^2 \qquad g'(x) = \cosh(x)$$
$$\text{compute} \rightarrow \quad f'(x) = 2x \qquad g(x) = \sinh(x)$$

so that with (7.2),

$$\int x^2 \cosh(x)\, dx = x^2 \sinh(x) - \int 2x \sinh(x)\, dx$$

A second round of integration by parts (can you figure out the details?) gives

$$\int x^2 \cosh(x)\, dx = x^2 \sinh(x) - \left(2x \cosh(x) - \int 2 \cosh(x)\, dx \right)$$
$$= x^2 \sinh(x) - 2x \cosh(x) + 2 \sinh(x) + C$$
$$= (x^2 + 2) \sinh(x) - 2x \cosh(x) + C \quad \blacksquare$$

(6) Evaluate $\int_0^1 \dfrac{x}{e^{2x}}\, dx$.

☐ Let's rewrite the integral

$$\int_0^1 \frac{x}{e^{2x}}\, dx = \int_0^1 x e^{-2x}\, dx$$

then go with

$$\text{choose} \rightarrow \quad f(x) = x \qquad g'(x) = e^{-2x}$$
$$\text{compute} \rightarrow \quad f'(x) = 1 \qquad g(x) = -\tfrac{1}{2} e^{-2x}$$

so

$$\int_0^1 \frac{x}{e^{2x}}\, dx = -\frac{1}{2} x e^{-2x} \Big|_0^1 + \int_0^1 \frac{1}{2} e^{-2x}\, dx = -\frac{1}{2} x e^{-2x} \Big|_0^1 - \frac{1}{4} e^{-2x} \Big|_0^1$$
$$= -\frac{1}{2} \left(e^{-2} - 0 \right) - \frac{1}{4} \left(e^{-2} - e^0 \right) = -\frac{1}{2} \left(e^{-2} \right) - \frac{1}{4} \left(e^{-2} - 1 \right)$$
$$= \frac{1}{4} - \frac{3}{4e^2} \quad \blacksquare$$

A.1.3 *Int. by Partial Fracs. — Practice — Solved*

(1) Find $\displaystyle\int \frac{1}{x^2 + 3x - 4}\, dx$.

☐ First, let's blow up the integrand:

$$\frac{1}{x^2 + 3x - 4} = \frac{1}{(x+4)(x-1)} \overset{kaboom!}{=} \frac{A}{x+4} + \frac{B}{x-1} \qquad \text{(A.1)}$$

so that we immediately know,

$$\int \frac{1}{x^2 + 3x - 4}\, dx = \int \frac{A}{x+4} + \frac{B}{x-1}\, dx$$

$$= A \ln |x+4| + B \ln |x-1| + C$$

Backing up to find A and B, we can multiply (A.1) by the denominator $(x+4)(x-1)$ to get:

$$1 = A(x-1) + B(x+4)$$

This must hold for ANY value of x, including $x = 1$ and $x = -4$ (wink, wink). With $x = 1$, this simplifies to $1 = 5B$ so that $B = 1/5$. With $x = -4$, this simplifies to $1 = -5A$ so that $A = -1/5$. Altogether,

$$\int \frac{1}{(x+4)(x-1)}\, dx = -\frac{1}{5} \ln |x+4| + \frac{1}{5} \ln |x-1| + C$$

$$= \frac{1}{5} \ln \frac{|x-1|}{|x+4|} + C \quad \blacksquare$$

(2) Find $\displaystyle\int \frac{x^2}{(x-3)(x+2)^2}\, dx$.

☐ First, let's disassemble the integrand. Remember to account for ALL possible denominators which could lead to the given one as a least common denominator:

$$\frac{x^2}{(x-3)(x+2)^2} = \frac{A}{x-3} + \frac{B}{x+2} + \frac{C}{(x+2)^2}$$

For variety, let's find A, B, C first. Multiplying by the denominator $(x-3)(x+2)^2$ gives

$$x^2 = A(x+2)^2 + B(x-3)(x+2) + C(x-3)$$

If we select $x = 3$, we find $A = 9/25$.

If we select $x - 2$, we find $C = -4/5$.

This ends our list of really clever values of x, so let's now select the next easiest value to deal with, $x = 0$. With the values of A and C we now have, this gives $0 = 36/25 - 6B + 12/5$ so that $B = 16/25$.

So,

$$\int \frac{x^2}{(x-3)(x+2)^2} \, dx = \int \left(\frac{9/25}{(x-3)} + \frac{16/25}{x+2} - \frac{4/5}{(x+2)^2} \right) dx$$
$$= \frac{9}{25} \ln|x-3| + \frac{16}{25} \ln|x+2| + \frac{4}{5(x+2)} + C \quad \blacksquare$$

---

(3) Find $\int \dfrac{x^2 - x + 6}{x^3 + 3x} \, dx$.

☐ First, let's disassemble the integrand. There is an irreducible quadratic in this denominator, so the numerator of its mystery term could be a linear function:

$$\frac{x^2 - x + 6}{x^3 + 3x} = \frac{x^2 - x + 6}{x(x^2 + 3)} = \frac{A}{x} + \frac{Bx + C}{x^2 + 3} \qquad \text{(A.2)}$$

Then,

$$\int \frac{x^2 - x + 6}{x^3 + 3x} \, dx = \int \frac{A}{x} + \frac{Bx + C}{x^2 + 3} \, dx$$
$$= \int \frac{A}{x} + \frac{Bx}{x^2 + 3} + \frac{C}{x^2 + 3} \, dx$$
$$= A \ln|x| + \frac{B}{2} \ln(x^2 + 3) + \frac{C}{\sqrt{3}} \tan^{-1}\left(\frac{x}{\sqrt{3}} \right) + D$$

To find A, B, C we can multiply (A.2) by the denominator $x(x^2 + 3)$ gives

$$x^2 - x + 6 = A(x^2 + 3) + (Bx + C)(x) = A(x^2 + 3) + Bx^2 + Cx$$

If we select $x = 0$, we get $A = 2$.
We still two other values of x, so how about $x = 1$ and $x = -1$. Neither will yield B or C immediately, but these produce two equations in two unknowns (oh no!! algebra!!): $x = 1$ gives $6 = 8 + B + C$ and $x = -1$ gives $8 = 8 + B - C$. Practice your algebra to get $B = -1$ and $C = -1$. All together,

$$\int \frac{x^2 - x + 6}{x^3 + 3x} \, dx = 2 \ln|x| - \frac{1}{2} \ln(x^2 + 3) - \frac{1}{\sqrt{3}} \tan^{-1} \frac{x}{\sqrt{3}} + C \quad \blacksquare$$

---

(4) Find $\displaystyle\int_0^1 \frac{x-1}{x^2+3x+2}\, dx$.

☐ We disassemble the integrand like so:

$$\frac{x-1}{x^2+3x+2} = \frac{x-1}{(x+2)(x+1)} = \frac{A}{x+2} + \frac{B}{x+1}$$

Multiplying by the denominator $(x+2)(x+1)$ gives

$$x - 1 = A(x+1) + B(x+2)$$

If we select $x = -1$, we get $B = -2$.

If we select $x = -2$, we get $A = 3$.

Now don't forget this is a definite integral, so keep those limits and apply the FTOC:

$$\int_0^1 \frac{x-1}{x^2+3x+2}\, dx = \int_0^1 \left(\frac{3}{x+2} + \frac{-2}{x+1}\right) dx$$

$$= \left(3\ln|x+2| - 2\ln|x+1|\right)\Big|_0^1$$

$$= (3\ln 3 - 2\ln 2) - (3\ln 2 - 2\ln 1)$$

$$= 3\ln 3 - 5\ln 2 = \ln\frac{27}{32} \quad \blacksquare$$

A.1.4 *Trig. Integrals — Practice — Solved*

---

(1) Find $\int \sin^6 x \cos^3 x \, dx$.

☐ The integrand has an odd power of cosine, so we borrow one to go with dx and convert the others to $\sin x$:

$$\int \sin^6 x \cos^3 x \, dx = \int \sin^6 x (1 - \sin^2 x) \cos x \, dx \ldots$$

which is ready for substitution:

$$\sin x = u \quad ; \quad \cos x \, dx = du$$

so that we now have

$$\cdots = \int u^6 (1 - u^2) \, du = \int (u^6 - u^8) \, du = \frac{1}{7} u^7 - \frac{1}{9} u^9 + C$$

$$= \frac{1}{7} \sin^7 x - \frac{1}{9} \sin^9 x + C \quad \blacksquare$$

---

(2) Evaluate $\int_0^{\pi/2} \sin^2 2\theta \, d\theta$.

☐ Since we have only an even power of sine, we need a half angle formula:

$$\int_0^{\pi/2} \sin^2 2\theta \, d\theta = \int_0^{\pi/2} \frac{1}{2}(1 - \cos 4\theta) \, d\theta$$

$$= \frac{1}{2} \left(\theta - \frac{1}{4} \sin 4\theta \right) \Big|_0^{\pi/2}$$

$$= \frac{\pi}{4} \quad \blacksquare$$

---

(3) Find $\int \sec^6 t \, dt$.

☐ Since there is an even power on the secant term, we save aside two of them with dx and convert the rest:

$$\int \sec^6 t \, dt = \int \sec^2 t \sec^4 t \, dt = \int \sec^2 t (1 + \tan^2 t)^2 \, dt \ldots$$

and we're ready for

$$\tan t = u \quad ; \quad \sec^2 t \, dt = du$$

leading to

$$\cdots = \int (1 + u^2)^2 \, du = \int (1 + 2u^2 + u^4) \, du = u + \frac{2}{3} u^3 + \frac{1}{5} u^5 + C$$

$$= \tan t + \frac{2}{3} \tan^3 t + \frac{1}{5} \tan^5 t + C \blacksquare$$

(4) Evaluate $\displaystyle\int_0^{\pi/4} \sec^4\theta \tan^4\theta\, d\theta$.

□ Since we have an even power of secant, we save two aside with dx and convert the rest:

$$\int_0^{\pi/4} \sec^4\theta \tan^4\theta\, d\theta = \int_0^{\pi/4} \sec^2\theta(1+\tan^2\theta)\tan^4\theta\, d\theta \dots$$

Now we're ready for

$$\tan\theta = u \quad ; \quad \sec^2\theta\, d\theta = du$$

and then,

$$\dots = \int_0^1 (1+u^2)u^4\, du = \int_0^1 (u^4+u^6)\, du = \left(\frac{1}{5}u^5 + \frac{1}{7}u^7\right)\Big|_0^1$$

$$= \frac{1}{5} + \frac{1}{7} = \frac{12}{35} \quad \blacksquare$$

(5) Use a strategy similar to that of EX 5 to find $\displaystyle\int \csc x\, dx$.

□ The trick in EX 5 when finding the antiderivative of $\sec x$ was to set the integral expression up so as to allow the substitution $\sec x + \tan x = u$ and $\sec x \tan x + \sec^2 x\, dx$
$= du$. Since we're dealing with $\csc x$ here, we'll likely have to introduce a mix of $\csc x$ and $\cot x$. Keeping in mind that

$$\frac{d}{dx}\csc x = -\csc x \cot x \quad \text{and} \quad \frac{d}{dx}\cot x = -\csc^2 x$$

we may want to modify the strategy of EX 5 as follows:

$$\int \csc x\, dx = \int \csc x \cdot \frac{\csc x + \cot x}{\csc x + \cot x}\, dx = \int \frac{\csc x \cot x + \csc^2 x}{\csc x + \cot x}\, dx$$

and this just happens to be an almost-perfect set up for $\csc x + \cot x = u$. That substitution directly gives $(-\csc x \cot x - \csc^2 x)\, dx = du$, or $\csc x \cot x + \csc^2 x = -du$. This allows us to reset the integral as:

$$\int \frac{\csc x \cot x + \csc^2 x}{\csc x + \cot x}\, dx = \frac{-du}{u} = -\ln|u| + C = -\ln|\csc x + \cot x| + C$$

Altogether,

$$\int \csc x\, dx = -\ln|\csc x + \cot x| + C \quad \blacksquare \qquad\qquad \text{(A.3)}$$

(6) Recall that hyperbolic trig functions come with identities quite similar to (but not always exactly the same as) regular trig functions. Find and use such an identity to solve $\int \sinh^3(x) \cosh^2(x) \, dx$.

□ The identity for hyperbolic trig functions most similar to $\sin^2 x + \cos^2 x = 1$ is $\cosh^2(x) - \sinh^2(x) = 1$. Or, rather, $\sinh^2(x) = \cosh^2(x) - 1$. We can use this to prepare the integral for a substitution:

$$\int \sinh^3(x) \cosh^2(x) \, dx = \int \sinh(x) \sinh^2(x) \cosh^2(x) \, dx$$

$$= \int (\cosh^2(x) - 1) \cosh^2(x) \sinh(x) \, dx$$

And now with $u = \cosh(x)$ and $du = \sinh(x) \, dx$, this integral becomes

$$\int (\cosh^2(x) - 1) \cosh^2(x) \sinh(x) \, dx = \int (u^2 - 1) u^2 \, du$$

$$= \int u^4 - u^2 \, du = \frac{1}{5} u^5 - \frac{1}{3} u^3 + C$$

Together,

$$\int \sinh^3(x) \cosh^2(x) \, dx = \frac{1}{5} \cosh^5(x) - \frac{1}{3} \cosh^3(x) + C \quad \blacksquare$$

A.1.5   *Trig. Substitution — Practice — Solved*

---

(1) Find $\int \sqrt{1 - 4x^2}\, dx$.

☐ Let's do some cosmetic changes first:

$$\int \sqrt{1 - 4x^2}\,dx = \int \sqrt{4\left(\frac{1}{4} - x^2\right)}\,dx = 2 \int \sqrt{\frac{1}{4} - x^2}\,dx$$

The square root term here now matches the form $\sqrt{a^2 - x^2}$ with $a = 1/2$, so we assign

$$x = \frac{1}{2}\sin\theta \quad ; \quad dx = \frac{1}{2}\cos\theta\, d\theta$$

and get started:

$$\int \sqrt{1 - 4x^2}\,dx = 2 \int \sqrt{\frac{1}{4} - x^2}\,dx$$

$$= 2 \int \sqrt{\frac{1}{4} - \frac{1}{4}\sin^2\theta} \cdot \frac{1}{2}\cos\theta\, d\theta$$

$$= \int \frac{1}{2}\sqrt{1 - \sin^2\theta}\cos\theta\, d\theta = \frac{1}{2}\int \cos^2\theta\, d\theta \quad (\dagger)$$

$$= \frac{1}{2}\left(\frac{1}{2}\theta + \frac{1}{4}\sin 2\theta\right) + C = \frac{1}{4}\theta + \frac{1}{8}\sin 2\theta + C$$

$$= \frac{1}{4}\theta + \frac{1}{4}\sin\theta\cos\theta + C$$

From the original substitution, we have $\sin\theta = 2x$; our favorite trig identity (you know which one this is, right?) gives $\cos\theta = \sqrt{1 - 4x^2}$. And so,

$$\int \sqrt{1 - 4x^2}\,dx = \frac{1}{4}\sin^{-1}(2x) + \frac{1}{4}(2x)\sqrt{1 - 4x^2} + C$$

$$= \frac{1}{4}\sin^{-1}(2x) + \frac{1}{2}x\sqrt{1 - 4x^2} + C$$

($\dagger$ The antiderivative of $\cos^2\theta$ has been done previously as part of You Try It 2 in Sec. 7.4). ∎

(2) Find $\displaystyle\int \frac{x}{\sqrt{x^2-7}}\,dx$.

☐ The square root term here matches the form $\sqrt{x^2-a^2}$ with $a=\sqrt{7}$, so we assign

$$x=\sqrt{7}\sec\theta \quad;\quad dx=\sqrt{7}\sec\theta\tan\theta\,d\theta$$

which leads to

$$\int \frac{x}{\sqrt{x^2-7}}\,dx = \int \frac{\sqrt{7}\sec\theta}{\sqrt{7\sec^2\theta-7}}\cdot\sqrt{7}\sec\theta\tan\theta\,d\theta$$

$$= \int \frac{7\sec^2\theta\tan\theta}{\sqrt{7}\tan\theta}\,d\theta = \sqrt{7}\int \sec^2\theta\,d\theta = \sqrt{7}\tan\theta + C$$

The original substitution shows $\sec\theta = \dfrac{x}{\sqrt{7}}$, which — when encoded in a right triangle (not shown, you can draw these by now, right?) — gives

$$\tan\theta = \frac{\sqrt{x^2-7}}{\sqrt{7}}$$

so that

$$\int \frac{x}{\sqrt{x^2-7}}\,dx = \sqrt{7}\tan\theta + C = \sqrt{x^2-7} + C$$

(Of course, using substitution with $u=x^2-7$ would have been easier!!) ∎

---

(3) Evaluate $\displaystyle\int_0^1 x\sqrt{x^2+4}\,dx$.

☐ The square root term here matches the form $\sqrt{a^2+x^2}$ with $a=2$, so we assign

$$x=2\tan\theta \quad;\quad dx=2\sec^2\theta\,d\theta$$

and change the limits:

$$x=0 \to \theta=0$$
$$x=1 \to \theta=\tan^{-1}(1/2)$$

so the integral becomes

$$\int_0^1 x\sqrt{x^2+4}\,dx = \int_0^{\tan^{-1}(1/2)} 2\tan\theta\sqrt{4\tan^2\theta+4}\cdot 2\sec^2\theta\,d\theta$$

$$= \int_0^{\tan^{-1}(1/2)} 2\tan\theta(2\sec\theta)\cdot 2\sec^2\theta\,d\theta$$

$$= 8\int_0^{\tan^{-1}(1/2)} \tan\theta\sec^3\theta\,d\theta$$

$$= 8\int_0^{\tan^{-1}(1/2)} \sec^2\theta(\sec\theta\tan\theta\,d\theta)$$

At this point, this trigonometric integral needs a new regular substitution:

$$u = \sec\theta \quad ; \quad du = \sec\theta\tan\theta\,d\theta$$

Changing limits (yes, again!), we have

$$\theta = 0 \to u = \sec 0 = 1$$

$$\theta = \tan^{-1}(1/2) \to u = \sec(\tan^{-1}(1/2)) = \frac{\sqrt{5}}{2}$$

After this substitution, the integral is now ... isn't this fun??? ...

$$8\int_0^{\tan^{-1}(1/2)} \sec^2\theta(\tan\theta\sec\theta\,d\theta) = 8\int_1^{\sqrt{5}/2} u^2\,du = \frac{8}{3}u^3\Big|_1^{\sqrt{5}/2}$$

$$= \frac{8}{3}\left(\frac{\sqrt{5}}{2}\right)^3 - \frac{8}{3}$$

$$= \frac{8}{3}\left(\frac{5\sqrt{5}}{8}\right) - \frac{8}{3} = \frac{5\sqrt{5}}{3} - \frac{8}{3} \quad \blacksquare$$

A.1.6 *Improper Integrals — Practice — Solved*

(1) Evaluate $\displaystyle\int_{-\infty}^{-1} e^{-2t}\, dt$.

☐ The endpoint at $-\infty$ is the only one we have to worry about:

$$\int_{-\infty}^{-1} e^{-2t}\, dt = \lim_{c \to -\infty} \int_{c}^{-1} e^{-2t}\, dt = \lim_{c \to -\infty} \left(-\frac{1}{2} e^{-2t} \right) \Big|_{c}^{-1}$$

$$= -\frac{1}{2} \lim_{c \to -\infty} \left(e^2 - e^{-2c} \right)$$

Since $\displaystyle\lim_{c \to -\infty} e^{-2c}$ diverges, the entire improper integral diverges. ∎

(2) Evaluate $\displaystyle\int_{0}^{\pi} \tan x\, dx$.

☐ This doesn't look like an improper integral until you remember that $\tan x$ has a discontinuity at $x = \pi/2$. It also helps to remember the general antiderivative of $\tan x$ is $\ln|\sec x| + C$ (found by writing $\tan x = \sin x / \cos x$, then applying $u = \cos x$).

$$\int_{0}^{\pi} \tan x\, dx = \lim_{c \to \pi/2-} \int_{0}^{c} \tan x\, dx + \lim_{d \to \pi/2+} \int_{0}^{c} \tan x\, dx$$

$$= \lim_{c \to \pi/2-} \ln|\sec x| \Big|_{0}^{c} + \lim_{d \to \pi/2+} \ln|\sec x| \Big|_{d}^{\pi}$$

$$= \lim_{c \to \pi/2-} \left(\ln|\sec c| - \ln \sec 0 \right)$$

$$+ \lim_{d \to \pi/2+} \left(\ln|\sec \pi| - \ln|\sec d| \right)$$

$$= \lim_{c \to \pi/2-} \left(\ln|\sec c| \right) + \lim_{d \to \pi/2+} \left(-\ln|\sec d| \right)$$

Since $\sec(x)$ has a vertical asymptote at $x = \pi/2$, both of these limits — and therefore the original integral — diverge. ∎

(3) Evaluate $\displaystyle\int_{4}^{20} \frac{1}{\sqrt{x-4}}\, dx$.

☐ The only problem spot is the left endpoint, where the integrand is undefined. And so, we have

$$\int_{4}^{20} \frac{1}{\sqrt{x-4}}\, dx = \lim_{c \to 4+} \int_{c}^{20} \frac{1}{\sqrt{x-4}}\, dx = \lim_{c \to 4+} 2\sqrt{x-4} \Big|_{c}^{20}$$

$$= \lim_{c \to 4+} \left(2\sqrt{20-4} - 2\sqrt{c-4} \right) = 2\sqrt{16} - 0 = 8 \quad ∎$$

---

(4) Evaluate $\displaystyle\int_{-2}^{2} \frac{1}{x^2 - 1}\, dx$.

☐ There are two discontinuities inside the interval of integration, $x = \pm 1$. So we need to chop up the interval $[-2, 2]$. When we cut at the discontinuities, we get intervals $[-2, -1)$, $(-1, 1)$, $(1, 2]$. The middle subinterval has two bad endpoints, so we have to split that one, too — say at $x = 0$. Then we have intervals $[-2, -1)$, $(-1, 0]$, $[0, 1)$, and $(1, 2]$. It also helps to recall that we developed the general antiderivative of the integrand at the beginning of Sec. 7.3:

$$\int \frac{1}{x^2 - 1}\, dx = \frac{1}{2}\ln\left|\frac{x-1}{x+1}\right| + C$$

Take a deep breath and let's get this started:

$$\int_{-2}^{2} \frac{1}{x^2 - 1}\, dx = \int_{-2}^{(-1)} \frac{1}{x^2 - 1}\, dx + \int_{(-1)}^{0} \frac{1}{x^2 - 1}\, dx$$

$$+ \int_{0}^{(1)} \frac{1}{x^2 - 1}\, dx + \int_{(1)}^{2} \frac{1}{x^2 - 1}\, dx$$

$$= \lim_{c_1 \to -1^-} \int_{-2}^{c_1} \frac{1}{x^2 - 1}\, dx + \lim_{c_2 \to -1^+} \int_{c_2}^{0} \frac{1}{x^2 - 1}\, dx$$

$$+ \lim_{c_3 \to 1^-} \int_{0}^{c_3} \frac{1}{x^2 - 1}\, dx + \lim_{c_4 \to 1^+} \int_{c_4}^{2} \frac{1}{x^2 - 1}\, dx$$

Let's examine just the first subintegral:

$$\lim_{c_1 \to -1^-} \int_{-2}^{c_1} \frac{1}{x^2 - 1}\, dx = \lim_{c_1 \to -1^-} \frac{1}{2}\ln\left|\frac{x-1}{x+1}\right|\Bigg|_{-2}^{c_1}$$

$$= \frac{1}{2}\lim_{c_1 \to -1^-}\left(\ln\left|\frac{c_1 - 1}{c_1 + 1}\right| - \ln\left|\frac{-2-1}{-2+1}\right|\right)$$

$$= \frac{1}{2}\lim_{c_1 \to -1^-}\left(\ln\left|\frac{c_1 - 1}{c_1 + 1}\right| - \ln 3\right)$$

Because $c_1 + 1 \to 0$ as $c_1 \to -1^-$, the given limit diverages, and so does the entire interval. Wow, that was a lot of work for nothing!

■

(5) Evaluate $\displaystyle\int_0^\infty \frac{1}{(x-1)^4}\,dx$.

□ The trouble spots are $x = 1$ and the infinite endpoint. We can partition the interval of integration into $[0, 1)$ and $(1, \infty)$. The latter interval needs to be split again, say at $x = 2$.

$$\int_0^\infty \frac{1}{(x-1)^4}\,dx$$

$$= \int_0^{(1)} \frac{1}{(x-1)^4}\,dx + \int_{(1)}^2 \frac{1}{(x-1)^4}\,dx + \int_2^\infty \frac{1}{(x-1)^4}\,dx$$

$$= \lim_{c_1 \to 1^-} \int_0^{c_1} \frac{1}{(x-1)^4}\,dx + \lim_{c_2 \to 1^+} \int_{c_2}^2 \frac{1}{(x-1)^4}\,dx$$

$$+ \lim_{c_3 \to \infty} \int_2^{c_3} \frac{1}{(x-1)^4}\,dx$$

Let's take a look at the first limit only; if it happens to converge, we'll do the others. If it diverges, we're done.

$$\lim_{c_1 \to 1^-} \int_0^{c_1} \frac{1}{(x-1)^4}\,dx = \lim_{c_1 \to 1^-} \left. \frac{-1}{3(x-1)^3} \right|_0^{c_1}$$

$$= -\frac{1}{3} \lim_{c_1 \to 1^-} \left(\frac{1}{(c_1-1)^3} - \frac{1}{(0-1)^3} \right)$$

$$= -\frac{1}{3} \lim_{c_1 \to 1^-} \left(\frac{1}{(c_1-1)^3} + 1 \right)$$

The limit term diverges, and so the entire improper integral diverges. ∎

(6) Evaluate $\displaystyle\int_{-\infty}^\infty \frac{1}{x^2+1}\,dx$.

□ The integrand has no discontinuities; let's split the integral at $x = 0$.

$$\int_{-\infty}^{\infty} \frac{1}{x^2+1}\,dx = \int_{-\infty}^{0} \frac{1}{x^2+1}\,dx + \int_{0}^{\infty} \frac{1}{x^2+1}\,dx$$

$$= \lim_{c\to-\infty} \int_{c}^{0} \frac{1}{x^2+1}\,dx + \lim_{d\to\infty} \int_{0}^{d} \frac{1}{x^2+1}\,dx$$

$$= \lim_{c\to-\infty} \tan^{-1}(x)\Big|_{c}^{0} + \lim_{d\to\infty} \tan^{-1}(x)\Big|_{0}^{d}$$

$$= \lim_{c\to-\infty} (\tan^{-1}(0) - \tan^{-1}(c))$$

$$+ \lim_{d\to\infty} (\tan^{-1}(d) - \tan^{-1}(0))$$

$$= \lim_{c\to-\infty} (-\tan^{-1}(c)) + \lim_{d\to\infty} (\tan^{-1}(d))$$

$$= -\left(-\frac{\pi}{2}\right) + \frac{\pi}{2} = \pi$$

Altogether, $\int_{-\infty}^{\infty} \frac{1}{x^2+1}\,dx = \pi$. (Isn't it fun when π pops up in unusual places? Why would π be the net total area underneath this function?) ■

A.2 Chapter 8: Practice Problem Solutions

A.2.1 *Area Betw. Curves — Practice — Solved*

---

(1) Find the area between $y = x^2$ and $y = x^4$.

☐ These curves intersect at $x = -1$ and $x = 1$. The curve $y = x^2$ is above $y = x^4$ on that interval. Plus, notice that by symmetry, the area is twice the area on $[0, 1]$. So, the area between them is:

$$A = 2 \int_0^1 [(x^2) - (x^4)]dx = 2 \left(\frac{1}{3}x^3 - \frac{1}{5}x^5 \right) \Big|_0^1 = \frac{4}{15} \quad \blacksquare$$

---

(2) Find the area between $y = 12 - x^2$ and $y = x^2 - 6$.

☐ These curves intersect at $x = -3$ and $x = 3$. The curve $y = 12 - x^2$ is above $y = x^2 - 6$ on that interval. Plus, notice that by symmetry, the area is twice the area on $[0, 3]$. So, the area between them is:

$$A = 2 \int_0^3 \left[(12 - x^2) - (x^2 - 6) \right] dx = 2 \int_0^3 (18 - 2x^2)dx$$

$$= 2 \left(18x - \frac{2}{3}x^3 \right) \Big|_0^3 = 2 (54 - 18) = 72 \quad \blacksquare$$

---

(3) Find the area between $y = x^3 - x$ and $y = 3x$.

☐ These curves intersect at $x = -2$, $x = 0$ and $x = 2$. The two subregions ($x = -2$ to $x = 0$ and $x = 0$ to $x = 2$) are symmetric, so we can just double the area of one of them. Using the latter, where $y = 3x$ is above $y = x^3 - x$, the total area between them is:

$$A = 2 \int_0^2 \left[(3x) - (x^3 - x) \right] dx = 2 \int_0^2 (4x - x^3)dx$$

$$= 2 \left(2x^2 - \frac{1}{4}x^4 \right) \Big|_0^2 = 2 (8 - 4) = 8 \quad \blacksquare$$

---

(4) Find (approximately) the area between $y = e^{-x}$ and $y = 2 - x$.

☐ Using tech, we can find that these curves intersect at $x \approx -1.15$ and $x \approx 1.84$. The curve $y = 2 - x$ is above the curve $y = e^{-x}$ in a north-south orientation, so:

$$A \approx \int_{-1.15}^{1.84} (2 - x) - e^{-x} \, dx \approx 1.95 \quad \blacksquare$$

---

(5) Write an expression that relates the net unsigned area between $\sin x$ and the x-axis on $[-12\pi, 12\pi]$ to the area under $\sin x$ on $[0, \pi]$. Then find that total net unsigned area.

☐ Consider that the area under one "lump"[1] of the graph of $\sin x$ on $[0, \pi]$ is

$$A_1 = \int_0^\pi \sin x \, dx = 2$$

The net unsigned area inside one complete cycle of $\sin x$ on $[0, 2\pi]$ is twice that, or 4. Let A_{tot} be the net unsigned area between $\sin x$ and the x-axis on the large interval $[-12\pi, 12\pi]$. There are twelve full cycles of $\sin x$ contributing to A_{tot} (six to the left of the origin, and six to the right), and so there are 24 repetitions of the net area on $[0, \pi]$. By symmetry, we have

$$A_{tot} = 24 \int_0^\pi \sin x \, dx = 24(2) = 48$$

Note that we cannot simply find A_{tot} with the integral

$$\int_{-12\pi}^{12\pi} \sin x \, dx$$

because that integral is zero. $\qquad\qquad\blacksquare$

---

(6) Find the area between $4x + y^2 = 12$ and $x = y$.

☐ This region is easier to deal with from the perspective of the y-axis. The curves intersect at $y = -6$ and $y = 2$. The curve $4x + y^2 = 12$ is "above" $x = y$ on that interval, and this curve is solved for x as $x = 3 - y^2/4$. So, the area between them is:

---

[1]This is very technical mathematics terminology.

$$A = \int_{-6}^{2} \left[\left(3 - \frac{y^2}{4} \right) - (y) \right] dy = \left(3y - \frac{1}{12}y^3 - \frac{1}{2}y^2 \right) \Big|_{-6}^{2}$$

$$= \left(6 - \frac{2}{3} - 2 \right) - (-18 + 18 - 18) = 22 - \frac{2}{3} = \frac{64}{3} \quad \blacksquare$$

(7) Find the area between $f(x) = \cosh(x)$ and $f(x) = \sinh(x)$ on $[0, \ln 2]$ and write that value as a rational number.

☐ It helps to remember that the graph of $\cosh(x)$ is above the graph of $\sinh(x)$ for all x (see Fig. 1.37). So, the area between them is:

$$A = \int_{0}^{\ln 2} \cosh(x) - \sinh(x)\, dx = (\sinh(x) - \cosh(x)) \Big|_{0}^{\ln 2}$$

$$= (\sinh(\ln 2) - \cosh(\ln 2)) - (\sinh(0) - \cosh(0))$$

$$= \sinh(\ln 2) - \cosh(\ln 2) + 1$$

But

$$\sinh(\ln 2) = \frac{e^{\ln 2} - e^{-\ln 2}}{2} = \frac{1}{2}\left(2 - \frac{1}{2} \right) = \frac{3}{4}$$

$$\cosh(\ln 2) = \frac{e^{\ln 2} + e^{-\ln 2}}{2} = \frac{1}{2}\left(2 + \frac{1}{2} \right) = \frac{5}{4}$$

so

$$A = \sinh(\ln 2) - \cosh(\ln 2) + 1 = \frac{3}{4} - \frac{5}{4} + 1 = \frac{1}{2} \quad \blacksquare$$

A.2.2　Arc Length — Practice — Solved

---

(1) Find the arc length of $y = 2(x+4)^{3/2}$ for $0 \leq x \leq 2$ and $y > 0$.

☐ We'll do a little prep-work for Eq. (8.7):

$$f(x) = 2(x+4)^{3/2}$$
$$f'(x) = 3(x+4)^{1/2} \quad 1 + [f'(x)]^2 \quad = 1 + 9(x+4) = 37 + 9x$$

Then,

$$L = \int_a^b \sqrt{1 + [f'(x)]^2}\, dx = \int_0^2 \sqrt{37 + 9x}\, dx$$

$$= \frac{2}{27}(37 + 9x)^{3/2}\Big|_0^2 = \frac{2}{27}\left((55)^{3/2} - (37)^{3/2}\right) \quad \blacksquare$$

---

(2) Find the arc length of $y = \ln \sec x$ for $0 \leq x \leq \pi/4$.

☐ We'll do a little prep-work for Eq. (8.7):

$$f(x) = \ln \sec x$$
$$f'(x) = \frac{1}{\sec x} \cdot \sec x \tan x = \tan x$$
$$1 + [f'(x)]^2 = 1 + \tan^2 x = \sec^2 x$$
$$\sqrt{1 + [f'(x)]^2} = \sqrt{\sec^2 x} = \sec x$$

and then, with the antiderivative of $\sec x$ known from Eq. (7.17),

$$L = \int_a^b \sqrt{1 + [f'(x)]^2}\, dx = \int_0^{\pi/4} \sec x\, dx = \ln|\sec x + \tan x|\Big|_0^{\pi/4}$$

$$= \ln\left|\sec\frac{\pi}{4} + \tan\frac{\pi}{4}\right| - \ln|\sec 0 + \tan 0| = \ln(\sqrt{2} + 1) \quad \blacksquare$$

---

(3) Write an integral that will give the arc length of $x = y^2 - y + 3$ for $-1 \leq y \leq 1$, but do not solve the integral. (Hint: How much do the letters matter?)

☐ Whether the curve is given as $y = f(x)$ for $a \leq x \leq b$ or $x = g(y)$ for $c \leq y \leq d$ doesn't really matter. In the latter case, which matches this problem, we'll have

$$L = \int_c^d \sqrt{1 + [g'(y)]^2}\, dy$$

Setting things up, we have:

$$g(y) = y^2 - y + 3$$
$$g'(y) = 2y - 1$$
$$1 + [g'(y)]^2 = 1 + (2y-1)^2 = 1 + (4y^2 - 4y + 1) = 4y^2 - 4y + 2$$
$$\sqrt{1 + [g'(y)]^2} = \sqrt{4y^2 - 4y + 2}$$

so that

$$L = \int_{-1}^{1} \sqrt{4y^2 - 4y + 2} \, dy$$

Good thing we don't have to solve that! ∎

---

(4) Find the arc length of $y = \cosh(x)$ for $0 \le x \le \ln 4$ and write that value as a rational number.

☐ First, a little prep-work. With $f(x) = \cosh(x)$, we have $f'(x) = \sinh(x)$, and then, with the identity $\cosh^2(x) - \sinh^2(x) = 1$, we get

$$\sqrt{1 + [f'(x)]^2} = \sqrt{1 + \sinh^2(x)} = \sqrt{1 + (\cosh^2(x) - 1)} = \cosh(x)$$

(Since $\cosh(x) > 0$ for all x, we don't have to worry about any $\pm$ business with the root.) So then, with Eq. (8.7)

$$L = \int_a^b \sqrt{1 + [f'(x)]^2} \, dx = \int_0^{\ln 4} \cosh(x) \, dx = \sinh(x) \Big|_0^{\ln 4} = \sinh(\ln 4)$$

Further,

$$\sinh(\ln 4) = \frac{e^{\ln 4} - e^{-\ln 4}}{2} = \frac{1}{2}\left(4 - \frac{1}{4}\right) = \frac{15}{8} \quad \blacksquare$$

A.2.3   *Average Value — Practice — Solved*

(1) Find the average of $g(x) = x^2\sqrt{1 + x^3}$ on $[0, 2]$.

$$\square\, g_{avg} = \frac{1}{b - a} \int_a^b g(x)\, dx = \frac{1}{2 - 0} \int_0^2 x^2\sqrt{1 + x^3}\, dx$$

$$= \frac{1}{2}\left(\frac{2}{9}(1 + x^3)^{3/2}\right)\Big|_0^2 = \frac{1}{9}\left(9^{3/2} - 1\right) = \frac{26}{9} \quad\blacksquare$$

(2) Find the average of $f(x) = \sqrt{x}$ on $[0, 4]$ and a value of c such that $f(c) = f_{avg}$.

$$\square\, f_{avg} = \frac{1}{b - a} \int_a^b f(x)\, dx = \frac{1}{4 - 0} \int_0^4 \sqrt{x}\, dx$$

$$= \frac{1}{4}\left(\frac{2}{3}(x)^{3/2}\right)\Big|_0^4 = \frac{1}{4} \cdot \frac{2}{3}(8) = \frac{4}{3}$$

Next, we need a c such that $f(c) = f_{avg}$ — this means we seek $\sqrt{c} = 4/3$, or $c = 16/9$. This point is indeed within the given interval $[0.4]$. $\quad\blacksquare$

(3) Suppose the water level in Lake Michigan over a 30 year period starting in 1990 has followed the function

$$M(t) = 580 + 2\sin\left(\frac{\pi}{12}t\right)$$

(where M is in feet). What has been the average water level in Lake Michigan for this period, i.e. for $0 \leq t \leq 30$?

$$\square\, M_{avg} = \frac{1}{b - a} \int_a^b M(t)\, dt = \frac{1}{30 - 0} \int_0^{30} 580 + 2\sin\left(\frac{\pi}{12}t\right) dt$$

$$= \frac{1}{30}\left(580t - \frac{24}{\pi}\cos\left(\frac{\pi}{12}t\right)\right)\Big|_0^{30}$$

$$= \frac{1}{30}\left\{\left(580 \cdot 30 - \frac{24}{\pi}\cos\frac{30\pi}{12}\right) - \left(580(0) - \frac{24}{\pi}\cos 0\right)\right\}$$

$$= \frac{1}{30}\left(580 \cdot 30 + \frac{24}{\pi}\right) = 580 + \frac{24}{30\pi} = 580 + \frac{4}{5\pi}$$

The average water level over the 30 year period is $M_{avg} = 580 + 4/(5\pi)\, ft$, or about $580.255\, ft$. $\quad\blacksquare$

A.2.4 Numerical Integration — Practice — Solved

(1) Decide in advance if you would expect each of the left and right hand rules to over- or under-estimate the area under $f(x) = x^4$ on $[0, 2]$. Then perform the left and right hand estimates using $n = 12$, and confirm your guess.

$f\!x$	=D5^4						
	A	B	C	D	E	F	G
1			i	x_i	f (x_i)		
2	a = 0		0	0.000000	0	LHR	
3	b = 2		1	0.166667	).0007716049383	sum:	30.8441358
4	n = 12		2	0.333333	0.01234567901	final:	5.1406893
5			3	0.500000	0.0625		
6	Delta-x =	0.1666(	4	0.666667	0.1975308642	RHR	
7			5	0.833333	0.4822530864	sum:	46.8441358
8			6	1.000000	1	final:	7.807355967
9			7	1.166667	1.852623457		
10			8	1.333333	3.160493827		
11			9	1.500000	5.0625		
12			10	1.666667	7.716049383		
13			11	1.833333	11.2970679		
14	PP 1		12	2.000000	16		

Fig. A.1   Left and right hand rule implementations, with PP 1.

☐ Because $f(x)$ is increasing, the left hand height of any partition leads to an underestimate of the area, and the right hand height leads to an overestimate. So the left hand rule will underestimate the total area, and the right hand rule will overestimate it. Figure A.1 shows the spreadsheet implementation. The left and right hand estimates are:

$$\text{LHR} \int_0^2 x^4 \, dx \approx 5.141 \qquad ; \qquad \text{RHR} \int_0^2 x^4 \, dx \approx 7.807$$

whereas the true area is

$$\int_0^2 x^4 \, dx = \frac{1}{5} x^5 \Big|_0^2 = \frac{32}{5} = 6.4$$

So the left hand estimate falls short, and the right hand estimate overshoots the true value, as predicted. Bonus fact: the trapezoid rule would have yielded the average of the left and right hand results, 6.474 ... not bad! ∎

(2) Find a trapezoid rule estimate for the area under $f(x) = e^{-\sqrt{x}}$ on $[0, 4]$ using $n = 6$, $n = 12$, and $n = 48$.

fx	=sum(G2:G50)							
	A	B	C	D	E	F	G	H	I
1			i	x_i	f (x_i)	c_i	c_i * f(x_i)		
2	a = 0		0	0.000000	1.00000	1	1.00000	TRAP RULE	
3	b = 4		1	0.083333	0.74926	2	1.49851	sum:	28.62453
4	n = 48		2	0.166667	0.66481	2	1.32963	multiple:	0.04166666667
5			3	0.250000	0.60653	2	1.21306	final:	1.192688838
6	Delta-x = 0.08333		4	0.333333	0.56138	2	1.12277		
7			5	0.416667	0.52440	2	1.04880		
8			6	0.500000	0.49307	2	0.98614		
9			7	0.583333	0.46591	2	0.93182		
10			8	0.666667	0.44198	2	0.88395		
11			9	0.750000	0.42062	2	0.84124		
12			10	0.833333	0.40137	2	0.80274		
13	PP 2		11	0.916667	0.38388	2	0.76776		
14			12	1.000000	0.36788	2	0.73576		
15			13	1.083333	0.35316	2	0.70632		

Fig. A.2   A trapezoid rule implementations, with PP 2.

☐ Figure A.2 shows the top portion of a spreadsheet implementation for $n = 48$, although the full extent of the sum is displayed. All three trapezoid rule estimates are:

- for $n = 6$: $\displaystyle\int_0^4 e^{-\sqrt{x}}\,dx \approx 1.283$

- for $n = 12$: $\displaystyle\int_0^4 e^{-\sqrt{x}}\,dx \approx 1.223$

- for $n = 48$: $\displaystyle\int_0^4 e^{-\sqrt{x}}\,dx \approx 1.193$

Fun fact: the true value, which cannot be obtained without some severe contortions of the integration process, is $2 - \dfrac{6}{e^2}$ or about 1.18. ∎

(3) Find a trapezoid rule estimate with $n = 12$ for $\int_0^\pi \cos(x^2)\,dx$.

	fx	=COS(D3^2)						
	A	B	C	D	E	F	G	H	I
1			i	x_i	f (x_i)	c_i	c_i * f(x_i)		
2	a = 0		0	0.000000	1.00000	1	1.00000	TRAP RULE	
3	b = 3.14159		1	0.261799	0.99765	2	1.99530	sum:	4.44310
4	n = 12		2	0.523599	0.96265	2	1.92531	multiple:	0.1308996939
5			3	0.785398	0.81570	2	1.63141	final:	0.5816002439
6	Delta-x = 0.26179		4	1.047198	0.45660	2	0.91321		
7			5	1.308997	-0.14219	2	-0.28439		
8			6	1.570796	-0.78121	2	-1.56242		
9			7	1.832596	-0.97659	2	-1.95318		
10			8	2.094395	-0.32016	2	-0.64032		
11			9	2.356194	0.74415	2	1.48830		
12			10	2.617994	0.84152	2	1.68304		
13	PP 3		11	2.879793	-0.42524	2	-0.85048		
14			12	3.141593	-0.90269	1	-0.90269		

Fig. A.3 A trapezoid rule implementations, with PP 3.

□ Figure A.3 shows the spreadsheet implementation which gives an estimate of $\int_0^\pi \cos(x^2)\,dx \approx 0.5816$. ∎

(4) Find a midpoint rule estimate with $n = 12$ for $\int_0^\pi \cos(x^2)\,dx$.

	fx	=cos(E4^2)						
	A	B	C	D	E	F	G	H
1			i	x_i	m_i	f (m_i)		
2	a = 0		0	0.0000000			MIDPT	
3	b = 3.14159265		1	0.2617994	0.13089969	0.9998532041	sum: 2.129749013	
4	n = 12		2	0.5235988	0.39269908	0.9881327882	final: 0.5575669877	
5			3	0.7853982	0.65449847	0.9096447696		
6	Delta-x = 0.26179939		4	1.0471976	0.91629786	0.6677593171		
7			5	1.3089969	1.17809725	0.1818654513		
8			6	1.5707963	1.43089663	-0.4816232411		
9			7	1.8325957	1.70169602	-0.9699372979		
10			8	2.0943951	1.96349541	-0.7559307878		
11			9	2.3561945	2.22529480	0.2372635088		
12			10	2.6179939	2.48709418	0.9952459823		
13	PP 4		11	2.8797933	2.74889357	0.2931938193		
14			12	3.1415927	3.01069296	-0.9357185014		

Fig. A.4 A midpoint rule implementations, with PP 4.

□ Figure A.4 shows the spreadsheet implementation which gives an estimate of $\int_0^\pi \cos(x^2)\,dx \approx 0.5576$. ∎

(5) Find a Simpson's rule estimate with $n = 12$ for $\int_0^\pi \cos(x^2)\,dx$. Of the three results in PP 3, PP 4, and PP 5, which one do you trust the most? The least?

f_x	=cos(D3^2)							
	A	B	C	D	E	F	G	H	I
1			i	x_i	f (x_i)	c_i	c_i * f(x_i)		
2	a = 0		0	0	1	1	1	SIMPSON'S RULE	
3	b = 3.14159		1	26179938	0.9976521276	4	3.99060851	sum:	6.470071931
4	n = 12		2	52359877	0.9626541277	2	1.925308255	multiple:	0.0872664626
5			3	78539816	0.8157045174	4	3.26281807	final:	**0.5646202902**
6	Delta-x = 0.26179		4	04719755	0.4566033934	2	0.9132067869		
7			5	30899693	-0.1421930827	4	-0.5687723306		
8			6	57079632	-0.7812118921	2	-1.562423784		
9			7	83259571	-0.9765876888	4	-3.906350755		
10			8	0943951C	-0.3201597757	2	-0.6403195514		
11			9	.3561944	0.7441513285	4	2.976605314		
12			10	61799387	0.8415194405	2	1.683038881		
13	PP 5		11	87979326	-0.4252405257	4	-1.700962103		
14			12	14159265	-0.9026853619	1	-0.9026853619		

Fig. A.5   A Simpson's rule implementations, with PP 5.

☐ Figure A.5 shows the spreadsheet implementation which gives an estimate of $\int_0^\pi \cos(x^2)\,dx \approx 0.5646$. Simpson's Rule estimates tend to be the most accurate (of the methods we've seen), so I trust this one the most. And so, I trust the trapezoid rule estimate the least, because it's farther away from this value than the midpoint estimate. ∎

(6) Find a Simpson's rule estimate with $n = 12$ for $\int_{-1}^{1} x^2 \cos(x^2)\, dx$. Then do an experiment to find how many partitions are required (i.e. determine n) for the trapezoid rule to give the same result to four places after the decimal, $m.mmmm$ (use increments of 2, so that your value of n is even).

fx	=sum(G2:G38)							
	A	B	C	D	E	F	G	H	I
1			i	x_i	f (x_i)	c_i	c_i * f(x_i)		
2	a = -1		0	-1.00000	0.5403023059	1	0.5403023059	**TRAP RULE**	
3	b = 1		1	-0.94444	0.5600497489	2	1.120099498	sum:	19.11217359
4	n = 36		2	-0.88889	0.5560553974	2	1.112110795	multiple:	0.02777777778
5			3	-0.83333	0.5336176062	2	1.067235212	final:	**0.5308937109**
6	Delta-x =	0.055	4	-0.77778	0.497584235	2	0.99516847		
7			5	-0.72222	0.4522422267	2	0.9044844535		
8			6	-0.66667	0.4012665197	2	0.8025330394		
9			7	-0.61111	0.3477150668	2	0.6954301336		
10			8	-0.55556	0.2940577064	2	0.5881154128		
11			9	-0.50000	0.2422281054	2	0.4844562109		
12			10	-0.44444	0.193689705	2	0.3873794099		
13	PP 6		11	-0.38889	0.1495083513	2	0.2990167026		
14			12	-0.33333	0.1104259454	2	0.2208518908		

Fig. A.6   A trapezoid rule implementations, with PP 6.

☐ The Simpson's rule estimate is (not shown) $\int_{-1}^{1} x^2 \cos(x^2)\, dx \approx$ 0.53090. As n increases, the trapezoid rule estimate approaches the Simpson's rule estimate from below, and we need to increase n until the trapezoid rule starts reporting an estimate of $0.5309m$. By trial and error, we can determine that at $n = 38$, the trapezoid rule reports an estimate of 0.53092. (At $n = 36$, the trapezoid rule reports 0.53089 which isn't quite enough. Figure A.6 shows the upper portion of the trapezoid rule implementation for $n = 36$.) As it turns out, the "true" answer for this integral is about 0.5312. So both methods are still off a little but, but the trapezoid rule had to work three times as hard just to get to the same amount of "wrong" as Simpson's rule.   ■

(7) Use the Trapezoid Rule to estimate $\int_a^b f(x)\,dx$ for the tabulated values shown in this table (which is broken into two parts):

x	-2	-1.6	-1.2	-0.8	-0.4
$g(x)$	-0.00134	-0.05323	-0.41329	-0.70549	-0.39918

0	0.4	0.8	1.2	1.6	2.0
0	-0.39918	-0.70549	-0.41329	-0.05323	-0.00134

	A	B	C	D	E	F	G	H	I
1			i	x_i	f (x_i)	c_i	c_i * f(x_i)		
2	a = -2		0	-2	-0.00134	1	-0.00134	TRAP RULE	
3	b = 2		1	-1.6	-0.05323	2	-0.10646	sum:	0.00000
4	n = 10		2	-1.2	-0.41329	2	-0.82659	multiple:	0.2
5			3	-0.8	-0.70549	2	-1.41098	final:	0
6	Delta-x = 0.4		4	-0.4	-0.39918	2	-0.79836		
7			5	0	0.00000	2	0.00000		
8			6	0.4	0.39918	2	0.79836		
9			7	0.8	0.70549	2	1.41098		
10			8	1.2	0.41329	2	0.82659		
11			9	1.6	0.05323	2	0.10646		
12			10	2	0.00134	1	0.00134		

Fig. A.7   Trapezoid rule implementation for tabulated data, with PP 7.

☐ Figure A.7 shows the spreadsheet implementation, which yields a result of $\int_{-2}^{2} f(x)\,dx \approx 0$. Based on the symmetry of equal but opposite values of $f(x)$ around $x = 0$, this result is not surprising. ∎

A.3 Chapter 9: Practice Problem Solutions

A.3.1 *Solids of Rev.: Sfc Area — Practice — Solved*

(1) Find the surface area of the solid formed when $y = \cos(2x), 0 \le x \le \pi/6$, is revolved around the x-axis.

$\square$ Getting ready for (9.2),

$$f'(x) = -2\sin(2x)$$
$$1 + (f'(x))^2 = 1 + 4\sin^2(2x)$$

So

$$S = \int_a^b 2\pi f(x)\sqrt{1 + [f'(x)]^2}\,dx = \int_0^{\pi/6} 2\pi \cos(2x)\sqrt{1 + 4\sin^2(2x)}\,dx$$

Substituting,

$$\sin(2x) = u \quad ; \quad 2\cos(2x)\,dx = du$$
$$x = 0 \to u = 0$$
$$x = \frac{\pi}{6} \to u = \frac{\sqrt{3}}{2}$$

Then,

$$S = \pi \int_0^{\sqrt{3}/2} \sqrt{1 + 4u^2}\,du = \pi \int_0^{\sqrt{3}/2} 2\sqrt{\frac{1}{4} + u^2}\,du$$

which matches Useful Fact 7.7, needing

$$u = \frac{1}{2}\tan\theta \quad ; \quad du = \frac{1}{2}\sec^2\theta\,d\theta$$
$$u = 0 \to \theta = 0$$
$$u = \frac{\sqrt{3}}{2} \to \tan\theta = \sqrt{3} \to \theta = \frac{\pi}{3}$$

And finishing up,

$$S = 2\pi \int_0^{\pi/3} \sqrt{\frac{1}{4} + \frac{1}{4}\tan^2\theta} \cdot \frac{1}{2}\sec^2\theta\,d\theta = \frac{\pi}{2}\int_0^{\pi/3} \sec^3\theta\,d\theta$$
$$= \frac{\pi}{2} \cdot \frac{1}{2}\left(\sec\theta\tan\theta + \ln|\sec\theta + \tan\theta|\right)\Big|_0^{\pi/3}$$
$$= \frac{\pi}{4}\left[\left(\sec\frac{\pi}{3}\tan\frac{\pi}{3} + \ln\left|\sec\frac{\pi}{3} + \tan\frac{\pi}{3}\right|\right)\right.$$
$$\left. - (\sec 0\tan 0 + \ln|\sec 0 + \tan 0|)\right]$$
$$= \frac{\pi}{4}\left(2\sqrt{3} + \ln|2 + \sqrt{3}|\right) \quad \blacksquare$$

(2) Find (estimate) the surface area of the solid formed when $y = x^3/6 + 1/2x$, for $1/2 \le x \le 1$, is revolved around the x-axis.

☐ Getting ready for (9.2),

$$f'(x) = \frac{x^2}{2} - \frac{1}{2x^2}$$

$$1 + (f'(x))^2 = 1 + \left(\frac{x^2}{2} - \frac{1}{2x^2}\right)^2$$

So

$$S = \int_a^b 2\pi f(x)\sqrt{1 + [f'(x)]^2}\, dx$$

$$= \int_{1/2}^1 2\pi \left(\frac{x^3}{6} + \frac{1}{2x}\right) \sqrt{1 + \left(\frac{x^2}{2}\frac{1}{2x^2}\right)^2}\, dx$$

This integral is too horrible to consider trying by hand. Using tech, we get

$$S = \frac{263\pi}{256} \approx 3.227 \quad \blacksquare$$

(3) Find the surface area of the solid formed when $y = 1 - x^2$, $0 \le x \le 1$, is revolved around the y-axis.

☐ Since we're revolving around the y-axis, let's rewrite the function as $x = \sqrt{1 - y}$. Also, the limits on x correspond to the y values $0 \le y \le 1$. So,

$$g'(y) = -\frac{1}{2\sqrt{1 - y}}$$

$$1 + (g'(y))^2 = 1 + \frac{1}{4(1 - y)} = \frac{5 - 4y}{4(1 - y)}$$

And

$$S = \int_a^b 2\pi g(y)\sqrt{1 + [g'(y)]^2}\, dy = \int_0^1 2\pi\sqrt{1 - y}\sqrt{\frac{5 - 4y}{4(1 - y)}}\, dy$$

$$= \pi \int_0^1 \sqrt{5 - 4y}\, dy$$

A brief substitution,

$$5 - 4y = u \quad ; \quad dy = -\frac{1}{4}\, du$$

$$x = 0 \to u = 5$$

$$x = 1 \to u = 1$$

gives

$$S = -\frac{\pi}{4}\int_5^1 \sqrt{u}\,du = \frac{\pi}{4}\int_1^5 \sqrt{u}\,du = \frac{\pi}{4}\cdot\frac{2}{3}u'^2\Big|_1^5 = \frac{\pi}{6}\left(5\sqrt{5}-1\right) \quad \blacksquare$$

(4) Find the surface area of the solid formed when $y = \cosh(x)$, $0 \le x \le \ln 3$, is revolved around the x-axis.

□ Since $f(x) = \cosh(x)$ then $f'(x) = \sinh(x)$ and

$$\sqrt{1+(f'(x))^2} = \sqrt{1+\sinh^2(x)} = \sqrt{1+(\cosh^2(x)-1)} = \cosh(x)$$

and then

$$S = \int_a^b 2\pi f(x)\sqrt{1+[f'(x)]^2}\,dx = \int_0^{\ln 3} 2\pi\cosh(x)\cdot\cosh(x)\,dx$$

$$= 2\pi \int_0^{\ln 3} \cosh^2(x)\,dx$$

$$= 2\pi\cdot\frac{1}{2}(x+\sinh(x)\cosh(x))\Big|_0^{\ln 3} = \pi\left(\frac{20}{9}+\ln 3\right)$$

Integration of $\cosh^2(x)$ can be done using identities or tech, and simplification of the result — if not done entirely with tech — uses the data $\sinh(0) = 0$, $\sinh(\ln 3) = 4/3$, and $\cosh(\ln 3) = 5/3$. ■

A.3.2   *Vol. by Discs — Practice — Solved*

(1) Find the volume of the solid generated by revolving the region between $y = \sqrt{x-1}$, $x = 2$, and $x = 5$ around the x-axis.

☐ This is an immediate application of (9.3):

$$V = \int_a^b \pi [f(x)]^2 \, dx = \int_2^5 \pi \left(\sqrt{x-1}\right)^2 dx$$

$$= \int_2^5 \pi (x-1) \, dx = \frac{15\pi}{2} \quad \blacksquare$$

(2) Find the volume of the solid generated by revolving the region between $y = x^{2/3}$, $x = 1$ and $y = 0$ around the y-axis.

☐ This is tricky for two reasons: (1) the function is written as $y = f(x)$, but we are revolving around the y-axis, so we need the function written as $x = g(y)$, i.e. $x = y^{3/2}$. (2) As described, the solid revolution of the region forms a hollowed solid about the y-axis. The outer solid is formed by $x = 1$, the inner one by $x = y^{3/2}$, so using (9.4),

$$V = \int_c^d \pi \left([g(y)]^2 - [h(y)]^2\right) dy = \int_0^1 \pi \left((1)^2 - (y^{3/2})^2\right) dy$$

$$= \int_0^1 \pi \left(1 - y^3\right) dy = \frac{3\pi}{4} \quad \blacksquare$$

(3) Find the volume of the solid generated by revolving the region between $y = x^2$ and $y = 4$ around the line $y = 4$.

☐ The curves intersect at $x = \pm 2$. We want to shift everything down to find an equivalent solid of revolution wound around the x-axis. Dropping by 4 units to move $y = 4$ down to the x-axis changes $y = x^2$ to $y = x^2 - 4$. We can also use symmetry to put in a limit of 0 on the integral, which is always nice:

$$V = 2 \int_a^b \pi [f(x)]^2 \, dx = 2 \int_0^2 \pi \left(x^2 - 4\right)^2 dx = \frac{512\pi}{15} \quad \blacksquare$$

(4) Find the volume of the solid generated by revolving the region between $y = 0$ and $y = \sin(x)$ from $x = 0$ to $x = \pi$ around the line $y = -2$.

☐ Note that as described, this gives us a hollowed out solid. We want to find the equivalent volume generated by revolving around the x-axis. If we move everything upward by 2 units, then the line of revolution,

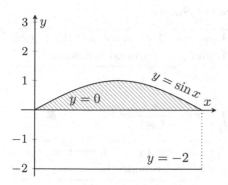

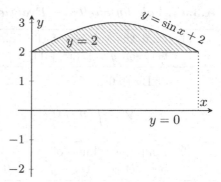

Fig. A.8 The area between $y = \sin x$ and $y = 0$, ready to be revolved around $y = -2$. (PP 4)

Fig. A.9 The equivalent area between $y = \sin x + 2$ and $y = 2$, ready to be revolved around $y = 0$. (PP 4)

$y = -2$ is moved up to the x-axis. This translates $y = 0$ to $y = 2$ and $y = \sin x$ to $y = \sin x + 2$. The volume we want is equivalent to revolving the region between $y = \sin x + 2$ and $y = 2$ around the x-axis from $x = 0$ to $x = \pi$. Figures A.8 and A.9 show the original region and equivalent region.

$$V = \int_a^b \pi \left([f(x)]^2 - [g(x)]^2\right) \, dx = \int_0^\pi \pi \left([\sin x + 2]^2 - [2]^2\right) \, dx$$

$$= \int_0^\pi \pi \left([\sin x + 2]^2 - 4\right) \, dx = 8\pi + \frac{\pi^2}{2} \quad \blacksquare$$

(5) Find the volume of the solid generated by revolving the region between $y = \sinh(x)$, $y = \cosh(x)$, $x = a$ and $x = b$, around the x-axis, for any a and b where $b > a$.

☐ The graph of $\cosh(x)$ is above $\sinh(x)$ (see Fig. 1.37), so this volume is

$$V = \int_a^b \pi \left(\cosh^2(x) - \sinh^2(x)\right) \, dx$$

Does it really get this easy? Yes! With the identity $\cosh^2(x) - \sinh^2(x) = 1$, we have

$$V = \int_a^b \pi \, dx = \pi(b - a)$$

That's crazy! ∎

A.3.3   *Vol. by Shells — Practice — Solved*

---

(1) Find the volume of the solid generated by revolving the region between $y = 1/(x^2 + 1)$, $x = 1$, and $x = 2$ around the y-axis.

☐ By (9.6),

$$V = \int_a^b 2\pi x f(x)\, dx = \int_1^2 2\pi x \cdot \frac{1}{x^2 + 1}\, dx = \pi \ln \frac{5}{2} \quad \blacksquare$$

---

(2) Find the volume of the solid generated by revolving the region between $x = \sqrt{1 - y^2}$, $y = 0$, and $y = 1$ around the x-axis.

☐ By (9.7), we get

$$V = \int_c^d 2\pi y g(y)\, dy = \int_0^1 2\pi y \sqrt{1 - y^2}\, dy = \frac{2\pi}{3} \quad \blacksquare$$

---

(3) Find the volume of the solid generated by revolving the region between $y = x$ and $y = x^2$ around the y-axis.

☐ These curves intersect at $(0,0)$ and $(1,1)$ so on the x-axis we use $x \in [0,1]$. Looking out from the y-axis, around which we're revolving, the curve $y = x$ generates the volume, so in preparation for (9.8), we assign $f(x) = x$ and $g(x) = x^2$, so:

$$V = \int_a^b 2\pi x \left(f(x) - g(x)\right) dx = \int_0^1 2\pi x \left(x - x^2\right) dx$$

$$= 2\pi \int_0^1 \left(x^2 - x^3\right) dx = \frac{\pi}{6} \quad \blacksquare$$

A.4   Chapter 10: Practice Problem Solutions

A.4.1   *Taylor Polynomials — Practice — Solved*

---

(1) Find the Taylor polynomial of order 3 for $f(x) = \ln(1 + 2x)$ at $x = 1$. Determine $R_3(x)$ for an interval of $d = 0.5$ centered at $x = 1$.

☐ We need $f(x)$ and its first three derivatives evaluated at $x = 1$. These are:

$$f(x) = \ln(1 + 2x) \rightarrow f(1) = \ln(3)$$

$$f'(x) = \frac{2}{1 + 2x} \rightarrow f'(1) = \frac{2}{3}$$

$$f''(x) = -\frac{4}{(1 + 2x)^2} \rightarrow f''(1) = -\frac{4}{9}$$

$$f^{(3)}(x) = \frac{16}{(1 + 2x)^3} \rightarrow f^{(3)}(1) = \frac{16}{27}$$

so

$$T_3(x) = \sum_{n=0}^{3} \frac{f^{(n)}(x_0)}{n!}(x - x_0)^n$$

$$= \frac{f^{(0)}(1)}{0!}(x - 1)^0 + \frac{f^{(1)}(1)}{1!}(x - 1)^1 + \frac{f^{(2)}(1)}{2!}(x - 1)^2$$

$$+ \frac{f^{(3)}(1)}{3!}(x - 1)^3$$

$$= \frac{\ln(3)}{0!}(x - 1)^0 + \frac{2/3}{1!}(x - 1)^1 + \frac{(-4/9)}{2!}(x - 1)^2$$

$$+ \frac{16/27}{3!}(x - 1)^3$$

$$= \ln(3) + \frac{2}{3}(x - 1) - \frac{2}{9}(x - 1)^2 + \frac{8}{81}(x - 1)^3$$

To get at $R_3(x)$ using Taylor's Inequality, we note that with $d = 0.5$, we will need an upper bound M of $|f^{(4)}(x)|$ on the interval $[0.5, 1.5]$. A graph of $|f^{(4)}(x)| = 96/(1 + 2x)^4$ is shown in Fig. A.10. Based on the graph, we'll assign an upper bound as $M = 6$. So,

$$R_3(x) \leq \frac{M}{(N + 1)!} \cdot d^{N+1} = \frac{6}{4!}(0.5)^4 \approx 0.015625 \quad \blacksquare$$

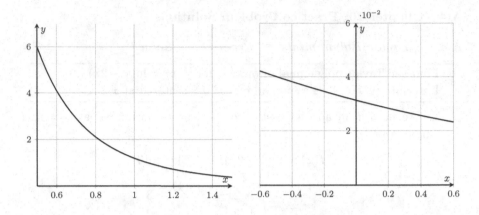

Fig. A.10   Looking for an upper bound of $|f^{(4)}(x)|$ on $[0.5, 1.5]$ (w/ PP 1).

Fig. A.11   Looking for an upper bound of $|f^{(5)}(x)|$ on $[-0.5, 0.5]$ (w/ PP 2).

---

(2) Find the Maclaurin polynomial of order 4 for for $f(x) = x + e^{-x/2}$. Determine $R_4(x)$ for the appropriate interval with $d = 0.5$.

□ We need $f(x)$ and its first four derivatives evaluated at $x = 0$. These are:

$$f(x) = x + e^{-x/2} \rightarrow f(0) = 1$$
$$f'(x) = 1 - \frac{1}{2}e^{-x/2} \rightarrow f'(0) = \frac{1}{2}$$
$$f''(x) = \frac{1}{4}e^{-x/2} \rightarrow f''(0) = \frac{1}{4}$$
$$f^{(3)}(x) = -\frac{1}{8}e^{-x/2} \rightarrow f^{(3)}(0) = -\frac{1}{8}$$
$$f^{(4)}(x) = \frac{1}{16}e^{-x/2} \rightarrow f^{(3)}(0) = \frac{1}{16}$$

so

$$M_4(x) = \sum_{n=0}^{4} \frac{f^{(n)}(x_0)}{n!}(x - 0)^n$$

$$= \frac{f^{(0)}(0)}{0!}(x)^0 + \frac{f^{(1)}(0)}{1!}(x)^1 + \cdots + \frac{f^{(4)}(0)}{4!}(x)^4$$

$$= \frac{1}{0!}(x)^0 + \frac{1/2}{1!}(x)^1 + \frac{1/4}{2!}(x)^2 + \frac{-1/8}{3!}(x)^3 + \frac{-1/16}{4!}(x)^4$$

$$= 1 + \frac{1}{2}x + \frac{1}{2^2 2!}x^2 - \frac{1}{2^3 3!}x^3 + \frac{1}{2^4 4!}x^4$$

To get at $R_4(x)$ using Taylor's Inequality, we note that with $d = 0.5$, we will need an upper bound M of $|f^{(5)}(x)|$ on the interval $[-0.5, 0.5]$. A graph of $|f^{(5)}(x)| = e^{-x/2}/32$ is shown in Fig. A.11; note the units on the y axis are given as $y \cdot 10^{-2}$. Based on the graph, we'll assign an upper bound as $M = 0.05$. So,

$$R_4(x) \leq \frac{M}{(N+1)!} \cdot d^{N+1} = \frac{0.05}{5!}(0.5)^5 \approx 1.3 \times 10^{-5} \quad \blacksquare$$

(3) Find the Taylor polynomial of order 3 for for $f(x) = x + e^{-x/2}$ at $x = 2$. Determine $R_3(x)$ for an interval of $d = 0.5$ centered at $x = 2$ and confirm this $R_3(x)$ holds at the test point $c = 2.3$.

□ We need $f(x)$ and its first three derivatives evaluated at $x = 2$. These are:

$$f(x) = x + e^{-x/2} \rightarrow f(2) = 2 + \frac{1}{e}$$

$$f'(x) = 1 - \frac{1}{2}e^{-x/2} \rightarrow f'(2) = 1 - \frac{1}{2e}$$

$$f''(x) = \frac{1}{4}e^{-x/2} \rightarrow f''(2) = \frac{1}{4e}$$

$$f^{(3)}(x) = -\frac{1}{8}e^{-x/2} \rightarrow f^{(3)}(2) = -\frac{1}{8e}$$

so

$$T_3(x) - \sum_{n=0}^{3} \frac{f^{(n)}(x_0)}{n!}(x - x_0)^n$$

$$= \frac{f^{(0)}(2)}{0!}(x-2)^0 + \frac{f^{(1)}(2)}{1!}(x-2)^1 + \frac{f^{(2)}(2)}{2!}(x-2)^2$$

$$+ \frac{f^{(3)}(2)}{3!}(x-2)^3$$

$$= \frac{2 + 1/e}{0!}(x-2)^0 + \frac{1 - 1/(2e)}{1!}(x-2)^1 + \frac{1/(4e)}{2!}(x-2)^2$$

$$+ \frac{-1/(8e)}{3!}(x-2)^3$$

$$= 2 + \frac{1}{e} + \frac{2e-1}{2e}(x-2) + \frac{1}{2^2 2!e}(x-2)^2 - \frac{1}{2^3 3!e}(x-2)^3$$

To get at $R_3(x)$ using Taylor's Inequality, we note that with $d = 0.5$, we will need an upper bound M of $|f^{(4)}(x)|$ on the interval $[1.5, 2.5]$. A graph of $|f^{(4)}(x)| = e^{-x/2}/16$ is shown in Fig. A.12; note the units on the y axis are given as $y \cdot 10^{-2}$. Based on the graph, we'll assign an upper bound as $M = 0.03$. So,

$$R_3(x) \leq \frac{M}{(N+1)!} \cdot d^{N+1} = \frac{0.03}{4!}(0.5)^4 \approx 7.8 \times 10^{-5}$$

At the test point $c = 2.3$, we have

$$f(2.3) = 2.3 + e^{-2.3/2} \approx 2.616636769$$
$$T_3(2.3) \approx 2.616629237$$

and $|f(2.3) - T_3(2.3)| \approx 7.5 \times 10^{-6}$, which is indeed less than $|R_3(x)| \approx 7.8 \times 10^{-5}$. ∎

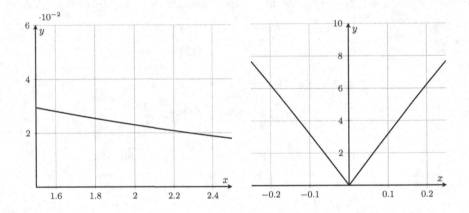

Fig. A.12   Looking for an upper bound of $|f^{(4)}(x)|$ on $[1.5, 2.5]$ (w/ PP 3).

Fig. A.13   Looking for an upper bound of $|f^{(4)}(x)|$ on $[-0.25, 0.25]$ (w/ PP 4).

(4) Find the Maclaurin polynomial of order 3 for for $f(x) = \sin(2x)$. Determine $R_3(x)$ for the appropriate interval of $d = 0.25$ and confirm this $R_3(x)$ holds at the test point $c = 0.1$.

☐ We need $f(x)$ and its first three derivatives evaluated at $x = 0$. These are:

$$f(x) = \sin(2x) \rightarrow f(0) = 0$$
$$f'(x) = 2\cos(2x) \rightarrow f'(0) = 2$$
$$f''(x) = -4\sin(2x) \rightarrow f''(0) = 0$$
$$f^{(3)}(x) = -8\cos(2x) \rightarrow f^{(3)}(0) = -8$$

so

$$M_3(x) = \sum_{n=0}^{3} \frac{f^{(n)}(x_0)}{n!}(x-0)^n$$

$$= \frac{f^{(0)}(0)}{0!}(x)^0 + \frac{f^{(1)}(0)}{1!}(x)^1 + \frac{f^{(2)}(0)}{2!}(x)^2 + \frac{f^{(3)}(0)}{3!}(x)^3$$

$$= \frac{0}{0!}(x)^0 + \frac{2}{1!}(x)^1 + \frac{0}{2!}(x)^2 + \frac{-8}{3!}(x)^3$$

$$= 2x - \frac{2^3}{3!}x^3$$

To get at $R_3(x)$ using Taylor's Inequality, we note that with $d = 0.25$, we will need an upper bound M of $|f^{(4)}(x)|$ on the interval $[-0.25, 0.25]$. A graph of $|f^{(4)}(x)| = |16\sin(2x)|$ is shown in Fig. A.13. Based on the graph, we'll assign an upper bound as $M = 8$. So,

$$R_3(x) \le \frac{M}{(N+1)!} \cdot d^{N+1} = \frac{8}{4!}(0.25)^4 \approx 0.001302$$

(Don't confuse the $M_3(x)$ meaning the third order Maclaurin series with the generic M that represents the upper bound used in the calculation of the remainder.) At the test point $c = 0.1$, we have

$$f(0.1) = \sin(0.2) \approx 0.1986693$$

$$M_3(0.1) \approx 0.1986667$$

With these results, we get $|f(0.1) - M_3(0.1)| \approx 2.6 \times 10^{-6}$, which is indeed less than $|R_3(x)| \approx 0.001302$. ∎

A.4.2   *Taylor Series Pt. 1 — Practice — Solved*

---

(1) Find the Maclaurin series for $f(x) = \cos \pi x$.

☐ The definition of Maclaurin series gives us:

$$M(x) = \frac{f(0)}{0!}x^0 + \frac{f'(0)}{1!}x^1 + \frac{f''(0)}{2!}x^2 + \frac{f^{(3)}(0)}{3!}x^3 + \cdots$$

But

$$f(x) = \cos \pi x \quad , \quad f'(x) = -\pi \sin \pi x \quad ,$$
$$f''(x) = -\pi^2 \cos \pi x \quad , \quad f^{(3)}(x) = \pi^3 \sin \pi x \quad \cdots$$

so

$$f(0) = 1 \quad f'(0) = 0 \quad f''(0) = -\pi^2 \quad f^{(3)}(0) = 0 \quad \cdots$$

$$M(x) = \frac{1}{0!}x^0 + 0 + \frac{-\pi^2}{2!}x^2 + 0 + \cdots$$
$$= x^0 - \frac{\pi^2}{2!}x^2 + \frac{\pi^4}{4!}x^4 \cdots = \sum_{n=0}^{\infty}(-1)^n \frac{\pi^{2n}}{(2n)!}x^{2n} \quad \blacksquare$$

---

(2) Find the Taylor series for $f(x) = \ln x$ around $x_0 = 2$.

☐ From the definition of Taylor series, we need to fill out

$$T(x) = \frac{f(2)}{0!}(x-2)^0 + \frac{f'(2)}{1!}(x-2)^1 + \frac{f''(2)}{2!}(x-2)^2 + \frac{f^{(3)}(2)}{3!}(x-2)^3 + \cdots$$

The necessary derivatives functions are:

$$f(x) = \ln x \quad , \quad f'(x) = \frac{1}{x} \quad , \quad f''(x) = -\frac{1}{x^2} \quad ,$$
$$f^{(3)}(x) = \frac{2!}{x^3} \quad , \quad f^{(4)}(x) = -\frac{3!}{x^4} \quad \cdots$$

so

$$f(2) = \ln 2 \quad , \quad f'(2) = \frac{1}{2} \quad , \quad f''(2) = -\frac{1}{2^2}$$
$$f^{(3)}(2) = \frac{2!}{2^3} \quad , \quad f^{(4)}(2) = -\frac{3!}{2^4} \quad \cdots$$

and

$$T(x) = \frac{\ln 2}{1}(x-2)^0 + \frac{1/2}{1!}(x-2) + \frac{-1/2^2}{2!}(x-2)^2$$
$$+ \frac{2!/2^3}{3!}(x-2)^3 + \cdots$$
$$= \frac{\ln 2}{1}(x-2)^0 + \frac{1}{2}(x-2) - \frac{1}{2 \cdot 2^2}(x-2)^2$$
$$+ \frac{1}{3 \cdot 2^3}(x-2)^3 + \cdots$$
$$= \ln 2 + \sum_{n=1}^{\infty} \frac{(-1)^{n-1}}{n2^n}(x-2)^n$$

(Note that since the first term, $\ln 2$, didn't fit a pattern with the rest of the terms, it was left out on its own.) ∎

---

(3) Find the Maclaurin series for $f(x) = 1/(1+x)^2$.

☐ The definition of Maclaurin series gives us:

$$M(x) = \frac{f(0)}{0!}x^0 + \frac{f'(0)}{1!}x^1 + \frac{f''(0)}{2!}x^2 + \frac{f^{(3)}(0)}{3!}x^3 + \cdots$$

But

$$f(x) = \frac{1}{(1+x)^2} \quad , \quad f'(x) = \frac{-2}{(1+x)^3} \quad ,$$
$$f''(x) = \frac{3!}{(1+x)^4} \quad , \quad f^{(3)}(x) = \frac{-4!}{(1+x)^5} \quad \cdots$$

so

$$f(0) = 1 \quad f'(0) = -2 \quad f''(0) = 3! \quad f^{(3)}(0) = -4! \quad \cdots$$

and

$$M(x) = \frac{1}{0!}x^0 + \frac{-2}{1!}x^1 + \frac{3!}{2!}x^2 + \frac{-4!}{3!}x^3 + \cdots$$
$$= 1 - 2x + 3x^2 - 4x^3 + \cdots$$
$$= \sum_{n=0}^{\infty} (-1)^n (n+1)x^n \quad ∎$$

A.4.3   *Infinite Sequences — Practice — Solved*

---

(1) Find a formula for the terms in the sequence $-\dfrac{1}{4}, \dfrac{2}{9}, -\dfrac{3}{16}, \dfrac{4}{25}, \cdots$

☐ Note some patterns: the numerators are increasing one by one, while the denominators are increasing as squares — and the number being squared in any one denominator is the numerator of the next term. Plus, the signs of each term alternate. We can put this together as:

$$a_n = (-1)^n \frac{n}{(n+1)^2} \text{ for } n \geq 1 \quad \blacksquare$$

---

(2) If it exists, determine the limit of the sequence $a_n = \dfrac{\sqrt{n}}{1+\sqrt{n}}$.

☐ If it's not clear that the limit is 1, we can multiply up and down by $\sqrt{n}$ to get

$$\frac{\sqrt{n}}{\sqrt{n}} \cdot \frac{\sqrt{n}}{1+\sqrt{n}} = \frac{n}{\sqrt{n}+n}$$

Then L-Hopital's Rule gets us to

$$\lim_{n\to\infty} a_n = \lim_{n\to\infty} \frac{1}{1/(2\sqrt{n})+1} = 1$$

The sequence converges to 1. $\blacksquare$

---

(3) If it exists, determine the limit of the sequence $a_n = \cos\left(\dfrac{2}{n}\right)$.

☐ We know that $2/n$ approaches 0 as $n \to \infty$, so a_n approaches $\cos 0$, or $a_n \to 1$. The sequence to 1. $\blacksquare$

---

(4) If it exists, determine the limit of the sequence $a_n = \dfrac{\ln n}{\ln 2n}$.

☐ Let's look at this one like a function and use L-Hopital's rule to see that

$$\lim_{n\to\infty} a_n = \lim_{n\to\infty} \frac{\frac{1}{n}}{2\frac{1}{2n}} = \lim_{n\to\infty} \frac{\frac{1}{n}}{\frac{1}{n}} = 1$$

Alternately, we could write $a_n = \ln n/(\ln 2 + \ln n)$, from which it's clear the limit is 1. Either way, the sequence converges to 1. $\blacksquare$

(5) Determine whether the sequence $a_n = ne^{-n}$ is increasing or decreasing, and if it is bounded.

□ Note that we can rewrite the formula as $a_n = n/e^n$. I have a suspicion this decreases since the denominator will quickly swamp the numerator. To show this, I need to show that $a_n > a_{n+1}$ for all $n > 0$.

$$a_n >? a_{n+1}$$
$$\frac{n}{e^n} >? \frac{n+1}{e^{n+1}}$$
$$\frac{n}{e^n} >? \frac{n+1}{e \cdot e^n}$$
$$n >? \frac{n+1}{e}$$
$$e \cdot n >? n+1$$

Since $e \approx 2.718$, this is true for all $n > 0$, so the sequence is indeed decreasing. The largest term in the sequence comes at $n = 1$, and is $a_1 = 1$. The sequence is bounded below by 0. Together, this is a decreasing sequence bounded above by 1 and below by 0. ∎

A.4.4   *Infinite Series — Practice — Solved*

---

(1) Investigate the convergence or divergence of $\displaystyle\sum_{n=1}^{\infty} \ln\left(\frac{n}{2n+5}\right)$.

☐ Are the terms trending towards zero? Let's see:

$$\lim_{n\to\infty} a_n = \lim_{n\to\infty} \ln\left(\frac{n}{2n+5}\right) = \ln\left(\frac{1}{2}\right) \neq 0$$

So by the divergence test, this series cannot converge. The series diverges. ∎

---

(2) Investigate the convergence or divergence of $\dfrac{1}{8} - \dfrac{1}{4} + \dfrac{1}{2} - 1 + \cdots$.

☐ This one's easy. The magnitude of the terms in the sequence is increasing, not approaching 0. Therefore, by the divergence test, the series diverges. ∎

---

(3) Investigate the convergence or divergence of $\displaystyle\sum_{n=0}^{\infty} \frac{\pi^n}{3^{n+1}}$.

☐ This series looks geometric-ish, let's pursue that:

$$\sum_{n=0}^{\infty} \frac{\pi^n}{3^{n+1}} = \sum_{n=0}^{\infty} \frac{1}{3} \cdot \left(\frac{\pi}{3}\right)^n$$

This is a geometric series in standard form with $a = 1/3$ and $r = \pi/3$. Since $|r| > 1$, the series diverges. ∎

---

(4) Write $6.2545454\ldots$ as a fraction.

☐

$$\square\, 6.2545454\ldots = 6.2 + \frac{54}{10^3} + \frac{54}{10^5} + \frac{54}{10^7} + \cdots$$

$$= 6.2 + \frac{54}{10^3}\left(1 + \frac{1}{10^2} + \frac{1}{10^4} + \cdots\right)$$

$$= 6.2 + \sum_{n=0}^{\infty} \frac{54}{10^3}\left(\frac{1}{10^2}\right)^n$$

This is a geometric series with $a = 54/10^3$ and $r = 1/10^2$. Since $|r| < 1$, the series converges, and the whole expression converges to

$$6.2545454\ldots = 6.2 + \frac{54/10^3}{1 - 1/10^2} = \frac{62}{10} + \frac{54}{990} = \frac{6192}{990} = \frac{344}{55} \quad ∎$$

(5) Find the values of x for which the series $\sum\limits_{n=1}^{\infty} (x-4)^n$ converges, and determine the algebraic form of the function the convergent series defines.

□ This is almost a geometric series, let's mess with it:

$$\sum_{n=1}^{\infty} (x-4)^n = \sum_{n=0}^{\infty} (x-4)^{n+1} \sum_{n=0}^{\infty} (x-4) \cdot (x-4)^n$$

Now it is a geometric series with $a = x - 4$ and $r = x - 4$. It will converge when $|r| < 1$, i.e. when $|x - 4| < 1$, i.e. when $3 < x < 5$. When it does converge, it converges to

$$\frac{a}{1-r} = \frac{x-4}{1-(x-4)} = \frac{x-4}{5-x}$$

So, the series $\sum\limits_{n=1}^{\infty} (x-4)^n$ is equivalent to the function $\dfrac{x-4}{5-x}$ for $3 < x < 5$. ∎

A.4.5   *Alternating Series — Practice — Solved*

(1) Investigate the convergence of these series:

$$(A)\ \sum_{n=1}^{\infty}(-1)^n\frac{\ln(n)}{\ln(2n)}\qquad ;\qquad (B)\ \sum_{n=1}^{\infty}(-1)^{n+1}\frac{n^{944}+\pi}{n^{946}}$$

☐ Series (A) cannot converge because the limit of its terms is not zero:

$$\lim_{n\to\infty}\frac{\ln(n)}{\ln(2n)}=\lim_{n\to\infty}\frac{\ln(n)}{\ln(2)+\ln(n)}=1$$

Series (B) converges.

$$\lim_{n\to\infty}\frac{n^{944}+\pi}{n^{946}}=0\quad\text{and}\quad\frac{(n+1)^{944}+\pi}{(n+1)^{946}}<\frac{n^{944}+\pi}{n^{946}}\quad\blacksquare$$

(2) Investigate the convergence of $\sum_{n=1}^{\infty}(-1)^n\dfrac{\sqrt{n}}{1+2\sqrt{n}}$. If this series converges, find how many terms are necessary for the error to be less than 0.0001.

☐ Since $a_n=\sqrt{n}/(1+2\sqrt{n})$, we have $\lim_{n\to\infty}a_n=1/2$, and so one of the conditions required for convergence of the alternating series is not met. We don't even have to check the other now. This series does not converge.   ∎

(3) Investigate the convergence of $\sum_{n=0}^{\infty}(-1)^{n+1}\dfrac{n^2+1}{2n^3+2}$. If this series converges, find the maximum error between the partial sum S_{40} and the true value of the series.

☐ With $a_n=(n^2+1)/(2n^3+2)$, then $\lim_{n\to\infty}a_n=0$. One of the conditions for convergence of this alternating series is met. For the other, we must compare a_{n+1} to a_n by:

$$a_{n+1}<a_n$$
$$\frac{(n+1)^2+1}{2(n+1)^3+2}<\frac{n^2+1}{2n^3+2}$$
$$(n^2+2n+1)(2n^3+2)<(n^2+1)(2n^3+6n^2+6n+4)$$
$$2n^5+4n^4+2n^3+2n^2+4n+2<2n^5+6n^4+8n^3+10n^2+6n+4$$
$$0<2n^4+6n^3+8n^2+2n+2$$

The final statement is true for all $n \geq 0$, so the equivalent first statement is also true. Now we know both conditions in Useful Fact 10.6 hold, and the series converges. Therefore, by Def. 10.3, the maximum error at S_{40} is equal to a_{41},

$$a_{41} = \frac{41^2 + 1}{2(41)^3 + 2} \approx 0.0122 \quad \blacksquare$$

---

(4) Investigate the convergence of $\displaystyle\sum_{n=0}^{\infty}(-1)^n \frac{e^{2n}}{e^{3n}}$. If this series converges, find how many terms are necessary for the error to be less than 0.0001.

☐ Note that the series term a_n can be simplified to $a_n = \dfrac{1}{e^n}$, so we'll clearly have both $a_n \to 0$ as $n \to \infty$ and $a_{n+1} < a_n$. The series converges. To find how far we have to dive into partial sums to find error less than 0.0001, we must find n such that $a_{n+1} < 0.0001$, or

$$\frac{1}{e^{n+1}} < 0.0001$$
$$e^{n+1} > 10^4$$
$$n + 1 > 4\ln(10)$$
$$n > 4\ln(10) - 1 \approx 8.21$$

Since n must be an integer, the first n we find that satisfies $n > 8.21$ is $n = 9$. It only takes a partial sum with $n = 9$ so that S_9 is within 0.0001 of the true value of the series. Not bad! $\quad \blacksquare$

---

(5) The series $\displaystyle\sum_{n=0}^{\infty}(-1)^n$ itself is an alternating series, but not much of one. Your somewhat distracted friend claims that the series obviously converges to 0 since it expands as $1 + (-1) + 1 + (-1) + \cdots$. Do you agree? Why or why not?

☐ My friend is a doofus. First of all, this series fails both of the conditions in Useful Fact 10.6. $\displaystyle\lim_{n \to \infty}(-1)^n$ is undefined. And, every other term is larger than the one before it, when we transition from -1

to 1. But even so, consider the sequence of partial sums:

$$s_0 = 1$$
$$s_1 = 1 + (-1) = 0$$
$$s_2 = 1 + (-1) + 1 = 1$$
$$s_3 = 1 + (-1) + 1 + (-1) = 0$$

$$\vdots \qquad\qquad \vdots$$

The sequence of partials sums is $1, 0, 1, 0, 1, 0, \ldots$. Since the fundamental definition of convergence of a series (whether an alternating series or not) is the limit of the sequence of its partial sums, and this sequence has no limit. So sure, maybe we want to say that $\sum_{n=0}^{\infty} (-1)^n$ converges to 0, but we can't.   ∎

A.5   Chapter 11: Practice Problem Solutions

A.5.1   *Int. Test & p-Series — Practice — Solved*

---

(1) Investigate the convergence of $\displaystyle\sum_{n=1}^{\infty} e^{-n}$.

□ Applying the integral test, this series will behave the same as the corresponding integral:

$$\int_1^{\infty} e^{-x}\, dx = \lim_{c\to\infty} -e^{-x}\Big|_1^c = \lim_{c\to\infty}\left(-e^{-c}+e^{-1}\right) = \lim_{c\to\infty}\left(\frac{1}{e}-\frac{1}{e^c}\right) = \frac{1}{e}$$

Since the integral converges, so does the series. However, remember that all we're doing is comparing convergence to convergence — the integral converged to $1/e$, but the series may converge to something different! ∎

---

(2) Investigate the convergence of $1 + \dfrac{1}{2\sqrt{2}} + \dfrac{1}{3\sqrt{3}} + \dfrac{1}{4\sqrt{4}} + \cdots$.

□ We can write this as

$$\sum_{n=1}^{\infty} \frac{1}{n\sqrt{n}} = \sum_{n=1}^{\infty} \frac{1}{n^{3/2}}$$

and recognize it as a p-series with $p = 3/2$; since $p > 1$, the series converges. ∎

---

(3) Investigate the convergence of $\displaystyle\sum_{n=1}^{\infty} \frac{1}{n^2 - 4n + 5}$.

□ Note that the denominator can be cleverly rewritten:

$$\frac{1}{n^2 - 4n + 5} = \frac{1}{n^2 - 4n + (4) - (4) + 5} = \frac{1}{(n-2)^2 + 1}$$

Applying the integral test, this series will behave the same as the corresponding integral:

$$\int_1^{\infty} \frac{1}{x^2 - 4x + 5}\, dx = \lim_{c\to\infty} \int_1^c \frac{1}{(x-2)^2 + 1}\, dx = \lim_{c\to\infty} \tan^{-1}(x-2)\Big|_1^c$$

$$= \lim_{c\to\infty} \tan^{-1}(c-2) - \tan^{-1}(-1) = \frac{\pi}{2} + \frac{\pi}{4} = \frac{3\pi}{4}$$

Since the (improper) integral converges, the series does, too. ∎

---

(4) Consider the series $\displaystyle\sum_{n=1}^{\infty} \frac{1}{n^2+1}$.

    (a) Estimate the 100th partial sum of this series.

    (b) Find an upper bound on the error between this result and the true value of the series.

    (c) Determine how many terms would be needed for an accuracy of 10^{-4}.

☐ (a) Using a spreadsheet (not shown), we can find $S_{100} \approx 1.067724$.

(b) The function $f(x) = 1/(x^2+1)$ is continuous, positive, and decreasing or $x > 1$, so the integral test would apply, and so the integral test remainder formula is appropriate. By Useful Fact 11.3, the remainder at S_{100} is bounded as follows:

$$R_{100} \le \int_{100}^{\infty} \frac{1}{x^2+1}\, dx \approx 0.01 = 10^{-2}$$

(c) To achieve an accuracy of 10^{-4}, we must find the value N for which

$$\int_{N}^{\infty} \frac{1}{x^2+1}\, dx = 10^{-4}$$

$$\frac{1}{2}(\pi - 2\tan^{-1} N) = 10^{-4}$$

$$N = \tan\left(\frac{\pi - 2\cdot 10^{-4}}{2}\right) \approx 10{,}000$$

The approximation for N is rounded up to $10{,}000$ from $9999.999967\ldots$ so not much rounding at all. It takes at least $10{,}000$ terms to achieve an accuracy of 10^{-4}. ■

---

(5) Investigate the convergence of $\displaystyle\sum_{n=0}^{\infty} \frac{1}{4}\cdot\left(\frac{2}{3} - \frac{1}{\pi}\right)^n$.

☐ Note that

$$\left(\frac{2}{3} - \frac{1}{\pi}\right)^n < \left(\frac{2}{3}\right)^n$$

for all $n \ge 1$. Since the series $\displaystyle\sum_{n=0}^{\infty} \frac{1}{4}\cdot\left(\frac{2}{3}\right)^n$ is a geometric series with $r = 2/3$, so that $|r| < 1$, we know it converges. Therefore, the given series we're testing also converges, by the direct comparison test. ■

(6) Investigate the convergence of $\displaystyle\sum_{n=1}^{\infty} \frac{1}{2^n - n}$.

☐ Golly, this looks a lot like the convergent geometric series with terms $1/2^n$. So a comparison test might be in order, but the direct comparison test won't work since $1/(2^n - n)$ is greater than $1/2^n$ for all $n \geq 1$. So let's try the limit comparison test:

$$\lim_{n\to\infty} \frac{a_n}{b_n} = \lim_{n\to\infty} \frac{2^n - n}{2^n} = \lim_{n\to\infty} \left(1 - \frac{n}{2^n}\right) = 1$$

Since the limit of the ratio of terms is a finite positive constant, then the two series will exhibit the same behavior. So the given series converges. ∎

(7) Investigate the convergence of $\displaystyle\sum_{n=3}^{\infty} \frac{5}{(n-2)^2}$.

☐ There are several ways to analyze this series. Since the function $f(x) = 5/(x-2)^2$ is continuous, positive, and decreasing for $x \geq 3$, the integral test would apply.

It looks like a p-series, but technically it is not one, because the denominator is $(n-2)^2$ rather than n^2. But also note the 1 series starts with $n = 3$, meaning the sequence of denominators is $1, 2, 3, \ldots$. So we can reindex the series: drop the starting value of the index down by 2 and add 2 back to all appearances of the index in the terms; we get

$$\sum_{n=3}^{\infty} \frac{5}{(n-2)^2} = \sum_{n=1}^{\infty} \frac{5}{n^2}$$

This is now definitely a p-series with $p = 2$; since $p > 1$, this series converges.

Since it "looks like" a p-series, perhaps we can use a comparison against $\displaystyle\sum_{n=3}^{\infty} \frac{5}{n^2}$, which is known to converge. The regular comparison test won't work since $5/(n-2)^2$ is not less than $5/n^2$. So how about the limit comparison test?

$$\lim_{n\to\infty} \frac{a_n}{b_n} = \lim_{n\to\infty} \frac{5/(n-2)^2}{5/n^2} = \lim_{n\to\infty} \frac{n^2}{(n-2)^2} = 1$$

Since the limit of the ratio of terms is a finite positive constant, then the two series will exhibit the same behavior. So the given series converges. ∎

A.5.2 Ratio Test — Practice — Solved

(1) Investigate the convergence of $\displaystyle\sum_{n=1}^{\infty}(-1)^{n-1}\frac{2^n}{n^4}$.

☐ The series contains factorials and powers of n, so the ratio test should work well.

$$L = \lim_{n\to\infty}\left|\frac{a_{n+1}}{a_n}\right| = \lim_{n\to\infty}\left|\frac{\left(\frac{2^{n+1}}{(n+1)^4}\right)}{\left(\frac{2^n}{n^4}\right)}\right|$$

$$= \lim_{n\to\infty}\left|\frac{2^{n+1}}{(n+1)^4}\frac{n^4}{2^n}\right| = 2\lim_{n\to\infty}\left(\frac{n}{n+1}\right)^4 = 2$$

Since $L > 1$ the series diverges. ∎

(2) Investigate the convergence of $\displaystyle\sum_{n=1}^{\infty}e^{-n}n! = \sum_{n=1}^{\infty}\frac{n!}{e^n}$.

☐ The series contains factorials and powers of n, so the ratio test should work well.

$$L = \lim_{n\to\infty}\left|\frac{a_{n+1}}{a_n}\right| = \lim_{n\to\infty}\left|\frac{\left(\frac{(n+1)!}{e^{n+1}}\right)}{\left(\frac{n!}{e^n}\right)}\right|$$

$$= \lim_{n\to\infty}\left|\frac{(n+1)!}{e^{n+1}}\frac{e^n}{n!}\right| = \lim_{n\to\infty}\left|\frac{n+1}{e}\right| = \infty$$

Since $L > 1$ the series diverges. ∎

(3) Investigate the convergence of $\displaystyle\sum_{n=1}^{\infty}\frac{1}{(2n)!}$.

☐ The series contains factorials, so the ratio test should work well:

$$L = \lim_{n\to\infty}\left|\frac{a_{n+1}}{a_n}\right| = \lim_{n\to\infty}\left|\frac{\left(\frac{1}{(2(n+1))!}\right)}{\left(\frac{1}{(2n)!}\right)}\right|$$

$$= \lim_{n\to\infty}\left|\frac{1}{(2n+2)(2n+1)(2n)!}\frac{(2n)!}{1}\right| = \lim_{n\to\infty}\left|\frac{1}{(2n+2)(2n+1)}\right| = 0$$

Since $L < 1$ the series converges. ∎

A.5.3   *Power Series — Practice — Solved*

---

(1) Find the interval and radius of convergence of the power series $\sum\limits_{n=0}^{\infty} \dfrac{x^n}{n!}$.

☐ Starting up the ratio test,

$$\left| \frac{a_{n+1}}{a_n} \right| = \left| \frac{\left(\frac{x^{n+1}}{(n+1)!} \right)}{\left(\frac{x^n}{n!} \right)} \right| \cdot \left| \frac{x^{n+1}}{(n+1)!} \cdot \frac{n!}{x^n} \right| = \left(\frac{1}{n+1} \right) |x|$$

To have convergence, we need the limit of this expression to be less than 1; that limit is

$$L = \lim_{n \to \infty} \left(\frac{1}{n+1} \right) |x| = |x| \lim_{n \to \infty} \frac{1}{n+1} = 0$$

Therefore, L is always less than 1, and the series converges for all values of x. The interval of convergence is $-\infty < x < \infty$; the radius of convergence is ∞.   ■

---

(2) Find the interval and radius of convergence of the power series $\sum\limits_{n=0}^{\infty} n^3 (x - 5)^n$.

☐ Starting up the ratio test,

$$\left| \frac{a_{n+1}}{a_n} \right| = \left| \frac{(n+1)^3 (x-5)^{n+1}}{n^3 (x-5)^n} \right| = |x - 5| \left(\frac{n+1}{n} \right)^3$$

To have convergence, we need the limit of this expression to be less than 1; that limit is

$$L = \lim_{n \to \infty} |x - 5| \left(\frac{n+1}{n} \right)^3 = |x - 5| \lim_{n \to \infty} \left(\frac{n+1}{n} \right)^3 = |x - 5|$$

To have $L < 1$, we need $|x - 5| < 1$, i.e. $4 < x < 6$. We also have to test the endpoints of this interval. When $x = 4$, we get

$$\sum_{n=1}^{\infty} (-1)^n n^3$$

which diverges by the divergence test. For $x = 6$, we get

$$\sum_{n=1}^{\infty} n^3$$

which also diverges. The interval of convergence is $4 < x < 6$; the radius of convergence is 1.   ■

(3) Find the interval and radius of convergence of the power series $\sum\limits_{n=1}^{\infty} \dfrac{(3x-2)^n}{n3^n}$.

☐ Starting up the ratio test,

$$\left|\frac{a_{n+1}}{a_n}\right| = \left|\frac{\frac{(3x-2)^{n+1}}{(n+1)3^{n+1}}}{\frac{(3x-2)^n}{n3^n}}\right| = \left|\frac{(3x-2)^{n+1}}{(n+1)3^{n+1}} \cdot \frac{n3^n}{(3x-2)^n}\right| = \frac{|3x-2|}{3}\left(\frac{n}{n+1}\right)$$

To have convergence, we need the limit of this expression to be less than 1; that limit is

$$L = \lim_{n\to\infty} \frac{|3x-2|}{3}\left(\frac{n}{n+1}\right) = \frac{|3x-2|}{3}\lim_{n\to\infty}\left(\frac{n}{n+1}\right) = \frac{|3x-2|}{3}$$

To have $L < 1$, we need $\dfrac{|3x-2|}{3} < 1$, or $|3x-2| < 3$. This opens up as:

$$-3 < 3x-2 < 3$$
$$-1 < 3x < 5$$
$$-\frac{1}{3} < x < \frac{5}{3}$$

We also have to test the endpoints of this interval. When $x = -\dfrac{1}{3}$, we have

$$\sum_{n=1}^{\infty} \frac{(-3)^n}{n3^n} = \sum_{n=1}^{\infty} \frac{(-1)^n}{n}$$

This is the alternating harmonic series, and converges. When $x = \dfrac{5}{3}$, we have

$$\sum_{n=1}^{\infty} \frac{3^n}{n3^n} = \sum_{n=1}^{\infty} \frac{1}{n}$$

This is the harmonic series, which diverges. Therefore the interval of convergence is $-1/3 \leq x < 5/3$. The radius of convergence is 1. ∎

A.5.4   *Funcs. in Disguise — Practice — Solved*

(1) Find the power series representation and interval of convergence of
$$f(x) = \frac{1}{1 + 9x^2}.$$

☐ With a little work, we can get this to look like the $a/(1-r)$ form of a geometric series:

$$f(x) = \frac{1}{1 + 9x^2} = \frac{1}{1 - (-9x^2)}$$

This matches with $a = 1$ and $r = -9x^2$. Thus a series form would be

$$\sum_{n=0}^{\infty} a(r)^n = \sum_{n=0}^{\infty} (1)(-9x^2)^n = \sum_{n=0}^{\infty} (-1)^n (3x)^{2n}$$

This will converge when $|r| = |9x^2| < 1$, i.e. $|x| < 1/3$, or $-1/3 < x < 1/3$.   ■

---

(2) Find the power series representation and interval of convergence of
$$f(x) = \frac{x}{9 + x^2}.$$

☐ With a little work, we can get this to look like the $a/(1-r)$ form of a geometric series:

$$f(x) = \frac{x}{9 + x^2} = \frac{x/9}{1 - (-x^2/9)}$$

This matches $a/(1-r)$ with $a = x/9$ and $r = -x^2/9$. Thus a series form would be

$$\sum_{n=0}^{\infty} a(r)^n = \sum_{n=0}^{\infty} \frac{x}{9} \left(-\frac{x^2}{9} \right)^n = \sum_{n=0}^{\infty} (-1)^n \frac{x}{9} \left(\frac{x}{3} \right)^{2n} = \sum_{n=0}^{\infty} (-1)^n \frac{x^{2n+1}}{3^{2n+2}}$$

This will converge when $|r| = |x^2/9| < 1$, i.e. when $|x| < 3$, or $-3 < x < 3$.   ■

---

(3) Find the power series representation and interval of convergence of
$$f(x) = \frac{1}{(1 + x)^3}.$$

☐ That form does not look like the $a/(1-r)$ form of a geometric series. However, note that

$$f(x) = \frac{1}{(1 + x)^3} = \frac{-1}{2} \frac{d}{dx} \frac{1}{(1 + x)^2}$$

and we can use the result from You Try It 6:

$$f(x) = \frac{-1}{2}\frac{d}{dx}\frac{1}{(1+x)^2} = -\frac{1}{2}\frac{d}{dx}\sum_{n=0}^{\infty}(-1)^n(n+1)x^n$$

$$= -\frac{1}{2}\sum_{n=1}^{\infty}(-1)^n n(n+1)x^{n-1} = \frac{1}{2}\sum_{n=1}^{\infty}(-1)^{n+1}n(n+1)x^{n-1}$$

$$= \frac{1}{2}\sum_{n=0}^{\infty}(-1)^{n+2}(n+1)(n+2)x^n$$

Wrapping up all that mess,

$$\frac{1}{(1+x)^3} = \frac{1}{2}\sum_{n=0}^{\infty}(-1)^n(n+1)(n+2)x^n$$

for $|x| < 1$. ∎

---

(4) Find the power series representation and interval of convergence of $f(x) = \dfrac{x^3}{(x-2)^2}$.

□ That form does not look like the $a/(1-r)$ form of a geometric series. However, note that

$$f(x) = x^3\frac{1}{(x-2)^2} = x^3\frac{d}{dx}\frac{-1}{x-2} = \frac{x^3}{2}\frac{d}{dx}\frac{1}{1-x/2}$$

The right side is a geometric series with $a = 1$ and $r = x/2$. So,

$$f(x) = \frac{x^3}{2}\frac{d}{dx}\sum_{n=0}^{\infty}\left(\frac{x}{2}\right)^n = \frac{x^3}{2}\sum_{n=1}^{\infty}\frac{nx^{n-1}}{2^n}$$

$$= \frac{x^3}{2}\sum_{n=0}^{\infty}\frac{(n+1)x^n}{2^{n+1}} = \sum_{n=0}^{\infty}\frac{(n+1)x^{n+3}}{2^{n+2}}$$

$$= \sum_{n=0}^{\infty}2(n+1)\left(\frac{x}{2}\right)^{n+3}$$

This will converge when $|r| = |x/2| < 1$, i.e. when $|x| < 2$, or $-2 < x < 2$. ∎

A.5.5   *Taylor Series Pt. 2 — Practice — Solved*

(1) Find the Maclaurin series and interval of convergence for $f(x) = \cos \pi x$.

☐ We found the series in Practice Problem 1 of Sec. 10.2:

$$\cos \pi x = \sum_{n=0}^{\infty} (-1)^n \frac{\pi^{2n}}{(2n)!} x^{2n}$$

The interval of convergence is determined by the ratio test:

$$\left| \frac{a_{n+1}}{a_n} \right| = \left| \frac{\pi^{2n+2} x^{2n+2}/(2n+2)!}{\pi^{2n} x^{2n}/(2n)!} \right| = \left| \frac{\pi^{2n+2} x^{2n+2}}{(2n+2)!} \cdot \frac{(2n)!}{\pi^{2n} x^{2n}} \right|$$

$$= \frac{\pi^2}{(2n+2)(2n+1)} |x^2|$$

In order to have convergence, we need the limit of this expression to be less than 1. But when we compute this limit, we get:

$$L = \lim_{n\to\infty} \frac{\pi^2}{(2n+2)(2n+1)} |x^2| = 0$$

Since $L < 1$ for all values of x, the interval of convergence is $-\infty < x < \infty$.   ■

---

(2) Find the Taylor series and interval of convergence for $f(x) = \ln x$ around $x_0 = 2$.

☐ We found the series in Practice Problem 2 of Sec. 10.2:

$$\ln x = \ln 2 + \sum_{n=1}^{\infty} \frac{(-1)^{n-1}}{n2^n} (x - 2)^n$$

The interval of convergence is determined by the ratio test (we don't need to consider the constant $\ln 2$, convergence is determined for the series itself):

$$\left| \frac{a_{n+1}}{a_n} \right| = \left| \frac{(x-2)^{n+1}/(n+1)2^{n+1}}{(x-2)^n/n2^n} \right| = \left| \frac{(x-2)^{n+1}}{(n+1)2^{n+1}} \cdot \frac{n2^n}{(x-2)^n} \right|$$

$$= \frac{n}{2(n+1)} |x - 2|$$

In order to have convergence, we need the limit of this expression to be less than 1. But when we compute this limit, we get:

$$L = \lim_{n \to \infty} \frac{n}{2(n+1)} |x - 2| = \frac{1}{2}|x - 2|$$

We get $L < 1$ when $|x - 2| < 2$ or $0 < x < 4$. At $x = 0$ the function is not defined, so the series cannot converge to the value; at $x = 4$, the series becomes:

$$\sum_{n=1}^{\infty} \frac{(-1)^{n-1}}{n2^n}(4-2)^n = \sum_{n=1}^{\infty} \frac{(-1)^{n-1}}{n}$$

This is the alternating harmonic series, and converges; therefore, $x = 4$ is in the interval of convergence, and that full interval is $0 < x \leq 4$, or $(0, 4]$. ∎

---

(3) Find the Maclaurin series and interval of convergence for $f(x) = \dfrac{1}{(1+x)^2}$.

□ We found the series in Practice Problem 3 of Sec. 10.2:

$$\frac{1}{(1+x)^2} = \sum_{n=0}^{\infty}(-1)^n n x^n$$

The interval of convergence is determined by the ratio test:

$$\left| \frac{a_{n+1}}{a_n} \right| = \left| \frac{(n+1)x^{n+1}}{nx^n} \right| = \frac{n+1}{n}|x|$$

In order to have convergence, we need the limit of this expression to be less than 1. But when we compute this limit, we get:

$$L = \lim_{n \to \infty} \frac{n+1}{n}|x| = |x|$$

We get $L < 1$ when $|x| < 1$ or $-1 < x < 1$. We still have to test the endpoints, too. At $x = -1$ the series becomes,

$$\sum_{n=0}^{\infty}(-1)^n n(-1)^n = \sum_{n=0}^{\infty} n$$

This will diverge (the divergence test). At $x = 1$, the series becomes

$$\sum_{n=0}^{\infty}(-1)^n n(1)^n = \sum_{n=0}^{\infty}(-1)^n n$$

This will also diverge (the divergence test). The interval of convergence remains $-1 < x < 1$, or $(-1, 1)$. ∎

(4) Find the Taylor series and interval of convergence for $f(x) = \dfrac{1}{x}$ around $x = -2$.

☐ Here's one we haven't done before! According to the definition of Taylor series, we need to fill out

$$T(x) = \sum_{n=0}^{\infty} \frac{f^{(n)}(-2)}{n!}(x - (-2))^n = \sum_{n=0}^{\infty} \frac{f^{(n)}(-2)}{n!}(x + 2)^n$$

But

$$f(x) = \frac{1}{x} \quad f'(x) = -\frac{1}{x^2} \quad f''(x) = \frac{2}{x^3} \quad f^{(3)}(x) = -\frac{3!}{x^4} \quad f^{(4)}(x) = \frac{4!}{x^5} \quad \cdots$$

Here come a bunch of values:

$$f(-2) = -\frac{1}{2} = -\frac{0!}{2^1} \quad ; \quad f'(-2) = -\frac{1}{(-2)^2} = -\frac{1!}{2^2} \quad ;$$

$$f''(-2) = \frac{2}{(-2)^3} = -\frac{2!}{2^3} \quad ; \quad f^{(3)}(-2) = -\frac{3!}{2^4} \quad ;$$

$$f^{(4)}(-2) = \frac{4!}{(-2)^5} = -\frac{4!}{2^5}$$

Using these first few values, we can generalize to: $f^{(n)}(-2) = -n!/2^{n+1}$. Therefore,

$$\frac{1}{x} = \sum_{n=0}^{\infty} \frac{f^{(n)}(-2)}{n!}(x + 2)^n = \sum_{n=0}^{\infty} \frac{-n!/2^{n+1}}{n!}(x + 2)^n = -\sum_{n=0}^{\infty} \frac{(x + 2)^n}{2^{n+1}}$$

The interval of convergence of this Taylor series is determined by the ratio test:

$$\left| \frac{a_{n+1}}{a_n} \right| = \left| \frac{(x + 2)^{n+1}/2^{n+2}}{(x + 2)^n/2^{n+1}} \right| = \left| \frac{(x + 2)^{n+1}}{2^{n+2}} \cdot \frac{2^{n+1}}{(x + 2)^n} \right| = \frac{1}{2}|x + 2|$$

In order to have convergence, we need the limit of this expression to be less than 1. But when we compute this limit, we get:

$$L = \lim_{n \to \infty} \frac{1}{2}|x + 2| = \frac{1}{2}|x + 2|$$

We get $L < 1$ when $|x + 2| < 2$ or $-4 < x < 0$. The function is not defined at $x = 0$ so the series cannot converge to a value there. At $x = -4$, the series becomes

$$-\sum_{n=0}^{\infty} \frac{(-4 + 2)^n}{2^{n+1}} = -\sum_{n=0}^{\infty} \frac{(-2)^n}{2^{n+1}} = \sum_{n=0}^{\infty} \frac{(-1)^{n+1}}{2}$$

This will diverge according the alternating series test, and so $x = -4$ is not in the interval of convergence. The interval of convergence remains $-4 < x < 0$, or $(-4, 0)$. ∎

A.6   Chapter 12: Practice Problem Solutions

A.6.1   *Intro to Param. Eqs. — Practice — Solved*

(1) Graph the parametric curve $x = \sqrt{t}, y = 1 - t$ for $0 \leq t \leq 4$. Eliminate the parameter to give the curve's $y = f(x)$ form.

☐ Let's pick some sample points:

t	0	1	2	3	4
x	0	1	$\sqrt{2}$	$\sqrt{3}$	2
y	1	0	-1	-2	-3

When we connect them, we see a portion of an an inverted parabola from $(0, 1)$ to $(2, -3)$. Solving the first equation for t, we get $t = x^2$. Plugging this into the second equation, we get $y = 1 - x^2$, which is an inverted parabola with vertex $(0, 1)$.   ■

(2) Eliminate the parameter to give the curve's $y = f(x)$ form for:

$$\begin{cases} x = -t^3 + 4 \\ y = \sqrt{t^6 + 1} \end{cases} \quad -\infty < t < \infty$$

☐ If we rearrange $x(t)$, we get $t^3 = 4 - x$. Now we can plug that into $y(t)$:

$$y = \sqrt{t^9 + 1} = \sqrt{(t^3)^2 + 1} = \sqrt{(4 - x)^2 + 1} = \sqrt{(16 - 8x + x^2) + 1}$$

So the curve is $y = \sqrt{x^2 - 8x + 17}$ ■

(3) Identify the curve $x = \sec\theta, y = \tan\theta$ for $-\pi/2 < \theta < \pi/2$ and state its orientation.

☐ If we square both equations, we get $x^2 = \sec^2\theta$ and $y^2 = \tan^2\theta$. Do you remember the trig identity $\sec^2\theta - \tan^2\theta = 1$? I hope so, because we used it a lot when doing trigonometric integrals! With our squared equations, this identity tells us we have the curve $x^2 - y^2 = 1$ — which is a hyperbola. As t increases from $-\pi/2$ to $\pi/2$, we traverse the right half of the hyperbola from the lower right, through the point $(1,0)$ at $t = 0$ and then to the upper right. Therefore, it's the right half of the unit hyperbola oriented clockwise.   ■

---

(4) Eliminate the parameter to identity the curve $x = 2\sinh(t)$, $y = 2\cosh(t)$ for $-\infty < t < \infty$.

□ We should plan to use the identity $\cosh^2(t) - \sinh(t) = 1$. So, let's rewrite the two parametric equations as

$$\frac{x}{2} = \sinh(t) \qquad \text{and} \qquad \frac{y}{2} = \cosh(t)$$

Then the identity gives us $(y/2)^2 - (x/2)^2 = 1$, or $y^2 - x^2 = 4$. This is a hyperbola. There are two branches: one with vertex $(0, 2)$ and opening upwards, the other with vertex $(0, -2)$ and downwards.   ■

A.6.2   Calc. of Param. Eqs. — Practice — Solved

(1) Find the equation of the line tangent to $x = e^{\sqrt{t}}; y = t - \ln t^2$ at $t = 1$.

☐ For the slope of the tangent line,

$$\frac{dy}{dx} = \frac{dy/dt}{dx/dt} = \frac{1 - 2/t}{e^{\sqrt{t}}/(2\sqrt{t})}$$

So at $t = 1$, we hit the point $(e, 1)$ and have a tangent slope of

$$m_{tan} = \frac{1 - 2}{e/2} = -\frac{2}{e}$$

and the equation of the tangent line is

$$y - (1) = -\frac{2}{e}(x - e)$$

$$y - 1 = -\frac{2}{e}x + 2$$

$$y = -\frac{2}{e}x + 3 \quad \blacksquare$$

---

(2) Find the locations of all horizontal and vertical tangent lines on the curve $x = 2t^3 + 3t^2 - 12t; y = 2t^3 + 3t^2 + 1$.

☐ Slopes of tangent lines are given by

$$\frac{dy}{dx} = \frac{dy/dt}{dx/dt} = \frac{6t^2 + 6t}{6t^2 + 6t - 12} = \frac{t^2 + t}{t^2 + t - 2} = \frac{t(t + 1)}{(t + 2)(t - 1)}$$

Tangent lines are horizontal when $dy/dx = 0$, i.e. when $t = -1, 0$. When $t = -1$, the point is $(13, 2)$; when $t = 0$, the point is $(0, 1)$.

Tangent lines are vertical when dy/dx is undefined, i.e. at $t = -2, 1$, which are the points $(20, -3)$ and $(-7, 6)$. ◼

---

(3) Find the point where the curve $x = \cos t; y = \sin t \cos t$ has more than one tangent line, and find the slopes of those tangent lines.

☐ This curve will have multiple tangent lines at any criss-cross point. The curve crosses itself where both x and y coordinates are the same, i.e. where $\cos t = \sin t \cos t$. This happens whenever $\cos t = 0$, and so there are intersections whenever t is an odd multiple of $\pi/2$. But these t values always give the point $(0,0)$. Thus, this curve intersects itself at $(0,0)$ and so has multiple tangents there. The number of tangents depends on the number of different tangent slopes we can find.

Slopes of tangent lines are given by

$$\frac{dy}{dx} = \frac{dy/dt}{dx/dt} = \frac{2\cos^2 t - 1}{-\sin t}$$

The numerator of this will always be the same, -1, for any of the above values of t. But the denominator could be either of ± 1. So tangent lines at $(0,0)$ have slopes of ± 1 and equations of $y = \pm x$.   ■

---

(4) Find the area bounded by the curve $x = t - 1/t;\, y = t + 1/t$ and the line $y = 2.5$.

□ First, we find where this curve intersects $y = 5/2$:

$i)$ $t + \dfrac{1}{t} = \dfrac{5}{2}$

$ii)$ $t^2 + 1 = \dfrac{5t}{2}$ $iv)$ $(t-2)\left(t - \dfrac{1}{2}\right) = 0$

$iii)$ $t^2 - \dfrac{5t}{2} + 1 = 0$ $v)$ $t = \dfrac{1}{2}, 2$

These endpoints are at $x = -3/2$ and $x = 3/2$. The region in question is above the parametric curve but below $y = 5/2$, so the area is

$$A = \int_{-3/2}^{3/2} \frac{5}{2}\, dx - \int_{1/2}^{2} y(t)x'(t)\, dt$$

$$= \int_{-3/2}^{3/2} \frac{5}{2}\, dx - \int_{1/2}^{2} (t + \frac{1}{t})(1 + \frac{1}{t^2})\, dt$$

$$= \frac{15}{2} - \left(4\ln 2 + \frac{15}{4}\right) = \frac{15}{4} - 4\ln 2 \quad ■$$

---

(5) Find the arc length of the curve $x = \cos t;\, y = \sin t$ for $0 \le t \le \pi$.

□ By Useful Fact 12.3,

$$L = \int_a^b \sqrt{[x'(t)]^2 + [y'(t)]^2}\, dt = \int_0^\pi \sqrt{(-\sin t)^2 + (\cos t)^2}\, dt$$

$$= \int_0^\pi \sqrt{1}\, dt = \pi$$

(This result should not be a surprise, since the curve is the upper half of the unit circle, and the arc length of that curve is half the circumference of the unit circle.)   ■

A.6.3 *Intro to Polar Coords.* — *Practice* — *Solved*

(1) Describe the region given by the polar bounds $2 < r \le 5$, $3\pi/4 < \theta < 5\pi/4$.

☐ This is a cheese-wedge shaped region in between circles of radius 2 and 5, starting at an angle of $\pi/4$ above the negative x-axis and sweeping counterclockwise to an angle of $\pi/4$ below the negative x-axis. The edge of the inner circle is not included and neither are the sides in QII, QIII. (See Fig. A.14.) ∎

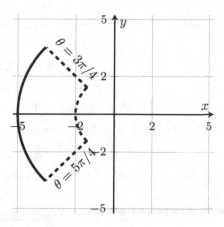

Fig. A.14　Polar wedge $2 < r \le 5$, $\dfrac{3\pi}{4} < \theta < \dfrac{5\pi}{4}$.

(2) Convert these points from polar to rectangular coordinates: (a) $(2, 2\pi/3)$, (b) $(4, 3\pi)$, (c) $(-2, -5\pi/6)$.

☐ (a) The Cartesian coordinates of $(2, 2\pi/3)$ are $(x, y) = (-1, \sqrt{3})$, because

$$x = r\cos\theta = 2\cos\frac{2\pi}{3} = -1$$

$$y = r\sin\theta = 2\sin\frac{2\pi}{3} = \sqrt{3}$$

(b) The Cartesian coordinates of $(4, 3\pi)$ are $(-4, 0)$. You should be able to tell that without using conversion equations.

(c) The Cartesian coordinates of $(-2, -5\pi/6)$ are $(x, y) = (\sqrt{3}, 1)$

because

$$x = r\cos\theta = -2\cos\frac{-5\pi}{6} = \sqrt{3}$$

$$y = r\sin\theta = -2\sin\frac{-5\pi}{6} = 1 \quad \blacksquare$$

---

(3) Convert $r = 2\sin\theta + 2\cos\theta$ to Cartesian form and identify it.

☐ Using the conversion equations in Useful Fact 12.4, we can replace $\sin\theta$ by y/r and s $\cos\theta$ by x/r, so

$$r = 2\sin\theta + 2\cos\theta = 2\cdot\frac{y}{r} + 2\frac{x}{r}$$

Multiplying both sides by r, then rearranging and completing the square gives:

$$r^2 = 2y + 2x$$
$$x^2 + y^2 = 2y + 2x$$
$$x^2 + y^2 - 2x - 2y = 0$$
$$x^2 - 2x + 1 + y^2 - 2y + 1 = 1 + 1$$
$$(x-1)^2 + (y-1)^2 = 2$$

So, our curve is a circle with center $(1,1)$ and radius $\sqrt{2}$.   ■

---

(4) Convert $x^2 + y^2 = 9$ to polar form.

☐ The circle $x^2 + y^2 = 9$ is the same as $r^2 = 9$, or $r = 3$, in polar form.
        ■

A.6.4 *Calc. of Polar Coords. — Practice — Solved*

(1) Find the slope of the line tangent to $r = 2 - \sin\theta$ at $\theta = \pi/3$.

□ Using Useful Fact 12.5,

$$
\begin{aligned}
\frac{dy}{dx} &= \frac{\frac{dr}{d\theta}\sin\theta + r\cos\theta}{\frac{dr}{d\theta}\cos\theta - r\sin\theta} \\
&= \frac{-\cos\theta\sin\theta + (2 - \sin\theta)\cos\theta}{-\cos\theta\cos\theta - (2 - \sin\theta)\sin\theta} \\
&= \frac{-2\cos\theta\sin\theta + 2\cos\theta}{-\cos^2\theta - 2\sin\theta + \sin^2\theta} \\
&= \frac{2\cos\theta - \sin 2\theta}{-2\sin\theta - \cos 2\theta}
\end{aligned}
$$

So when $\theta = \pi/3$,

$$
\frac{dy}{dx} = \frac{2\cos\frac{\pi}{3} - \sin\frac{2\pi}{3}}{-2\sin\frac{\pi}{3} - \cos\frac{2\pi}{3}} = \frac{1}{11}\left(4 - 3\sqrt{3}\right) \approx -0.1
$$

This polar curve is shown in Fig. A.15. At the point on the curve located by $\theta = \pi/3$, a slope of -0.1 (slightly downhill, almost flat) is reasonable. ∎

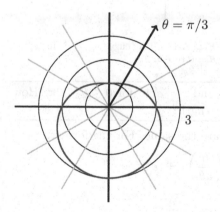

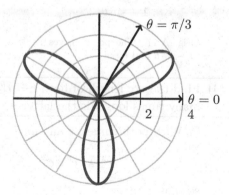

Fig. A.15 The polar curve $r = 2 - \sin\theta$.

Fig. A.16 $r = 4\sin 3\theta$ and bounding θ values for one loop.

(2) Find the area inside one loop of $r = 4\sin 3\theta$.

□ This polar curve is shown in Fig. A.16. One loop is delineated by two consecutive θ values which produce $r = 0$. Here, $\theta = 0$ and $\theta = \pi/3$

are the first two values which yield $r = 0$. So by Useful Fact 12.6,

$$A = \int_a^b \frac{1}{2} r^2 \, d\theta = \int_0^{\pi/3} \frac{1}{2} (4 \sin 3\theta)^2 \, d\theta = \frac{4\pi}{3} \quad \blacksquare$$

(3) Find the area inside $r = 1 - \sin \theta$ and outside $r = 1$.

☐ These curves are shown in Fig. A.17. We must find the θ values which mark the intersections. But $1 - \sin \theta = 1$ when $\sin \theta = 0$, so we have intersections when $\theta = 0, \pi$. But looking at the graphs, π and 2π are better bounds (left to right). So,

$$A \int_a^b \frac{1}{2} [(r_{out})^2 - (r_{in})^2] \, d\theta = \int_\pi^{2\pi} \frac{1}{2} \left[(1 - \sin \theta)^2 - (1)^2 \right] \, d\theta = \frac{\pi}{4} + 2 \quad \blacksquare$$

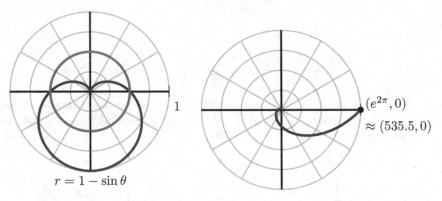

$r = 1 - \sin \theta$

1

$(e^{2\pi}, 0)$

$\approx (535.5, 0)$

Fig. A.17   The polar curves $r = 1 - \sin \theta$ and $r = 1$.

Fig. A.18   The curve $r = e^\theta$ for $\theta = 0$ to $\theta = 2\pi$.

(4) Find the arc length of both $r = e^\theta$ and $r = e^{\theta+1}$ for $0 \le \theta \le 2\pi$. How many times longer is the latter arc length?

☐ With both curves, r and $dr/d\theta$ are the same. For the former,

$$r^2 = (e^\theta)^2 e^{2\theta} \quad ; \quad \left(\frac{dr}{d\theta} \right)^2 = (e^\theta)^2 = e^{2\theta}$$

so that

$$\sqrt{r^2 + \left(\frac{dr}{d\theta} \right)^2} = \sqrt{2e^{2\theta}} = \sqrt{2} e^\theta$$

and by Useful Fact 12.7,

$$L_1 = \int_a^b \sqrt{r^2 + \left(\frac{dr}{d\theta} \right)^2} \, d\theta = \int_0^{2\pi} \sqrt{2} e^\theta \, d\theta = \sqrt{2} \left(e^{2\pi} - 1 \right)$$

For the second curve, $e^{\theta+1} = e^1 e^{\theta}$, so

$$r^2 = e^2 e^{2\theta} \qquad ; \qquad \left(\frac{dr}{d\theta}\right)^2 = e^2 e^{2\theta}$$

so that

$$\sqrt{r^2 + \left(\frac{dr}{d\theta}\right)^2} = \sqrt{2e^2 e^{2\theta}} = \sqrt{2} e^1 e^{\theta}$$

and

$$L_2 = \int_a^b \sqrt{r^2 + \left(\frac{dr}{d\theta}\right)^2} \, d\theta = \int_0^{2\pi} \sqrt{2} e^1 e^{\theta} \, d\theta = \sqrt{2} e \left(e^{2\pi} - 1\right) d\theta$$

Note that $L_2/L_1 = e$, so the second curve is longer than the first by a multiple of e. The pattern is evident; the arc length of $e^{\theta+a}$ would be e^a times longer than e^{θ}.

Figure A.18 shows one of these curves. Note that to see the terminal point of the curve, we need to zoom out so far, we cannot see the full spiral starting from $\theta = 0$ in the center of the diagram! ∎

Appendix B

Solutions to All Challenge Problems

B.1 Chapter 7: Challenge Problem Solutions

B.1.1 *Int. by Parts — Challenge — Solved*

(1) Find $\int e^{2x} \sin 3x \, dx$.

□ The initial assignment of parts can go either way, so let's try

$$\begin{array}{llll} \text{choose} & \rightarrow & f(x) = \sin 3x & g'(x) = e^{2x} \\ \text{compute} & \rightarrow & f'(x) = 3\cos 3x & g(x) = \frac{1}{2}e^{2x} \end{array}$$

so that with (7.2),

$$\int e^{2x} \sin 3x \, dx = \frac{1}{2}e^{2x}\sin 3x - \frac{3}{2}\int e^{2x}\cos 3x dx$$

Using integration by parts again for the new integral,

$$\begin{array}{llll} \text{choose} & \rightarrow & f(x) = \cos 3x & g'(x) = e^{2x} \\ \text{compute} & \rightarrow & f'(x) = -3\sin 3x & g(x) = \frac{1}{2}e^{2x} \end{array}$$

so that (ignoring the $+C$ for the moment),

$$\int e^{2x} \cos 3x \, dx = \frac{1}{2}e^{2x}\cos 3x + \frac{3}{2}\int e^{2x}\sin 3x \, dx$$

and going back to the original integral,

$$\int e^{2x} \sin 3x \, dx = \frac{1}{2} e^{2x} \sin 3x - \frac{3}{2} \left(\frac{1}{2} e^{2x} \cos 3x + \frac{3}{2} \int e^{2x} \sin 3x \, dx \right)$$

$$\frac{13}{4} \int e^{2x} \sin 3x \, dx = \frac{1}{2} e^{2x} \sin 3x - \frac{3}{4} e^{2x} \cos 3x + C$$

$$\int e^{2x} \sin 3x \, dx = \frac{4}{13} \left(\frac{1}{2} e^{2x} \sin 3x - \frac{3}{4} e^{2x} \cos 3x + C \right)$$

$$= \frac{2}{13} e^{2x} \sin 3x - \frac{3}{13} e^{2x} \cos 3x + C \quad \blacksquare$$

---

(2) Evaluate $\int_0^{1/2} \cos^{-1} x \, dx$.

☐ We'll get the general antiderivative first, then apply the limits at the end. There are not two parts to choose from, so

$$\begin{array}{llll} \text{choose} & \to & f(x) = \cos^{-1}(x) & g'(x) = 1 \\ \text{compute} & \to & f'(x) = -\frac{1}{\sqrt{1-x^2}} & g(x) = x \end{array}$$

Then, using (7.2),

$$\int \cos^{-1} x \, dx = x \cos^{-1} x + \int \frac{x}{\sqrt{1-x^2}} \, dx$$

For the new integral, consider the substitution $u = 1 - x^2$, then $x dx = -du/2$, so we have (not worrying about the $+C$ since this will be part of a definite integral),

$$\int \frac{x}{\sqrt{1-x^2}} \, dx = -\frac{1}{2} \int \frac{1}{\sqrt{u}} \, du = -\frac{1}{2} \left(2\sqrt{u} \right) + C = -\sqrt{1-x^2} + C$$

Finally, then, collecting the antiderivative and applying limits of integration,

$$\int_0^{1/2} \cos^{-1} x \, dx = \left(x \cos^{-1} x - \sqrt{1-x^2} \right) \Big|_0^{1/2}$$

$$= \frac{1}{2} \cos^{-1} \frac{1}{2} - \sqrt{1 - (1/2)^2} - (-1)$$

$$= \frac{1}{2} \cdot \frac{\pi}{3} - \sqrt{3/4} + 1 = \frac{\pi}{6} - \frac{\sqrt{3}}{2} + 1 \quad \blacksquare$$

(3) A reduction formula is a formula that reduces the complexity of an integral by reducing the power on one of its terms. Use integration by parts to derive the reduction formula

$$\int \cos^n x \, dx = \frac{1}{n} \cos^{n-1} x \sin x + \frac{n-1}{n} \int \cos^{n-2} x \, dx$$

□ choose → $f(x) = \cos^{n-1} x$ $\qquad\qquad\qquad\qquad$ $g'(x) = \cos x \, dx$

$\quad$ compute → $f'(x) = -(n-1) \cos^{n-2} x \sin x$ $\qquad$ $g(x) = \sin x$

Then, using (7.2),

$$\int \cos^n x \, dx = \sin x \cos^{n-1} x + (n-1) \int \sin^2 x \cos^{n-2} x \, dx$$

$$= \sin x \cos^{n-1} x + (n-1) \int (1 - \cos^2 x) \cos^{n-2} x \, dx$$

$$= \sin x \cos^{n-1} x + (n-1) \int \cos^{n-2} x \, dx - (n-1) \int \cos^n x \, dx$$

Combining both appearances of $\int \cos^n x \, dx$ on the left hand side,

$$(1 + (n-1)) \int \cos^n x \, dx = \sin x \cos^{n-1} x + (n-1) \int \cos^{n-2} x \, dx$$

$$n \int \cos^n x \, dx = \sin x \cos^{n-1} x + (n-1) \int \cos^{n-2} x \, dx$$

$$\int \cos^n x \, dx = \frac{1}{n} \sin x \cos^{n-1} x + \frac{n-1}{n} \int \cos^{n-2} x \, dx$$

which is the reduction formula given. $\qquad\qquad\qquad\qquad\qquad$ ■

B.1.2   *Int. by Partial Fracs. — Challenge — Solved*

---

(1) Find $\displaystyle\int \frac{1}{x^2(x-1)^2}\,dx$.

☐ First, let's disassemble the integrand. Remember to account for ALL possible denominators which could lead to the given one as a least common denominator. This leads to four possible denominators:

$$\frac{1}{x^2(x-1)^2} = \frac{A}{x} + \frac{B}{x^2} + \frac{C}{x-1} + \frac{D}{(x-1)^2} \qquad (B.1)$$

And so we already know the overall outcome:

$$\int \frac{1}{x^2(x-1)^2}\,dx = A\ln|x| - \frac{B}{x} + C\ln|x-1| - \frac{D}{x-1} + E$$

To find A, B, C, D, we multiply (B.1) by the denominator $x^2(x-1)^2$ to find,

$$1 = Ax(x-1)^2 + B(x-1)^2 + Cx^2(x-1) + Dx^2$$

If we select $x = 0$, we find $B = 1$.

If we select $x = 1$, we find $D = 1$.

This ends our list of really clever values of x, so let's now select ... how about $x = -1$ and $x = 2$? Why not. Neither one will yield A or C immediately, but these produce two equations in two unknowns (oh no!! algebra!!): $x = -1$ gives (when simplified) $2A + C = 2$ and $x = 2$ gives (when simplified) $A + 2C = -2$. Practice your algebra to get $A = 2$ and $C = -2$. Along with $B = D = 1$, we now have the true result:

$$\int \frac{1}{x^2(x-1)^2}\,dx = 2\ln|x| - \frac{1}{x} - 2\ln|x-1| - \frac{1}{x-1} + E$$

$$= 2\ln\frac{|x|}{|x-1|} - \frac{2x-1}{x(x-1)} + E \quad \blacksquare$$

(2) Find $\displaystyle\int \frac{x^2 - 2x - 1}{(x-1)^2(x^2+1)}\, dx.$

☐ First, let's disassemble the integrand. Remember to account for ALL possible denominators which could lead to the given one as a least common denominator. This leads two four possible denominators:

$$\frac{x^2 - 2x - 1}{(x-1)^2(x^2+1)} = \frac{A}{x-1} + \frac{B}{(x-1)^2} + \frac{Cx+D}{x^2+1} \qquad \text{(B.2)}$$

so

$$\int \frac{x^2 - 2x - 1}{(x-1)^2(x^2+1)}\, dx$$

$$= \int \frac{A}{x-1} + \frac{B}{(x-1)^2} + \frac{Cx}{x^2+1} + \frac{D}{x^2+1} + E$$

$$= A \ln|x-1| - \frac{B}{x-1} + \frac{C}{2}\ln(x^2+1) + D\tan^{-1}(x) + E$$

To find A, B, C, D, we can multiply (B.2) by the denominator $(x-1)^2(x^2+1)$ to get:

$$x^2 - 2x - 1 = A(x-1)(x^2+1) + B(x^2+1) + Cx(x-1)^2 + D(x-1)^2$$

If we select $x = 1$, we find $B = -1$.

If we select $x = 0$, we find $A = D$.

If we select $x = -1$, we get $1 = -A - C + D$, and since $A = D$, this means $C = -1$.

If we select $x = 2$, we get $4 = 5A - 2 + D$, and since $A = D$, this means $A = D = 1$.

So overall, with $A = 1, B = -1, C = -1, D = 1$, we have:

$$\int \frac{x^2 - 2x - 1}{(x-1)^2(x^2+1)}\, dx$$

$$= A \ln|x-1| - \frac{B}{x-1} + \frac{C}{2}\ln(x^2+1) + D\tan^{-1}(x) + E$$

$$= \ln|x-1| + \frac{1}{x-1} - \frac{1}{2}\ln(x^2+1) + \tan^{-1}(x) + E \quad \blacksquare$$

---

(3) Find $\displaystyle\int \frac{e^{2x}}{e^{2x}+3e^x+2}\,dx$.

□ This problem requires the substitution $u = e^x$, which gives $du = e^x\,dx$. Note that in the numerator, we can write $e^{2x}\,dx = e^x \cdot e^x\,dx$, so that

$$\int \frac{e^{2x}}{e^{2x}+3e^x+2}\,dx = \int \frac{u}{u^2+3u+2}\,du = \int \frac{u}{(u+2)(u+1)}\,du$$

The partial fraction decomposition of the integrand is pretty simple,

$$\frac{u}{(u+2)(u+1)} = \frac{A}{u+2} + \frac{B}{u+1}$$

We might as well find A and B right away. Multiplying by the full denominator, we get

$$u = A(u+1) + B(u+2)$$

Selecting $u = -1$ gives $B = -1$ and $u = -2$ gives $A = 2$. So,

$$\int \frac{u}{(u+2)(u+1)}\,du = \int \left(\frac{2}{u+2} - \frac{1}{u+1} \right)\,du$$
$$= 2\ln|u+2| - \ln|u+1| + C$$

Reversing the substitution,

$$\int \frac{e^{2x}}{e^{2x}+3e^x+2}\,dx = 2\ln|e^x+2| - \ln|e^x+1| + C$$
$$= \ln\frac{(e^x+2)^2}{e^x+1} + C \quad \blacksquare$$

B.1.3 *Trig. Integrals — Challenge — Solved*

---

(1) Evaluate $\displaystyle\int_0^{\pi/2} \cos^5 x \, dx$.

☐ Let's set aside one $\cos x$ to go with dx and convert the rest:
$$\int_0^{\pi/2} \cos^5 x \, dx = \int_0^{\pi/2} \cos^4 x \cos x \, dx = \int_0^{\pi/2} (1 - \sin^2 x)^2 \cos x \, dx \ldots$$
which sets up the substitution
$$\sin x = u \quad ; \quad \cos x \, dx = du$$
giving
$$\ldots = \int_0^1 (1 - u^2)^2 \, du = \int_0^1 (1 - 2u^2 + u^4) \, du = \left(u - \frac{2}{3}u^3 + \frac{1}{5}u^5 \right)\Big|_0^1$$
$$= 1 - \frac{2}{3} + \frac{1}{5} = \frac{8}{15} \quad \blacksquare$$

---

(2) Find $\displaystyle\int x \cos^2 x \, dx$.

☐ We must start this one with integration by parts; we identify and compute
$$f(x) = x \qquad g'(x) = \cos^2 x$$
$$f'(x) = 1 \qquad g(x) = \tfrac{1}{2}x + \tfrac{1}{4}\sin 2x$$
(The antiderivative of $\cos^2 x$ was found in You Try It 2 of this section.)
With these parts,
$$\int x \cos^2 x \, dx = \frac{1}{2}x^2 + \frac{1}{4}x \sin 2x - \int \left(\frac{1}{2}x + \frac{1}{4}\sin 2x \right) dx$$
$$= \frac{1}{2}x^2 + \frac{1}{4}x \sin 2x - \frac{1}{4}x^2 + \frac{1}{8}\cos 2x + C$$
$$= \frac{1}{4}x^2 + \frac{1}{4}x \sin 2x + \frac{1}{8}\cos 2x + C \quad \blacksquare$$

---

(3) Find $\displaystyle\int \tan^3(2x) \sec^5(2x) \, dx$.

☐ We start with a quick $2x = u$ so that $dx = du/2$:
$$\int \tan^3(2x) \sec^5(2x) \, dx = \frac{1}{2} \int \tan^3(u) \sec^5(u) \, du$$
$$= \frac{1}{2} \int \tan^2 u \sec^4 u \sec u \tan u \, du$$
$$= \frac{1}{2} \int (\sec^2 u - 1) \sec^4 u \sec u \tan u \, du$$

which then needs another substitution that we'll call w, since u is already taken:

$$\sec u = w \quad ; \quad \sec u \tan u \, du = dw$$

so carrying on,

$$\frac{1}{2} \int (\sec^2 u - 1) \sec^4 u \sec u \tan u \, du = \frac{1}{2} \int (w^2 - 1) w^4 \, dw$$

$$= \frac{1}{2} \int (w^6 - w^4) \, dw = \frac{1}{14} w^7 - \frac{1}{10} w^5 + C$$

$$= \frac{1}{14} \sec^7 u - \frac{1}{10} \sec^5 u + C$$

$$= \frac{1}{14} \sec^7 (2x) - \frac{1}{10} \sec^5 (2x) + C \quad \blacksquare$$

B.1.4   *Trig. Substitution — Challenge — Solved*

(1) Find $\displaystyle\int \frac{x^5}{\sqrt{x^2+2}}dx$.

☐ The square root term here matches the form $\sqrt{a^2+x^2}$ with $a = \sqrt{2}$, so we assign

$$x = \sqrt{2}\tan\theta \qquad ; \qquad dx = \sqrt{2}\sec^2\theta\,d\theta$$

and then,

$$\int \frac{x^5}{\sqrt{x^2+2}}\,dx = \int \frac{4\sqrt{2}\tan^5\theta}{\sqrt{2\tan^2\theta+2}}\cdot\sqrt{2}\sec^2\theta\,d\theta$$

$$= \int \frac{4\sqrt{2}\tan^5\theta}{\sqrt{2}\sec\theta}\cdot\sqrt{2}\sec^2\theta\,d\theta$$

$$= 4\sqrt{2}\int \tan^5\theta\sec\theta\,d\theta$$

$$= 4\sqrt{2}\int \tan^4\theta\sec\theta\tan\theta\,d\theta$$

$$= 4\sqrt{2}\int (\sec^2\theta-1)^2\sec\theta\tan\theta\,d\theta\ldots$$

We pause here for a "regular" substitution,

$$u = \sec\theta \qquad ; \qquad du = \sec\theta\tan\theta d\theta$$

and carry on,

$$\cdots = 4\sqrt{2}\int (u^2-1)^2\,du = 4\sqrt{2}\int (u^4-2u^2+1)\,du$$

$$= 4\sqrt{2}\left(\frac{1}{5}u^5 - \frac{2}{3}u^3 + u\right) + C$$

$$= 4\sqrt{2}\left(\frac{1}{5}\sec^5\theta - \frac{2}{3}\sec^3\theta + \sec\theta\right) + C$$

Now we're ready to reverse the substitution. With the original, $\tan\theta = \frac{x}{\sqrt{2}}$, we get via a right triangle (not shown),

$$\sec\theta = \frac{\sqrt{x^2+2}}{\sqrt{2}}$$

so that

$$\int \frac{x^5}{\sqrt{x^2+2}}\,dx$$

$$= 4\sqrt{2}\left[\frac{1}{5}\left(\frac{\sqrt{x^2+2}}{\sqrt{2}}\right)^5 - \frac{2}{3}\left(\frac{\sqrt{x^2+2}}{\sqrt{2}}\right)^3 + \left(\frac{\sqrt{x^2+2}}{\sqrt{2}}\right)\right] + C$$

$$= 4\sqrt{x^2+2}\left[\frac{1}{5}\left(\frac{(x^2+2)^2}{4}\right) - \frac{2}{3}\left(\frac{x^2+2}{2}\right) + 1\right] + C$$

$$= \sqrt{x^2+2}\left[\frac{1}{5}(x^2+2)^2 - \frac{4}{3}(x^2+2) + 4\right] + C \quad \blacksquare$$

---

(2) Find $\displaystyle\int \frac{x}{\sqrt{25-x^2}}\,dx$.

□ The square root term here matches the form $\sqrt{a^2-x^2}$ with $a = 5$, so we assign

$$x = 5\sin\theta \quad ; \quad dx = 5\cos\theta\,d\theta$$

and begin:

$$\int \frac{x}{\sqrt{25-x^2}}\,dx = \int \frac{5\sin\theta}{\sqrt{25-25\sin^2\theta}}\cdot 5\cos\theta\,d\theta$$

$$= \int \frac{5\sin\theta}{5\cos\theta}\cdot 5\cos\theta\,d\theta$$

$$= 5\int \sin\theta\,d\theta = -5\cos\theta + C$$

From the original substitution $\sin\theta = x/5$, we find via right triangle (not shown)

$$\cos\theta = \frac{\sqrt{25-x^2}}{5}$$

and thus,

$$\int \frac{x}{\sqrt{25-x^2}}\,dx = -5\cdot\frac{\sqrt{25-x^2}}{5} + C = -\sqrt{25-x^2} + C$$

Don't you wish you had used integration by substitution instead? $\quad\blacksquare$

(3) Evaluate $\displaystyle\int_0^1 \sqrt{x^2 + 1}\, dx$.

☐ The square root term here matches the form $\sqrt{a^2 + x^2}$ with $a = 1$, so we assign

$$x = \tan\theta \quad ; \quad dx = \sec^2\theta\, d\theta$$

and change limits,

$$x = 0 \to \theta = 0$$
$$x = 1 \to \theta = \tan^{-1}(1) = \frac{\pi}{4}$$

so the integral becomes

$$\int_0^1 \sqrt{x^2 + 1}\, dx = \int_0^{\pi/4} \sqrt{\tan^2\theta + 1} \cdot \sec^2\theta\, d\theta = \int_0^{\pi/4} \sec^3\theta\, d\theta$$

$$= \frac{1}{2}\left(\sec\theta\tan\theta + \ln|\sec\theta + \tan\theta|\right)\Big|_0^{\pi/4}$$

$$= \frac{1}{2}\left(\sec\frac{\pi}{4}\tan\frac{\pi}{4} + \ln\left|\sec\frac{\pi}{4} + \tan\frac{\pi}{4}\right|\right)$$

$$\quad - \frac{1}{2}\left(\sec 0\tan 0 + \ln|\sec 0 + \tan 0|\right)$$

$$= \frac{1}{2}\left((\sqrt{2})(1) + \ln|\sqrt{2} + 1|\right) - \frac{1}{2}\left(0 + \ln|1 + 0|\right)$$

$$= \frac{1}{2}\left((\sqrt{2})(1) + \ln|\sqrt{2} + 1|\right) - \frac{1}{2}\left(0 + \ln|1 + 0|\right)$$

$$= \frac{1}{2}\left(\sqrt{2} + \ln(\sqrt{2} + 1)\right) \quad \blacksquare$$

B.1.5   *Improper Integrals — Challenge — Solved*

---

(1) Evaluate $\displaystyle\int_0^\infty f(x)\,dx$, where $f(x) = \dfrac{1}{x^2 + x - 6}$.

The integrand can be written as

$$f(x) = \frac{1}{(x+3)(x-2)}$$

It has two discontinuities, and one of them ($x = 2$) is within the interval of integration. Plus, we have the infinite endpoint. Let's get to chopping up the overall interval of integration to isolate these trouble spots, one per sub-integral. A first cut at the discontinuity $x = 2$ gives subintervals $[0, 2)$ and $(2, \infty)$. The latter has two bad endpoints, so we still need another cut. I will choose to make my extra cut at $x = 4$.

$$\int_0^\infty f(x)\,dx = \int_0^{(2)} f(x)\,dx + \int_{(2)}^4 f(x)\,dx + \int_4^\infty f(x)\,dx$$

$$= \lim_{c \to 2^-} \int_0^c f(x)\,dx + \lim_{d \to 2^-} \int_d^4 f(x)\,dx + \lim_{s \to \infty} \int_4^s f(x)\,dx$$

Now just taking the first subintegral, we have

$$\lim_{c \to 2^-} \int_0^c \frac{1}{(x+3)(x-2)}\,dx$$

$$= \lim_{c \to 2^-} \frac{1}{5}\left(\ln|x - 2| - \ln|x + 3|\right)\Big|_0^c$$

$$= \lim_{c \to 2^-} \frac{1}{5}\left(\ln|c - 2| - \ln|c + 3|\right) - \frac{1}{5}\left(\ln 2 - \ln 3\right)$$

Since $\displaystyle\lim_{c \to 2^-} \frac{1}{5}\ln|c - 2|$ diverges, then this sub-integral (as well as the whole improper intergal) diverges as well. ∎

---

(2) Evaluate $\displaystyle\int_{-1}^\infty f(x)\,dx$ where $f(x) = \dfrac{e^{1/x}}{x^2}$.

□ First, note that the general antiderivative of $f(x)$ is easy to find as $-e^{1/x} + C$ (use $u = 1/x$ if needed). Of course, this only applied where the function and antiderivative are defined, which is a good reason to avoid the obvious discontinuity $x = 0$.

So let's start chopping up our interval of integration to isolate $x = 0$ and ∞ as endpoints (one per subintegral). If we first subdivide into $[-1, 0)$ and $(0, \infty)$, the latter interval has two bad endpoints, so we need to chop it again. How about at $x = 1$?

$$\int_{-1}^{\infty} f(x)\,dx = \int_{-1}^{(0)} f(x)\,dx + \int_{(0)}^{1} f(x)\,dx + \int_{1}^{\infty} f(x)\,dx$$

$$= \lim_{c_1 \to 0^-} \int_{-1}^{c_1} f(x)\,dx + \lim_{c_2 \to 0^+} \int_{c_2}^{1} f(x)\,dx$$

$$+ \lim_{c_3 \to \infty} \int_{1}^{c_3} f(x)\,dx$$

The first and third subintervals converge (try them!), but the second one diverges. Therefore the entire interval diverges. ∎

---

(3) If the area under $f(x) = \dfrac{\ln x}{x^2}$ on $[1, \infty)$ is finite, find the coordinate $x = A$ ($A > 1$) at which the area under $f(x)$ on $[1, A]$ is half of the total area over $[1, \infty)$. (You should use tech strategically.)

☐ First, let's see if the area under $f(x)$ is finite. The only endpoint of integration to worry about is ∞, but before we worry about that, let's find the general antiderivative of the function — because it's going to take integration by parts. So here we go. We assign

$$f(x) = \ln x \qquad \text{and} \qquad g'(x) = \frac{1}{x^2}$$

and then compute

$$f'(x) = \frac{1}{x} \qquad \text{and} \qquad g(x) = -\frac{1}{x}$$

so that

$$\int \frac{\ln x}{x^2}\,dx = -\frac{1}{x} \cdot \ln x - \int \frac{1}{x}\left(-\frac{1}{x}\right)\,dx$$

$$= -\frac{\ln x}{x} + \int \frac{1}{x^2}\,dx = -\frac{\ln x}{x} - \frac{1}{x} + C$$

$$= -\frac{1}{x}(\ln x + 1) + C$$

Choosing the antiderivative for which $C = 0$, we can implement this in the definite integral. (Can you spot the step where L-Hopital's Rule is used?)

$$\int_1^\infty \frac{\ln x}{x^2}\, dx = \lim_{c \to \infty} \int_1^c \frac{\ln x}{x^2}\, dx = \lim_{c \to \infty} \left(-\frac{1}{x}(\ln x + 1) \right) \Big|_1^c$$

$$= \lim_{c \to \infty} \left(-\frac{\ln c + 1}{c} + 1 \right) = \lim_{c \to \infty} \left(-\frac{1/c}{1} \right) + 1 = 1$$

Well, that's a nice number for the total area. To solve for where we get half of that total area, we look for A such that

$$\int_1^A \frac{\ln x}{x^2}\, dx = \frac{1}{2}$$

We can borrow the general antiderivative again,

$$\int_1^A \frac{\ln x}{x^2}\, dx = -\frac{1}{x}(\ln x + 1) \Big|_1^A$$

$$= -\frac{1}{A}(\ln A + 1) + \frac{1}{1}(\ln 1 + 1) = -\frac{1}{A}(\ln A + 1) + 1$$

Solving

$$-\frac{1}{A}(\ln A + 1) + 1 = \frac{1}{2}$$

is not something we can do by hand. Using tech, we can estimate the solution as about $A \approx 5.36$.   ∎

B.2 Chapter 8: Challenge Problem Solutions

B.2.1 *Area Betw. Curves — Challenge — Solved*

---

(1) Find the area between $y = 1/x$, $y = 1/x^2$ and $x = 2$.

☐ These curves intersect at $x = 1$. The line $y = 1/x$ is above $y = 1/x^2$ on the interval from $x = 1$ to $x = 2$. So, the area between them is:

$$A = \int_1^2 \left[\frac{1}{x} - \frac{1}{x^2} \right] dx = \left(\ln x + \frac{1}{x} \right) \Big|_1^2$$

$$= \left(\ln 2 + \frac{1}{2} - 1 \right) = \ln 2 - \frac{1}{2} \quad \blacksquare$$

---

(2) Find the net unsigned area between $y = 8 - x^2$, $y = x^2$, $x = -3$ and $x = 3$.

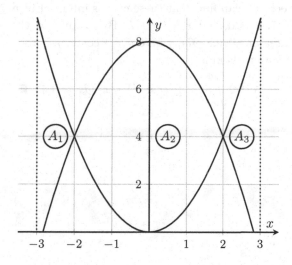

Fig. B.1   Using symmetry for areas between curves (with CP 2).

☐ These curves intersect at $x = -2$ and $x = 2$. Since limits of $x = -3$ and $x = 3$ are given in the problem, we must integrate farther out. The contributing areas are shown in Fig. B.1 as A_1, A_2, A_3. The curve $y = 8 - x^2$ is above $y = x^2$ for $-2 < x < 2$ but x^2 is on top for $x < -2$ and $x > 2$. Notice that by symmetry, we can write the total area

$$A_1 + A_2 + A_3 = 2 \left(\frac{1}{2} A_2 + A_3 \right)$$

So, the net unsigned area between the two curves is:

$$A = 2 \left(\int_0^2 \left[(8 - x^2) - (x^2) \right] dx + \int_2^3 \left[(x^2) - (8 - x^2) \right] dx \right)$$

$$= 2 \left(\int_0^2 (8 - 2x^2) dx + \int_2^3 (2x^2 - 8) dx \right)$$

$$= 2 \left[\left(8x - \frac{2}{3}x^3 \right) \Big|_0^2 + \left(\frac{2}{3}x^3 - 8x \right) \Big|_2^3 \right]$$

$$= 2 \left[\left(16 - \frac{16}{3} \right) + (18 - 24) - \left(\frac{16}{3} - 16 \right) \right]$$

$$= 2 \left(26 - \frac{32}{3} \right) = \frac{92}{3} \quad \blacksquare$$

---

(3) Find (approximately) the area between $y = x \cos(x^2)$ and $y = x^3$.

☐ Using tech, we can find that these curves intersect at approximately $x = -0.8597$, $x = 0$, and $x = 0.8957$. By symmetry, the total area is twice the area on $(0, 0.8597)$, where $y = x \cos(x^2)$ is above $y = x^3$. So, the area between them is:

$$A \approx 2 \int_0^{0.8957} (x \cos(x^2) - x^3) dx \approx 0.4 \quad \blacksquare$$

B.2.2 *Arc Length — Challenge — Solved*

(1) Find the arc length of $y = \ln x$ for $1 \le x \le \sqrt{3}$.

□ We'll do a little prep-work to get ready for (8.7) with $f(x) = \ln x$:

$$f'(x) = \frac{1}{x}$$

$$1 + [f'(x)]^2 = 1 + \frac{1}{x^2} = \frac{x^2 + 1}{x^2}$$

$$\sqrt{1 + [f'(x)]^2} = \sqrt{\frac{x^2 + 1}{x^2}} = \frac{\sqrt{x^2 + 1}}{x}$$

The arc length is then

$$L = \int_a^b \sqrt{1 + [f'(x)]^2}\, dx = \int_1^{\sqrt{3}} \frac{\sqrt{x^2 + 1}}{x}\, dx$$

which, sadly, looks like it requires trigonometric substitution, and seems to match the case posed in Useful Fact 7.7.

With $x = \tan\theta$ we have $dx = \sec^2\theta\, d\theta$ and new endpoints:

$$x = 1 \to \theta = \pi/4$$

$$x = \sqrt{3} \to \theta = \pi/3$$

and here we go:

$$L = \int_{\pi/4}^{\pi/3} \frac{\sqrt{\tan^2\theta + 1}}{\tan\theta} \sec^2\theta\, d\theta = \int_{\pi/4}^{\pi/3} \frac{\sec\theta}{\tan\theta} \sec^2\theta\, d\theta$$

$$= \int_{\pi/4}^{\pi/3} \frac{\sec\theta}{\tan\theta}(1 + \tan^2\theta)\, d\theta = \int_{\pi/4}^{\pi/3} \frac{\sec\theta}{\tan\theta} + \sec\theta\tan\theta\, d\theta$$

$$= \int_{\pi/4}^{\pi/3} \csc\theta + \sec\theta\tan\theta\, d\theta$$

The antiderivative of $\csc\theta$ was determined in Practice Problem 5 of Sec. 7.4, and the antiderivative of $\sec x \tan x$ should be well known by now. And so, continuing,

$$L = (-\ln|\csc\theta + \cot\theta| + \sec\theta)\Big|_{\pi/4}^{\pi/3}$$

$$= \left(-\ln\left|\csc\frac{\pi}{3} + \cot\frac{\pi}{3}\right| + \sec\frac{\pi}{3}\right) - \left(-\ln\left|\csc\frac{\pi}{4} + \cot\frac{\pi}{4}\right| + \sec\frac{\pi}{4}\right)$$

We need a whole bunch of trig values now, and here they are:

$$\csc\frac{\pi}{3} = \frac{2}{\sqrt{3}} \quad ; \quad \csc\frac{\pi}{4} = \sqrt{2}$$

$$\cot\frac{\pi}{3} = \frac{1}{\sqrt{3}} \quad ; \quad \cot\frac{\pi}{4} = 1$$

$$\sec\frac{\pi}{3} = 2 \quad ; \quad \sec\frac{\pi}{4} = \sqrt{2}$$

So altogether,

$$L = \left(-\ln\left|\frac{2}{\sqrt{3}} + \frac{1}{\sqrt{3}}\right| + 2\right) - \left(-\ln|\sqrt{2} + 1| + \sqrt{2}\right)$$

$$= -\ln\sqrt{3} + 2 + \ln|\sqrt{2} + 1| - \sqrt{2}$$

$$= 2 - \sqrt{2} - \frac{1}{2}\ln 3 + \ln|\sqrt{2} + 1|$$

Who would have thought that such an innocent sounding problem like "the arc length of $y = \ln x$ on $[1, \sqrt{3}]$" could have turned out to be so hideous! ∎

---

(2) Find the arc length of $y = x^2/2 + 6\pi$ for $0 \le x \le 1/\sqrt{3}$.

☐ A little prep work:

$$f(x) = x^2/2 + 6\pi$$

$$f'(x) = x$$

$$1 + [f'(x)]^2 = 1 + x^2$$

so that by (8.7),

$$L = \int_a^b \sqrt{1 + [f'(x)]^2}\, dx = \int_0^{1/\sqrt{3}} \sqrt{1 + x^2}\, dx$$

Gosh, another trig substitution problem! It's of the form indicated in Useful Fact 7.7 of Sec. 7.5), in which we will set

$$x = \tan\theta \quad ; \quad dx = \sec^2\theta\, d\theta$$

The endpoints follow as:

$$x = 0 \to \theta = 0$$

$$x = 1/\sqrt{3} \to \theta = \pi/6$$

So that

$$L = \int_0^{\pi/6} \sqrt{1 + \tan^2 \theta} \cdot \sec^2 \theta \, d\theta$$

$$L = \int_0^{\pi/6} \sec^3 \theta \, d\theta$$

$$= \left(\frac{1}{2} \sec \theta \tan \theta + \frac{1}{2} \ln |\sec \theta + \tan \theta| \right) \Big|_0^{\pi/6}$$

$$= \frac{1}{2} \left(\sec(\pi/6) \tan(\pi/6) + \ln |\sec(\pi/6) + \tan(\pi/6)| \right)$$

$$= \frac{1}{2} \left((2/\sqrt{3})(1/\sqrt{3}) + \ln |(2/\sqrt{3} + 1/\sqrt{3}| \right)$$

$$= \frac{1}{2} \left((\frac{2}{3} + \ln |\sqrt{3}| \right) \quad \blacksquare$$

---

(3) You have a skeptical friend who does not believe the circumference of a circle is $2\pi r$. You now have the tools to conclusively prove him wrong. Do it, for his own sake.

□ The equation of a circle is $x^2 + y^2 = r^2$, but we can't work with it like that. Rather, let's find the arc length of a quarter of the circle, from $x = 0$ to $x = r$; then, four times that length will be the circumference of the whole circle. Getting ready,

$$f(x) = \sqrt{r^2 - x^2}$$

$$f'(x) = \frac{-x}{\sqrt{r^2 - x^2}}$$

$$1 + [f'(x)]^2 = 1 + \frac{x^2}{r^2 - x^2} = \frac{(r^2 - x^2) + x^2}{r^2 - x^2} = \frac{r^2}{r^2 - x^2}$$

$$\sqrt{1 + [f'(x)]^2} = \sqrt{\frac{r^2}{r^2 - x^2}} = \frac{r}{\sqrt{r^2 - x^2}}$$

so that

$$L = \int_a^b \sqrt{1 + [f'(x)]^2} \, dx = \int_0^r \frac{r}{\sqrt{r^2 - x^2}} \, dx$$

This is a trig substitution problem of the form indicated in Useful Fact 7.6 of Sec. 7.5), in which we will set

$$x = r \sin \theta \quad ; \quad dx = r \cos \theta \, d\theta$$

Endpoints are updated as follows:

$$x = 0 \rightarrow \theta = 0$$

$$x = r \rightarrow \theta = \frac{\pi}{2}$$

And then

$$L = \int_0^{\pi/2} \frac{r}{r\cos\theta} \cdot r\cos\theta \, d\theta = \int_0^{\pi/2} r \, d\theta = r\theta \Big|_0^{\pi/2} = \frac{\pi r}{2}$$

Since this is the arc length of a quarter of the circle, the whole circumference is

$$C = 4L = 4\left(\frac{\pi r}{2}\right) = 2\pi r \quad \blacksquare$$

B.2.3   *Average Value — Challenge — Solved*

---

(1) Find the average of $f(\theta) = \sec\theta\tan\theta$ on $[0, \pi/4]$.

$$\square \; f_{avg} = \frac{1}{b-a}\int_a^b f(\theta)d\theta = \frac{1}{\pi/4 - 0}\int_0^{\pi/4}\sec\theta\tan\theta d\theta = \frac{4}{\pi}\left.(\sec\theta)\right|_0^{\pi/4}$$

$$= \frac{4}{\pi}\left(\sec\frac{\pi}{4} - \sec 0\right) = \frac{4}{\pi}\left(\sqrt{2} - 1\right) \quad \blacksquare$$

---

(2) Find the average of $f(x) = 2\sin x - \sin 2x$ on $[0, \pi]$ and a estimate value of c such that $f(c) = f_{avg}$.

$$\square \quad f_{avg} = \frac{1}{b-a}\int_a^b f(x)dx = \frac{1}{\pi - 0}\int_0^\pi (2\sin x - \sin 2x)dx$$

$$= \frac{1}{\pi}\left.\left(-2\cos x + \frac{1}{2}\cos 2x\right)\right|_0^\pi = \frac{1}{\pi}\left(2 + \frac{1}{2} + 2 - \frac{1}{2}\right) = \frac{4}{\pi}$$

Next, we need a c such that $f(c) = f_{avg}$, i.e. $2\sin c - \sin 2c = 4/\pi$. This equation must be solved numerically. So, using tech, we can find two solutions within our interval $[0, \pi]$ — they are

$$c \approx 2.808 \quad \text{or} \quad c \approx 1.238$$

They are both equally good values of c. $\quad\blacksquare$

---

(3) Consider the arc of the parabola $y = x^2$ on the interval $x \in [0, 2]$. What is the average distance to the origin of all the points on this arc?

□ The distance from *any* point (x, y) to the origin is

$$d = \sqrt{(x-0)^2 + (y-0)^2} = \sqrt{x^2 + y^2}$$

The distance from any point on the parabola $y = x^2$, i.e. any point (x, x^2) to the origin is then

$$d = \sqrt{x^2 + (x^2)^2} = \sqrt{x^2 + x^4} = x\sqrt{1 + x^2}$$

The average value of this distance over $[0, 2]$ is given by this integral:

$$d_{avg} = \frac{1}{b-a}\int_a^b f(x)dx = \frac{1}{2}\int_0^2 x\sqrt{1 + x^2}\,dx$$

We need a substitution (the good old fashioned kind, not a trigonometric substitution),

$$1 + x^2 = u \quad \text{and} \quad x\,dx = \frac{1}{2}\,du$$

The endpoints change as follows:

$$x = 0 \to u = 1 \quad ; \quad x = 2 \to u = 5$$

And now we get

$$d_{avg} = \frac{1}{2} \int_1^5 \frac{1}{2} \sqrt{u}\, du = \frac{1}{4} \cdot \frac{2}{3} u^{3/2} \Big|_1^5 = \frac{1}{6}(5\sqrt{5} - 1) \quad \blacksquare$$

B.2.4 *Numerical Int. — Challenge — Solved*

(1) Do an experiment to find how big n must get for the midpoint rule to get a 5 decimal place match (within rounding) $(n.nnnnn)$ to Wolfram Alpha's super-accurate answer for $\displaystyle\int_0^2 \frac{1}{\sqrt{x^3+1}}\, dx$. Use increments of 10.

☐ Wolfram Alpha reports that $\displaystyle\int_0^2 \frac{1}{\sqrt{x^3+1}}\, dx \approx 1.402182$, so we can round this to 1.40218. Thus, we are looking for the first n (to an increment of 10) for which the midpoint rule also gives an approximation that is rounded to 1.40218. At $n = 110$, we have a midpoint approximation of 1.4021851, which is barely too large. But at $n = 120$, we get a midpoint approximation of 1.4021847, which now rounds to 1.40129, to "match" the best value. ∎

(2) Do an experiment to find how big n must get for the trapezoid rule to get a 5 decimal place match (within rounding) $(n.nnnnn)$ to Wolfram Alpha's super-accurate answer for $\displaystyle\int_0^2 \frac{1}{\sqrt{x^3+1}}\, dx$. Use increments of 10.

f_x	=1/sqrt(D3^3 + 1)							
	A	B	C	D	E	F	G	H	I
1			i	x_i	f(x_i)	c_i	c_i * f(x_i)		
2	a = 0		0	0.000000	1.00000	1	1.00000	TRAP RULE	
3	b = 2		1	0.018182	1.00000	2	1.99999	sum:	154.23936
4	n = 110		2	0.036364	0.99998	2	1.99995	multiple:	0.00909090909091
5			3	0.054545	0.99992	2	1.99984	final:	1.4021760
6	Delta-x = 0.01818		4	0.072727	0.99981	2	1.99962		
7			5	0.090909	0.99962	2	1.99925		
8			6	0.109091	0.99935	2	1.99870		
9			7	0.127273	0.99897	2	1.99794		
10			8	0.145455	0.99846	2	1.99693		
11			9	0.163636	0.99782	2	1.99563		
12			10	0.181818	0.99701	2	1.99402		
13	CP 2		11	0.200000	0.99602	2	1.99205		
14			12	0.218182	0.99485	2	1.98969		

Fig. B.2 A trapezoid rule implementation, with CP 2.

☐ Per CP 1, we want to match (within rounding) a value of 1.40218. With the trapezoid rule, $n = 100$ gives 1.4021747, which isn't quite ready to round up. But at $n = 110$, we get 1.402176, which is ready to be rounded up to 1.40218, and "matches" the best value. Figure B.2 shows the upper portion of this trapezoid rule implementation. ∎

(3) Do an experiment to find how big n must get for the trapezoid rule to get a 5 decimal place match (within rounding) ($n.nnnnn$) to Wolfram Alpha's super-accurate answer for $\int_0^2 \dfrac{1}{\sqrt{x^3+1}}\,dx$. Use increments of 10. Compare to CP 1 and CP 2.

☐ As in CP 1 and CP 2, we are looking to match (within rounding) a value of 1.40218. With Simpson's rule, $n = 10$ gives 1.4022, but at $n = 20$, we get 1.4021836, which rounds to 1.40218, and "matches" the best value. (Figure B.3 shows this Simpson's rule implementation.) Whereas the midpoint and trapezoid rules took over 100 partitions to achieve the desired accuracy, Simpson's rule did it somewhere between only $n = 10$ and $n = 20$.   ■

f_x | =1/sqrt(D4^3 + 1)

	A	B	C	D	E	F	G	H	I
1			i	x_i	f (x_i)	c_i	c_i * f(x_i)		
2	a =	0	0	0	1	1	1	SIMPSON'S RULE	
3	b =	2	1	0.1	0.9995003747	4	3.998001499	sum:	42.06550883
4	n =	20	2	0.2	0.9960238411	2	1.992047682	multiple:	0.03333333333
5			3	0.3	0.9867673659	4	3.947069464	final:	1.402183628
6	Delta-x =	0.1	4	0.4	0.9694584179	2	1.938916836		
7			5	0.5	0.9428090416	4	3.771236166		
8			6	0.6	0.9068453126	2	1.813690625		
9			7	0.7	0.8629030294	4	3.451612118		
10	CP 3		8	0.8	0.8132500608	2	1.626500122		
11			9	0.9	0.7605057524	4	3.04202301		
12			10	1	0.7071067812	2	1.414213562		
13			11	1.1	0.6549812432	4	2.619924973		
14			12	1.2	0.6054493496	2	1.210898699		
15			13	1.3	0.559279218	4	2.237116872		
16			14	1.4	0.5168113941	2	1.033622788		
17			15	1.5	0.4780914437	4	1.912365775		
18			16	1.6	0.442981195	2	0.8859623899		
19			17	1.7	0.4112406723	4	1.644962689		
20			18	1.8	0.3825833549	2	0.7651667097		
21			19	1.9	0.3567108798	4	1.426843519		
22			20	2	0.3333333333	1	0.3333333333		

Fig. B.3   A Simpson's rule implementation, with CP 3.

B.3   Chapter 9: Challenge Problem Solutions

B.3.1   *Solids of Rev.: Sfc. Area — Challenge — Solved*

---

(1) Find the surface area of the solid formed when $9x = y^2 + 18, 2 \le x \le 6$, is revolved around the x-axis.

☐ Rewrite the function as $y = \sqrt{9x - 18} = 3\sqrt{x - 2}$. Then

$$f'(x) = \frac{3}{2\sqrt{x - 2}}$$

$$1 + (f'(x))^2 = 1 + \frac{9}{4(x - 2)} = \frac{4x + 1}{4(x - 2)}$$

So

$$S = \int_a^b 2\pi f(x)\sqrt{1 + [f'(x)]^2}\, dx = \int_2^6 2\pi(3\sqrt{x - 2})\sqrt{\frac{4x + 1}{4(x - 2)}}\, dx$$

$$= 3\pi \int_2^6 \sqrt{4x + 1}\, dx$$

A quick substitution,

$$4x + 1 = u \quad ; \quad dx = \frac{1}{4}\, du$$

$$x = 2 \rightarrow u = 9$$

$$x = 6 \rightarrow u = 25$$

Gives

$$S = \frac{3\pi}{4} \int_9^{25} \sqrt{u}\, du = \frac{3\pi}{4} \cdot \frac{2}{3} u^{3/2} \Big|_9^{25} = \frac{\pi}{2}(25^{3/2} - 9^{3/2})$$

$$= \frac{\pi}{2}(125 - 27) = 49\pi \quad \blacksquare$$

---

(2) Your skeptical friend from the Challenge Problems of Sec. 8.2 is at it again. Now he's claiming that the surface area of a sphere isn't really $4\pi r^2$, and that quantity is just a conspiracy spread by the fake media. Prove him wrong.

☐ A sphere is formed by revolving $y = \sqrt{r^2 - x^2}$ around the x-axis over the interval is $[-r, r]$. Just to keep the limits of the integral a bit simpler, we can also note that the surface area of the sphere on $[-r, r]$ is twice the surface area on $[0, r]$. With this info to get us started,

$$f'(x) = -\frac{2x}{2\sqrt{r^2 - x^2}} = -\frac{x}{\sqrt{r^2 - x^2}}$$

$$1 + (f'(x))^2 = 1 + \frac{x^2}{r^2 - x^2} = \frac{r^2}{r^2 - x^2}$$

$$\sqrt{1 + (f'(x))^2} = \frac{r}{\sqrt{r^2 - x^2}}$$

And so

$$S = \int_a^b 2\pi f(x)\sqrt{1 + [f'(x)]^2}\,dx = 2\int_0^r 2\pi\sqrt{r^2 - x^2}\cdot\frac{r}{\sqrt{r^2 - x^2}}\,dx$$

$$= 4\pi\int_0^r r\,dx = 4\pi rx\Big|_0^r = 4\pi r^2 \quad \blacksquare$$

---

(3) Show that the area of the surface formed when $y = 1/x$ is revolved around the x-axis, for $x \geq 1$, is infinite. (Hint: The fact that $1 + 1/x^4 > 1$ will be useful.)

□ As usual, we get ready with:

$$f'(x) = -\frac{1}{x^2}$$

$$1 + (f'(x))^2 = 1 + \frac{1}{x^4}$$

So

$$S = \int_a^b 2\pi f(x)\sqrt{1 + [f'(x)]^2}\,dx$$

$$= \int_1^\infty 2\pi\frac{1}{x}\sqrt{1 + \frac{1}{x^4}}\,dx \geq \int_1^\infty 2\pi\frac{1}{x}\,dx$$

$$= \lim_{c\to\infty}\int_1^c 2\pi\frac{1}{x}\,dx = 2\pi\lim_{c\to\infty}\ln x\Big|_1^c$$

$$= 2\pi\lim_{c\to\infty}(\ln c - \ln 1) = \infty$$

So the integral defining our surface area is greater than an integral that is infinite. Fun fact: while we have shown the surface area of this surface of revolution is infinite, we will show later that the *volume* of this same solid is finite!   ∎

B.3.2   *Vol. by Discs — Challenge — Solved*

---

(1) Find the volume of the solid generated by revolving the region between $y = \sec x$, $y = 1$, $x = -1$, and $x = 1$ around the x-axis.

☐ This is a hollowed solid, with $y = \sec x$ generating the outside and $y = 1$ generating the inside. We can use symmetry to tidy one of the limits of integration:

$$V = \int_a^b \pi \left([f(x)]^2 - [g(x)]^2\right) \, dx = \int_{-1}^1 \pi \left([\sec x]^2 - [1]^2\right) \, dx$$

$$= 2 \int_0^1 \pi \left([\sec x]^2 - [1]^2\right) \, dx = 2 \int_0^1 \pi \left(\sec^2 x - 1\right) \, dx$$

$$= 2\pi(\tan(1) - 1) \quad \blacksquare$$

---

(2) Find the volume of the solid generated by revolving the region between $y = 1/x$, $y = 0$, $x = 1$ and $x = 3$ around the line $y = -1$.

☐ As described, this produces a hollowed solid, with the function $y = 1/x$ generating the outer solid, and $y = 0$ generating the inner one. We can move everything up by 1 to get an equivalent volume — the volume we want is the same as when the region between $y = 1/x + 1$ (outer) and $y = 1$ (inner) is revolved around the x-axis. And so using (9.5),

$$V = \int_a^b \pi \left([f(x)]^2 - [g(x)]^2\right) \, dx = \int_1^3 \pi \left([\frac{1}{x} + 1]^2 - [1]^2\right) \, dx$$

$$= \int_0^1 \pi \left(\frac{1}{x^2} + \frac{2}{x}\right) \, dx = 2\pi(\ln 3 + \frac{1}{3}) \quad \blacksquare$$

---

(3) Approximate the volume of the solid generated by revolving the region between $y = x^2$ and $y = \ln(x + 1)$ around the x-axis.

☐ One point of intersection of these curves is $x = 0$ (hopefully that's fairly obvious). Using tech, we can determine a second intersection at $x \approx 0.747$. The revolution of the region between the curves produces a hollowed solid, with $y = \ln(x + 1)$ generating the outside and $y = x^2$ generating the inside. With our estimated point of intersection

$$V = \int_a^b \pi \left([f(x)]^2 - [g(x)]^2\right) \, dx \approx \int_0^{0.747} \pi \left([\ln(x + 1)]^2 - [x^2]^2\right) \, dx$$

$$\approx \int_0^{0.747} \pi \left([\ln(x + 1)]^2 - x^4\right) \, dx \approx 0.132 \quad \blacksquare$$

B.3.3   *Vol. by Shells — Challenge — Solved*

---

(1) Find the volume of the solid generated by revolving the region between $y = \ln x$, $x = 1$, and $x = 2$ around the y-axis.

☐ Using (9.6), we get

$$V = \int_a^b 2\pi x f(x)\, dx = \int_1^2 2\pi x (\ln x)\, dx = 4\pi \ln(2) - \frac{3\pi}{2} \quad \blacksquare$$

---

(2) Find the volume of the solid generated by revolving the region between $x = \sin(y)$, $y = 0$, and $y = \pi$ around the x-axis.

☐ By (9.7), we get

$$V = \int_c^d 2\pi y g(y)\, dy = \int_0^\pi 2\pi y (\sin y)\, dy = 2\pi^2 \quad \blacksquare$$

---

(3) Remember your skeptical friend from the Challenge Problems of the previous two sections? He has a sister! She claims to not believe that the volume of a sphere of radius r is $V = 4\pi r^2/3$. Use the method of cylindrical shells to set her straight.

☐ We can form a hemisphere of radius r by revolving $y = \sqrt{r^2 - x^2}$ around the y-axis. The volume of this is:

$$V = \int_0^r 2\pi x \sqrt{r^2 - x^2}\, dx = \frac{2\pi}{3}(r^3)$$

This solid is half of a full sphere, so the volume of a full sphere is twice this, or

$$V = \frac{4}{3}\pi r^3 \quad \blacksquare$$

B.4   Chapter 10: Challenge Problem Solutions

B.4.1   *Taylor Poly. — Challenge — Solved*

---

(1) Find the Maclaurin polynomial of order 3 for $f(x) = \sqrt{1 - x^2}$.

□ We need $f(x)$ and its first three derivatives evaluated at $x = 0$. These are:

$$f(x) = \sqrt{1 - x^2} \rightarrow f(0) = 1$$
$$f'(x) = \frac{-x}{\sqrt{1 - x^2}} \rightarrow f'(0) = 0$$
$$f''(x) = -\frac{1}{(1 - x^2)^{3/2}} \rightarrow f''(0) = -1$$
$$f^{(3)}(x) = -\frac{3x}{(1 - x^2)^{5/2}} \rightarrow f^{(3)}(0) = 0$$

so

$$M_3(x) = \sum_{n=0}^{3} \frac{f^{(n)}(x_0)}{n!}(x - 0)^n$$
$$= \frac{f^{(0)}(0)}{0!}(x)^0 + \frac{f^{(1)}(0)}{1!}(x)^1 + \frac{f^{(2)}(0)}{2!}(x)^2 + \frac{f^{(3)}(0)}{3!}(x)^3$$
$$= \frac{1}{0!}(x)^0 + \frac{0}{1!}(x)^1 + \frac{-1}{2!}(x)^2 + \frac{0}{3!}(x)^3$$
$$= 1 - \frac{1}{2}x^2$$

(Taylor polynomials for this function at this point will have only even powers.) ∎

---

(2) Find the Taylor polynomial of order 2 for $\tan^{-1}(x)$ at $x = 1$. Determine $R_2(x)$ for an interval of $d = 0.5$ centered at $x = 1$ and confirm this $R_2(x)$ holds at the test point $c = 0.8$.

□ We need $f(x)$ and its first two derivatives evaluated at $x = 1$. These are:

$$f(x) = \tan^{-1}(x) \rightarrow f(1) = \frac{\pi}{4}$$
$$f'(x) = \frac{1}{1 + x^2} \rightarrow f'(1) = \frac{1}{2}$$
$$f''(x) = -\frac{2x}{(1 + x^2)^2} \rightarrow f''(1) = -\frac{1}{2}$$

so

$$T_2(x) = \sum_{n=0}^{2} \frac{f^{(n)}(x_0)}{n!}(x - x_0)^n$$

$$= \frac{f^{(0)}(1)}{0!}(x - 1)^0 + \frac{f^{(1)}(1)}{1!}(x - 1)^1 + \frac{f^{(2)}(1)}{2!}(x - 1)^2$$

$$= \frac{\pi/4}{0!}(x - 1)^0 + \frac{1/2}{1!}(x - 1)^1 + \frac{-1/2}{2!}(x - 1)^2$$

$$= \frac{\pi}{4} + \frac{1}{2}(x - 1) - \frac{1}{4}(x - 1)^2$$

To get at $R_2(x)$ using Taylor's Inequality, we note that with $d = 0.5$, we will need an upper bound M of $|f^{(3)}(x)|$ on the interval $[0.5, 1.5]$. A graph of

$$|f^{(3)}(x)| = \left| \frac{6x^2 - 2}{(x^2 + 1)^3} \right|$$

is shown in Fig. B.4. Based on the graph, we'll assign an upper bound as $M = 0.5$. So,

$$R_2(x) \leq \frac{M}{(N+1)!} \cdot d^{N+1} = \frac{0.5}{3!}(0.5)^3 \approx 0.010417$$

At the test point $c = 0.8$, we have

$$f(0.8) = \tan^{-1}(0.8) \approx 0.674741$$

$$T_2(0.8) \approx 0.675398$$

and $|f(0.8) - T_2(0.8)| \approx 0.000657$, which is indeed less than $|R_2(x)| \approx 0.010417$. ∎

---

(3) Find the Taylor polynomial of order 3 for $f(x) = \ln(1 + x^2)$ at $x = 1$. Determine $R_3(x)$ for an interval of $d = 1$ centered at $x = 1$.

□ We need $f(x)$ and its first three derivatives evaluated at $x = 1$. These are:

$$f(x) = \ln(1 + x^2) \to f(1) = \ln(2)$$

$$f'(x) = \frac{2x}{1 + x^2} \to f'(1) = 1$$

$$f''(x) = -\frac{2(-1 + x^2)}{(1 + x^2)^2} \to f''(1) = 0$$

$$f^{(3)}(x) = -\frac{4x(-3 + x^2)}{(1 + x^2)^3} \to f^{(3)}(1) = -1$$

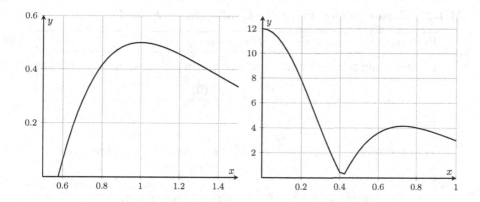

Fig. B.4 Looking for an upper bound of $|f^{(3)}(x)|$ on $[0.5, 1.5]$ (w/ CP 2).

Fig. B.5 Looking for an upper bound of $|f^{(4)}(x)|$ on $[0, 1]$ (w/ CP 3).

so

$$T_3(x) = \sum_{n=0}^{3} \frac{f^{(n)}(x_0)}{n!}(x - x_0)^n$$

$$= \frac{f^{(0)}(1)}{0!}(x - 1)^0 + \frac{f^{(1)}(1)}{1!}(x - 1)^1 + \frac{f^{(2)}(1)}{2!}(x - 1)^2$$

$$+ \frac{f^{(3)}(1)}{3!}(x - 1)^3$$

$$= \frac{\ln 2}{0!}(x - 1)^0 + \frac{1}{1!}(x - 1)^1 + \frac{0}{2!}(x - 1)^2 + \frac{-1}{3!}(x - 1)^3$$

$$= \ln 2 + (x - 1) - \frac{1}{6}(x - 1)^3$$

To get at $R_3(x)$ using Taylor's Inequality, we note that with $d = 1$, we will need an upper bound M of $|f^{(4)}(x)|$ on the interval $[0, 1]$. A graph of

$$|f^{(4)}(x)| = \left| \frac{12(x^4 - 6x^2 + 1)}{(x^2 + 1)^4} \right|$$

is shown in Fig. B.5. Based on the graph, we'll assign an upper bound as $M = 12$. So,

$$R_3(x) < \frac{M}{(N + 1)!} \cdot d^{N+1} - \frac{12}{4!}(1)^4 = 0.5 \quad \blacksquare$$

B.4.2   Taylor Series Pt. 1 — Challenge — Solved

---

(1) Find the Maclaurin series for $f(x) = \cosh x$.

☐ According to the definition of Maclaurin series, we need to fill out

$$M(x) = \frac{f(0)}{0!}x^0 + \frac{f'(0)}{1!}x^1 + \frac{f''(0)}{2!}x^2 + \frac{f^{(3)}(0)}{3!}x^3 + \cdots$$

But

$$f(x) = \cosh x \quad f'(x) = \sinh x \quad f''(x) = \cosh x \quad f^{(3)}(x) = \sinh x \quad \cdots$$

so

$$f(0) = \cosh(0) = 1 \quad , \quad f'(0) = \sinh(0) = 0 \quad ,$$
$$f''(0) = \cosh(0) = 1 \quad , \quad f^{(3)}(x) = \sinh(0) = 0 \quad \cdots$$

Inserting these values into the template for $M(x)$,

$$M(x) = \frac{1}{0!}x^0 + 0 + \frac{1}{2!}x^2 + 0 + \cdots$$

$$= x^0 + \frac{x^2}{2!} + \frac{x^4}{4!} + \cdots$$

$$= \sum_{n=0}^{\infty} \frac{x^{2n}}{(2n)!} \quad \blacksquare$$

---

(2) Find the Taylor series for $f(x) = \sin x$ around $x_0 = \pi/2$.

☐ According to the definition of Taylor series, we need to fill out

$$T(x) = \frac{f(\pi/2)}{0!}\left(x - \frac{\pi}{2}\right)^0 + \frac{f'(\pi/2)}{1!}\left(x - \frac{\pi}{2}\right)^1$$
$$+ \frac{f''(\pi/2)}{2!}\left(x - \frac{\pi}{2}\right)^2 + \frac{f^{(3)}(\pi/2)}{3!}\left(x - \frac{\pi}{2}\right)^3 + \cdots$$

But

$$f(x) = \sin x \quad , \quad der f(x) = \cos x \quad ,$$
$$f''(x) = -\sin x \quad , \quad f^{(3)}(x) = -\cos x \quad , \quad f^{(4)}(x) = \sin x \quad \cdots$$

so

$$f\left(\frac{\pi}{2}\right) = 1 \quad , \quad f'\left(\frac{\pi}{2}\right) = 0 \quad ,$$
$$f''\left(\frac{\pi}{2}\right) = -1 \quad , \quad f^{(3)}\left(\frac{\pi}{2}\right) = 0 \quad ,$$
$$f^{(4)}\left(\frac{\pi}{2}\right) = 1 \quad \cdots$$

and

$$T(x) = \frac{1}{1}\left(x - \frac{\pi}{2}\right)^0 + 0 + \frac{-1}{2!}\left(x - \frac{\pi}{2}\right)^2 + 0 + \frac{1}{4!}\left(x - \frac{\pi}{2}\right)^4 - \cdots$$

$$= \left(x - \frac{\pi}{2}\right)^0 - \frac{1}{2!}\left(x - \frac{\pi}{2}\right)^2 + \frac{1}{4!}\left(x - \frac{\pi}{2}\right)^4 - \cdots$$

$$= \sum_{n=0}^{\infty} \frac{(-1)^n}{(2n)!}\left(x - \frac{\pi}{2}\right)^{2n} \quad \blacksquare$$

---

(3) Find the Taylor series for $f(x) = e^{-2x}$ around $x_0 = -\frac{1}{2}$.

☐ Note that since $x_0 = -1/2$, and term $(x - x_0)^n$ will appear as $(x + 1/2)^n$. According to the definition of Taylor series, we need to fill out

$$T(x) = \frac{f(-1/2)}{0!}\left(x + \frac{1}{2}\right)^0 + \frac{f'(-1/2)}{1!}\left(x + \frac{1}{2}\right)^1$$

$$+ \frac{f''(-1/2)}{2!}\left(x + \frac{1}{2}\right)^2 + \frac{f^{(3)}(-1/2)}{3!}\left(x + \frac{1}{2}\right)^3 + \cdots$$

But

$$f(x) = e^{-2x} \quad f'(x) = -2e^{-2x} \quad f''(x) = 2^2e^{-2x} \quad f^{(3)}(x) = -2^3e^{-2x} \quad \cdots$$

so

$$f(-1/2) = e \quad f'(-1/2) = -2e \quad f''(-1/2) = 2^2e \quad f^{(3)}(-1/2) = -2^3e \quad \cdots$$

and

$$T(x) = \frac{e}{0!}\left(x + \frac{1}{2}\right)^0 - \frac{2e}{1!}\left(x + \frac{1}{2}\right)^1 + \frac{2^2e}{2!}\left(x + \frac{1}{2}\right)^2$$

$$- \frac{2^3e}{3!}\left(x + \frac{1}{2}\right)^3 + \cdots$$

$$= \frac{2^0e}{0!}\left(x + \frac{1}{2}\right)^0 - \frac{2^1e}{1!}\left(x + \frac{1}{2}\right)^1 + \frac{2^2e}{2!}\left(x + \frac{1}{2}\right)^2$$

$$- \frac{2^3e}{3!}\left(x + \frac{1}{2}\right)^3 + \cdots$$

$$= \sum_{n=0}^{\infty} (-1)^n \frac{2^n e}{n!}\left(x + \frac{1}{2}\right)^n \quad \blacksquare$$

B.4.3   *Infinite Sequences — Challenge — Solved*

---

(1) If it exists, determine the limit of the sequence $a_n = \dfrac{(-1)^n n^3}{n^3 + 2n^2 + 1}$.

☐ Let's factor n^3 from the denominator and get

$$a_n = \frac{(-1)n^3}{n^3(1 + \frac{2}{n} + \frac{1}{n^3})} = \frac{(-1)^n}{1 + \frac{2}{n} + \frac{1}{n^3}}$$

so

$$\lim_{n \to \infty} a_n = \lim_{n \to \infty} (-1)^n$$

and the sequence diverges.   ■

---

(2) If it exists, determine the limit of the sequence $a_n = \ln(n+1) - \ln(n)$.

☐ We can use properties of logarithms to write

$$a_n = \ln(n+1) - \ln(n) = \ln\left(\frac{n+1}{n}\right) = \ln\left(1 + \frac{1}{n}\right)$$

so

$$\lim_{n \to \infty} a_n = \ln(1) = 0$$

and the sequence converges to 0.   ■

---

(3) Determine whether the sequence $a_n = n + (1/n)$ is increasing or decreasing, and if it is bounded.

☐ I have a suspicion this increases. To show this, I need to show that $a_n < a_{n+1}$ for all $n > 0$.

$$a_n <? a_{n+1}$$
$$n + \frac{1}{n} <? (n+1) + \frac{1}{n+1}$$
$$\frac{1}{n} <? 1 + \frac{1}{n+1}$$
$$1 <? n + \frac{n}{n+1}$$

This is surely true for all $n > 1$, so the sequence is indeed increasing. It is unbounded above, and bounded below by 2.   ■

B.4.4   *Infinite Series — Challenge — Solved*

---

(1) Investigate the convergence of $\displaystyle\sum_{n=1}^{\infty} \frac{e^n}{3^{n-1}}$.

☐ This series looks geometric-ish, let's pursue that:

$$\sum_{n=1}^{\infty} \frac{e^n}{3^{n-1}} = \sum_{n=0}^{\infty} \frac{e^{n+1}}{3^n} \sum_{n=0}^{\infty} e \cdot \left(\frac{e}{3}\right)^n$$

This is a geometric series with $a = e$ and $r = e/3$. Since $|r| < 1$, the series converges to

$$\sum_{n=1}^{\infty} \frac{e^n}{3^{n-1}} = \frac{a}{1-r} = \frac{e}{1 - e/3} = \frac{3e}{3-e} \quad \blacksquare$$

---

(2) Write $0.123456456\ldots$ as a fraction.

☐ $0.123456456\ldots = 0.123 + \dfrac{456}{10^6} + \dfrac{456}{10^9} + \dfrac{456}{10^{12}} + \cdots$

$$= 0.123 + \frac{456}{10^6}\left(1 + \frac{1}{10^3} + \frac{1}{10^6} + \cdots\right)$$

$$= 0.123 + \sum_{n=0}^{\infty} \frac{456}{10^6}\left(\frac{1}{10^3}\right)^n$$

This is a geometric series with $a = 456/10^6$ and $r = 1/10^3$. Since $|r| < 1$, the series converges, and the whole expression converges to

$$0.123456456\ldots = 0.123 + \sum_{n=0}^{\infty} \frac{456}{10^6}\left(\frac{1}{10^3}\right)^n = \frac{123}{1000} + \frac{456/10^6}{1 - 1/10^3}$$

$$= \frac{123}{1000} + \frac{456}{999000} = \frac{41111}{333000} \quad \blacksquare$$

---

(3) Find the values of x for which the series $\displaystyle\sum_{n=0}^{\infty} \frac{(x+3)^n}{2^n}$ converges, and determine the algebraic form of the function the convergent series defines.

☐ Let's get it in a slightly better form:

$$\sum_{n=0}^{\infty} \frac{(x+3)^n}{2^n} = \sum_{n=0}^{\infty} \left(\frac{x+3}{2}\right)^n$$

This is a geometric series with $a = 1$ and $r = (x+3)/2$. It will converge when $|r| < 1$, i.e. when $|x + 3| < 2$, i.e. when $-5 < x < -1$. When it does converge, it converges to

$$\frac{a}{1-r} = \frac{1}{1-(x+3)/2} = \frac{2}{-1-x} = \frac{-2}{x+1}$$

So, the series $\displaystyle\sum_{n=0}^{\infty} \frac{(x+3)^n}{2^n}$ is equivalent to the function $-\dfrac{2}{x+1}$ for $-5 < x < 1.$ ∎

B.4.5 *Alternating Series — Challenge — Solved*

---

(1) Investigate the convergence of $\sum_{n=0}^{\infty} \frac{(-1)^n}{4^n n!}$. If this series converges, find how many terms are necessary for the error to be less than 0.0001.

$\square$ The easy part is the convergence; $a_n = 1/(4^n n!)$, so clearly, $a_n \to 0$ as $n \to \infty$. Also, since the denominator for $n+1$ will be larger than the denominator for n, then $a_{n+1} < a_n$, and both conditions of the alternating series test are met. This is a convergent series. For the error to be less than 0.0001, we must find n such that $a_{n+1} < 0.0001 = 10^{-4}$, i.e.

$$\frac{1}{4^{n+1}(n+1)!} < 10^{-4} \quad \text{or, inverted} \quad 4^{n+1}(n+1)! > 10^4$$

This is not solvable algebraically, so we'll need numerical help. Let's just find where $4^{n+1}(n+1)! = 10^4$, then set n to be the next integer up. By Wolfram Alpha, the solution to $4^{n+1}(n+1)! = 10^4$ is $n = 3.17$, which means we need 4 terms for the remainder between the partial sums and the true value of the series to be less than 10^{-4}. ∎

---

(2) Investigate the convergence of $\sum_{n=0}^{\infty}(-1)^n \cos\left((2n+1)\frac{\pi}{2}\right)$ and $\sum_{n=0}^{\infty}(-1)^n \cos\left((2n+1)\frac{\pi}{2}\right)$.

$\square$ Let's take a look at the first few terms in this series:

$$n = 0 : a_0 = (-1)^0 \cos\left(\frac{\pi}{2}\right) = 0$$

$$n = 1 : a_1 = (-1)^1 \cos\left(\frac{3\pi}{2}\right) = 0$$

$$n = 2 : a_2 = (-1)^2 \cos\left(\frac{5\pi}{2}\right) = 0$$

$$\vdots \qquad \vdots$$

This pattern will continue. Each term in the series is actually just 0, so the series converges to 0.

$$n = 0 : a_0 = (-1)^0 \sin\left(\frac{\pi}{2}\right) = (1)(1) = 1$$

$$n = 1 : a_1 = (-1)^1 \sin\left(\frac{3\pi}{2}\right) =_{(} -1)(-1) = 1$$

$$n = 2 : a_2 = (-1)^2 \sin\left(\frac{5\pi}{2}\right) = (1)(1) = 1$$

$$\vdots \qquad\qquad \vdots$$

This pattern will also continue. Each term in the series is equal to 1, so the series diverges. I guess one could argue whether or not these series are really alternating series even though there are a multiple of $(-1)^n$ in the recipes. ∎

---

(3) Investigate the convergence of $\displaystyle\sum_{n=0}^{\infty}(-1)^n \frac{|\sin(n+1)|}{n+1}$.

☐ The varying size of the numerator $\sin(n+1)$ throws a wrench into the comparison of a_{n+1} vs a_n that is required by the regular alternating series test. So instead, how about we use a comparison? We know that the alternating harmonic series, $\displaystyle\sum_{n=0}^{\infty}(-1)^n \frac{1}{n+1}$, converges. And each term in the given series is going to be less than the corresponding term in the alternating harmonic series; that is,

$$\frac{|\sin(n+1)|}{n+1} \leq \frac{1}{n+1} \qquad \text{for each } n$$

And so it stands to reason that

$$\sum_{n=0}^{\infty}(-1)^n \frac{|\sin(n+1)|}{n+1} \leq \sum_{n=0}^{\infty}(-1)^n \frac{1}{n+1}$$

The series on the right converges, so the series on the left must converge, too. ∎

B.5   Chapter 11: Challenge Problem Solutions

B.5.1   *Int. Test & p-Series — Challenge — Solved*

---

(1) Investigate the convergence of $\displaystyle\sum_{n=1}^{\infty} ne^{-n}$.

□ Since $f(x) = xe^{-x} = x/e^x$ is continuous, positive, and decreasing for $x \geq 1$, then the integral test applies, and this series will behave the same as the corresponding integral:

$$\int_{1}^{\infty} xe^{-x}\,dx = \lim_{c\to\infty} -e^{-x}(1+x)\Big|_{1}^{c}$$

$$= \lim_{c\to\infty}\left(-\frac{1+c}{e^c} + \frac{2}{e}\right) = \frac{2}{e}$$

Since the integral converges, so does the series. Note that in this solution, there are two "hidden" steps: (1) the integral requires integration by parts ($f(x) = x, g'(x) = e^{-x}$); (2) evaluation of the limit term involving c requires L-Hopital's rule. ∎

---

(2) Investigate the convergence of $\displaystyle\sum_{n=1}^{\infty} \frac{3n^2 + 4n}{\sqrt{n^5 + 2}}$.

□ Eventually, the $+4n$ in the numerator and the $+2$ in the denominator will get dwarfed by the other terms, and so intuition says that this series should behave like

$$\sum_{n=1}^{\infty} \frac{3n^2}{\sqrt{n^5}} = \sum_{n=1}^{\infty} \frac{3n^2}{n^{5/2}} = \sum_{n=1}^{\infty} \frac{3}{n^{1/2}}$$

which is a divergent p-series. We can use the limit comparison test to see if these two series are alike enough for divergence of the comparison p-series to force divergence of the given series.

$$\lim_{n\to\infty} \frac{a_n}{b_n} = \lim_{n\to\infty} \frac{(3n^2 + 4n)/(\sqrt{n^5 + 2})}{(3n^2)/(\sqrt{n^5})} = \lim_{n\to\infty} \frac{(3n^2 + 4n)(\sqrt{n^5})}{(3n^2)\sqrt{n^5 + 2}}$$

$$= \lim_{n\to\infty} \frac{3n^2 + 4n}{3n^2} \cdot \frac{\sqrt{n^5}}{\sqrt{n^5 + 2}} = 1$$

Since the limit of the ratio of terms is a finite positive constant, then the two series will exhibit the same behavior. So the given series diverges. ∎

(3) Investigate the convergence of $\displaystyle\sum_{n=1}^{\infty} \frac{\ln n}{n^2 + 1}$.

□ First, note that since $\ln n/(n^2 + 1) < \ln n/n^2$ for each $n \geq 1$, then the convergence of the given series will be determined if we can show that the partner comparison series $\displaystyle\sum_{n=1}^{\infty} \frac{\ln n}{n^2}$ converges. The latter can be determined by the integral test; this series will behave the same as the corresponding integral:

$$\int_1^{\infty} \frac{\ln x}{x^2}\, dx = \lim_{c \to \infty} \left(-\frac{1}{x}(\ln x + 1) \right) \Big|_1^c$$

$$= \lim_{c \to \infty} \left(-\frac{\ln c + 1}{c} + 1 \right) = \lim_{c \to \infty} \left(-\frac{1/c}{1} + 1 \right) = 1$$

Since the integral converges, so does the temporary series. And since the comparison series converges, so does the given series — which is bounded above by the comparison series.

Note that in this solution, there are two "hidden" steps: (1) the demonstrated integral requires integration by parts ($f(x) = \ln x, g'(x) = 1/x^2\, dx$); (2) the limit of the c-term at the very end of the integral solution requires L-Hopital's rule. ∎

B.5.2   *Ratio Test — Challenge — Solved*

(1) Investigate the convergence of $\sum\limits_{n=1}^{\infty}(-1)^{n+1}\dfrac{n^2 2^n}{n!}$.

☐ The series contains factorials and powers of n, so the ratio test should work well.

$$L = \lim_{n\to\infty}\left|\frac{a_{n+1}}{a_n}\right| = \lim_{n\to\infty}\left|\frac{\left(\frac{(n+1)^2 2^{n+1}}{(n+1)!}\right)}{\left(\frac{n^2 2^n}{n!}\right)}\right|$$

$$= \lim_{n\to\infty}\left|\frac{(n+1)^2 2^{n+1}}{(n+1)!}\frac{n!}{n^2 2^n}\right| = \lim_{n\to\infty}\frac{2(n+1)}{n^2} = 0$$

Since $L < 1$ the series converges.   ■

(2) Investigate the convergence of $\sum\limits_{n=1}^{\infty}\dfrac{n!}{n^n}$.

☐ By the ratio test,

$$L = \lim_{n\to\infty}\left|\frac{a_{n+1}}{a_n}\right| = \lim_{n\to\infty}\left|\frac{\left(\frac{(n+1)!}{(n+1)^{n+1}}\right)}{\left(\frac{n!}{n^n}\right)}\right|$$

$$= \lim_{n\to\infty}\left(\frac{(n+1)!}{(n+1)^{n+1}}\right)\cdot\frac{n^n}{n!} = \lim_{n\to\infty}\left(\frac{(n+1)!}{(n+1)(n+1)^n}\right)\cdot\frac{n^n}{n!}$$

$$= \lim_{n\to\infty}\left(\frac{n!}{(n+1)^n}\right)\cdot\frac{n^n}{n!} = \lim_{n\to\infty}\frac{n^n}{(n+1)^n}$$

$$= \lim_{n\to\infty}\left(\frac{n}{n+1}\right)^n = \frac{1}{e}$$

Since $L = 1/e < 1$, the series converges.   ■

(3) Investigate the convergence of $\sum\limits_{n=1}^{\infty}\dfrac{3^n n!}{(2n)!}$.

☐ The series contains factorials and powers of n, so the ratio test should work well.

$$L = \lim_{n \to \infty} \left| \frac{a_{n+1}}{a_n} \right| = \lim_{n \to \infty} \left| \frac{\left(\frac{3^{n+1}(n+1)!}{(2(n+1))!} \right)}{\left(\frac{3^n n!}{(2n)!} \right)} \right|$$

$$= \lim_{n \to \infty} \left| \frac{3^{n+1}(n+1)!}{(2n+2)!} \frac{(2n)!}{3^n n!} \right| = \lim_{n \to \infty} \frac{3(n+1)}{(2n+1)(2n+2)}$$

$$= \lim_{n \to \infty} \frac{3(n+1)}{2(2n+1)(n+1)} = \frac{3}{2} \lim_{n \to \infty} \frac{1}{2n+1} = 0$$

Since $L < 1$ the series converges. ∎

B.5.3   *Power Series — Challenge — Solved*

---

(1) Find the interval and radius of convergence of the power series $\displaystyle\sum_{n=1}^{\infty} \frac{x^n}{n3^n}$.

☐ Starting up the ratio test,

$$\left| \frac{a_{n+1}}{a_n} \right| = \left| \frac{\frac{x^{n+1}}{(n+1)3^{n+1}}}{\frac{x^n}{n3^n}} \right| = \left| \frac{x^{n+1}}{(n+1)3^{n+1}} \cdot \frac{n3^n}{x^n} \right| = \frac{|x|}{3} \left(\frac{n}{n+1} \right)$$

To have convergence, we need the limit of this expression to be less than 1; that limit is

$$L = \lim_{n \to \infty} \frac{|x|}{3} \left(\frac{n}{n+1} \right) = \frac{|x|}{3}$$

To have $L < 1$, we need $\dfrac{|x|}{3} < 1$, i.e. $-3 < x < 3$. We also have to test the endpoints of this interval. When $x = -3$, we have

$$\sum_{n=1}^{\infty} \frac{(-3)^n}{n3^n} = \sum_{n=1}^{\infty} \frac{(-1)^n}{n}$$

This is the alternating harmonic series, and converges. When $x = 3$, we have

$$\sum_{n=1}^{\infty} \frac{3^n}{n3^n} = \sum_{n=1}^{\infty} \frac{1}{n}$$

This is the harmonic series, which diverges. Therefore the interval of convergence is $-3 \leq x < 3$. The radius of convergence is 3.   ∎

---

(2) Find the interval and radius of convergence of the power series $\displaystyle\sum_{n=1}^{\infty} \frac{(-2)^n}{\sqrt{n}} (x + 3)^n$.

☐ Starting up the ratio test,

$$\left| \frac{a_{n+1}}{a_n} \right| = \left| \frac{\frac{(-2)^{n+1}}{\sqrt{n+1}} (x+3)^{n+1}}{\frac{(-2)^n}{\sqrt{n}} (x+3)^n} \right| = \left| \frac{(-2)^{n+1}}{\sqrt{n+1}} (x+3)^{n+1} \cdot \frac{\sqrt{n}}{(-2)^n (x+3)^n} \right|$$

$$= 2|x+3| \sqrt{\frac{n}{n+1}}$$

To have convergence, we need the limit of this expression to be less than 1; that limit is

$$L = \lim_{n \to \infty} 2|x+3| \sqrt{\frac{n}{n+1}} = 2|x+3|$$

To have $L < 1$, we need $|x + 3| < \dfrac{1}{2}$, which means:

$$-\frac{1}{2} < x + 3 < \frac{1}{2}$$

$$-\frac{7}{2} < x < -\frac{5}{2}$$

We also have to test the endpoints of this interval. When $x = -\dfrac{7}{2}$, we have

$$\sum_{n=1}^{\infty} \frac{(-2)^n}{\sqrt{n}} \left(-\frac{1}{2}\right)^n = \sum_{n=1}^{\infty} \frac{1}{\sqrt{n}}$$

This is a p-series with $p = \dfrac{1}{2}$ and so diverges. When $x = -\dfrac{5}{2}$, we have

$$\sum_{n=1}^{\infty} \frac{(-2)^n}{\sqrt{n}} \left(+\frac{1}{2}\right)^n = \sum_{n=1}^{\infty} \frac{(-1)^n}{\sqrt{n}}$$

This converges by the alternating series test. Therefore the interval of convergence is $-7/2 < x \leq -5/2$. The radius of convergence is $1/2$.

■

---

(3) The following series defines a *Bessel Function*. What is its interval of convergence?

$$\sum_{n=0}^{\infty} \frac{(-1)^n x^{2n+1}}{n!(n+1)!2^{2n+1}}$$

□ Starting up the ratio test,

$$\left|\frac{a_{n+1}}{a_n}\right| = \left|\frac{x^{2(n+1)+1}}{(n+1)!((n+1)+1)!2^{2(n+1)+1}} \cdot \frac{n!(n+1)!2^{2n+1}}{x^{2n+1}}\right|$$

$$= \left|\frac{x^{2n+3}}{(n+1)!(n+2)!2^{2n+3}} \cdot \frac{n!(n+1)!2^{2n+1}}{x^{2n+1}}\right|$$

$$= \left|\frac{x^2}{(n+1)(n+2)2^2}\right|$$

To have convergence, we need the limit of this expression to be less than 1; that limit is

$$L = \lim_{n \to \infty} \left|\frac{x^2}{(n+1)(n+2)2^2}\right| = x^2 \lim_{n \to \infty} \left|\frac{1}{4(n+1)(n+2)}\right| = 0$$

Therefore, L is always less than 1 and the series converges for all values of x. The interval of convergence is $-\infty < x < \infty$.

■

B.5.4 *Funcs. in Disguise — Challenge — Solved*

(1) Find the power series representation and interval of convergence of $f(x) = \dfrac{x}{4x+1}$.

□ With a little work, we can get this to look like the $a/(1-r)$ form of a geometric series:

$$\frac{x}{4x+1} = \frac{x}{1-(-4x)}$$

This matches with $a = x$ and $r = -4x$. So $f(x)$ is

$$\sum_{n=0}^{\infty} a(r)^n = \sum_{n=0}^{\infty} x(-4x)^n = \sum_{n=0}^{\infty} (-1)^n 4^n x^{n+1}$$

This will converge when $|r| = |4x| < 1$, i.e. when $|x| < 1/4$, or $-1/4 < x < 1/4$. ∎

(2) Find the power series representation and interval of convergence of $f(x) = \dfrac{x^2}{(1+x)^3}$.

□ There's a hard way and an easy way. The easy way is to borrow the result of Practice Problem 3:

$$f(x) = x^2 \cdot \frac{1}{(1+x)^3} = x^2 \cdot \frac{1}{2} \sum_{n=0}^{\infty} (-1)^n (n+1)(n+2)x^n$$

$$= \frac{1}{2} \sum_{n=0}^{\infty} (-1)^n (n+1)(n+2)x^{n+2}$$

Together,

$$\frac{x^2}{(1+x)^3} = \frac{1}{2} \sum_{n=2}^{\infty} (-1)^n (n)(n-1)x^n$$

for $|x| < 1$. ∎

(3) Find the power series representation and interval of convergence of $f(x) = \tan^{-1}\left(\dfrac{x}{3}\right)$.

□ There's a hard way and an easy way. The easy way is to use the now known series for $\tan^{-1}(x)$ developed in EX 5 of this section:

$$\tan^{-1}(x) = \sum_{n=0}^{\infty} (-1)^n \frac{x^{2n+1}}{2n+1}$$

so

$$\tan^{-1}\left(\frac{x}{3}\right) = \sum_{n=0}^{\infty}(-1)^n\frac{(x/3)^{2n+1}}{2n+1} = \sum_{n=0}^{\infty}(-1)^n\frac{1}{2n+1}\left(\frac{x}{3}\right)^{2n+1}$$

The original series for $\tan^{-1}(x)$ converges for $|x| < 1$ so this one will converge for $|x/3| < 1$, i.e. $|x| < 3$. ∎

B.5.5 *Taylor Series Pt. 2 — Challenge — Solved*

(1) Find the Taylor series and interval of convergence for $f(x) = \cos x$ around $x_0 = \pi/2$.

☐ According to the definition of Taylor series, we need to fill out

$$T(x) = \sum_{n=0}^{\infty} \frac{f^{(n)}(\pi/2)}{n!}(x - \pi/2)^n$$

But

$$f(x) = \cos x \quad, \quad f'(x) = -\sin x \quad,$$
$$f''(x) = -\cos x \quad, \quad f^{(3)}(x) = \sin x \quad, \quad f^{(4)}(x) = \cos x \quad \dots$$

so

$$f(\pi/2) = 0 \quad, \quad f'(\pi/2) = -1 \quad, \quad f''(\pi/2) = 0 \quad,$$
$$f^{(3)}(\pi/2) = 1 \quad, \quad f^{(4)}(\pi/2) = 0 \quad \dots$$

(we can expect the pattern to continue). Expanding the Taylor series formula,

$$T(x) = 0 + \frac{-1}{1}(x - \pi/2)^1 + 0 + \frac{1}{3!}(x - \pi/2)^3 + 0 + \frac{-1}{5!}(x - \pi/2)^5 - \cdots$$
$$= -(x - \pi/2)^1 + \frac{1}{3!}(x - \pi/2)^3 - \frac{1}{5!}(x - \pi/2)^5 + \cdots$$
$$= \sum_{n=0}^{\infty} \frac{(-1)^{n+1}}{(2n+1)!}(x - \pi/2)^{2n+1}$$

The interval of convergence of this Taylor series is determined by the ratio test:

$$\left| \frac{a_{n+1}}{a_n} \right| = \left| \frac{(x - \pi/2)^{2n+3}/(2n+3)!}{(x - \pi/2)^{2n+1}/(2n+1)!} \right|$$
$$= \left| \frac{(x - \pi/2)^{2n+3}}{(2n+3)!} \cdot \frac{(2n+1)!}{(x - \pi/2)^{2n+1}} \right|$$
$$= \frac{1}{(2n+3)(2n+2)}|x - \pi/2|^2$$

In order to have convergence, we need the limit of this expression to be less than 1. But when we compute this limit, we get:

$$L = \lim_{n \to \infty} \frac{1}{(2n+3)(2n+2)}|x - \pi/2|^2 = 0$$

Since $L < 0$ for all x, the interval of convergence is $-\infty < x < \infty$. ∎

(2) Find the Maclaurin series and interval of convergence for $f(x) = x^2 \tan^{-1}(x^2)$.

□ We already know the Maclaurin series

$$\tan^{-1}(x) = \sum_{n=0}^{\infty} \frac{(-1)^n}{2n+1} x^{2n+1} \quad \text{for} \quad -1 \leq x \leq 1$$

Therefore

$$\tan^{-1}(x^2) = \sum_{n=0}^{\infty} \frac{(-1)^n}{2n+1}(x^2)^{2n+1} = \sum_{n=0}^{\infty} \frac{(-1)^n}{2n+1} x^{4n+2} \quad \text{for} \quad -1 \leq x \leq 1$$

and multiplying by x^2, we see that for $-1 < x < 1$,

$$x^2 \tan^{-1}(x^2) = x^2 \cdot \sum_{n=0}^{\infty} \frac{(-1)^n}{2n+1} x^{4n+2} = \sum_{n=0}^{\infty} \frac{(-1)^n}{2n+1} x^{4n+4} \quad \blacksquare$$

(3) Find the (fully simplified!) Maclaurin series and its interval of convergence for $f(x) = e^x + 1/e^x$.

□ Writing the function as $f(x) = e^x + e^{-x}$ and then expanding each part as its own Maclaurin series, we get

$$e^x + e^{-x} = \sum_{n=0}^{\infty} \frac{x^n}{n!} + \sum_{n=0}^{\infty}(-1)^n \frac{x^n}{n!} = \sum_{n=0}^{\infty} \left(\frac{x^n}{n!} + (-1)^n \frac{x^n}{n!} \right)$$

$$= \sum_{n=0}^{\infty} \frac{(1 + (-1)^n)x^n}{n!}$$

Now here's the sneaky part: $1 + (-1)^n$ is 2 for even n and 0 for odd n. So only the even terms $n = 0, 2, 4, \ldots$ survive in the series, as shown:

$$e^x + e^{-x} = \sum_{n=0}^{\infty} \frac{2x^{2n}}{(2n)!}$$

Since the interval of convergence for both original series was $-\infty < x < \infty$, the interval for this combined series is the same.   ∎

B.6 Chapter 12: Challenge Problem Solutions

B.6.1 *Intro to Param. Eqs. — Challenge — Solved*

---

(1) Eliminate the parameter and identify the curve $x = 1 + 3t, y = 2 - t^2$.

□ Solving the first equation for t, we get $t = (x - 1)/3$. Plugging this into the second equation gives

$$y = 2 - \left(\frac{x}{3} - \frac{1}{3}\right)^2 = 2 - \frac{1}{9}(x - 1)^2$$

This is an inverted parabola with vertex $(1, 2)$. Wow, we're seeing a lot of parabolas, aren't we? ■

---

(2) Eliminate the parameter and identify the curve $x = e^t + 1, y = e^{2t}$.

□ First, note that the second equation can be written $y = (e^t)^2$. We can identify e^t from the first equation as $e^t = x - 1$, and therefore we have $y = (x - 1)^2$. This is an upright parabola with vertex $(1, 0)$. Another parabola? Good grief! ■

---

(3) Identify the curve $x = 4\cos\theta, y = 5\sin\theta$ for $-\pi/2 < \theta < \pi/2$ and state its orientation.

□ Let's rearrange the parametric equations slightly:

$$\frac{x}{4} = \cos\theta \quad ; \quad \frac{y}{5} = \sin\theta$$

and square both:

$$\frac{x^2}{16} = \cos^2\theta \quad ; \quad \frac{y^2}{25} = \sin^2\theta$$

so that we're ready to use our favorite trig identity:

$$\frac{x^2}{16} + \frac{y^2}{25} = \cos^2\theta + \sin^2\theta = 1$$

Our curve is also known as $x^2/16 + y^2/25 = 1$, which is an ellipse. Due to the restrictions on θ, though, we only get the right half of the ellipse. At $t = -\pi/2$, we're at the point $(0, -5)$; at $t = \pi/2$, we're at the point $(0, 5)$ — so we traverse the right half of the ellipse counterclockwise. ■

B.6.2   *Calc. of Param. Eqs. — Challenge — Solved*

(1) Find the locations on the parametric curve $x = a\cos^3\theta, y = a\sin^3\theta$ where (a) tangent lines are only horizontal, (b) tangent lines are only vertical, and (c) where the curve forms a cusp, and therefore has no unique tangent line.

☐ First, note these derivative expressions:

$$\frac{dx}{d\theta} = -3a\cos^2\theta\sin\theta$$

$$\frac{dy}{d\theta} = 3a\sin^2\theta\cos\theta$$

Since the parameter used for this curve is θ, we adapt Useful Fact 12.1 to write

$$\frac{dy}{dx} = \frac{dy/d\theta}{dx/d\theta} = -\tan\theta$$

Note: for one circuit of this curve, it is sufficient to use $0 \le \theta < 2\pi$.

(a) Tangent lines are horizontal where $dy/dx = 0$; on this curve, that will be where $\tan\theta = 0, \pi$. This yields the points $(x, y) = (a, 0)$ and $(x, y) = (-a, 0)$.

(b) Tangent lines are vertical when $dx/dt = 0$; on this curve, that will be where $\tan\theta = \pi/2, 3\pi/2$. This yields the points $(x, y) = (0, a)$ and $(x, y) = (0, -a)$.

So far, we've come up with four points at which tangent lines are horizontal or vertical, right? Wrong! Those results just above are total nonsense! Why are they still there, then? Because they give an important lesson: when you're simplifying things, don't go on autopilot.

The story in this problem is that because dy/dx is measured parametrically as $(dy/d\theta)/(dx/d\theta)$, then dy/dx will be zero at a location where $dy/d\theta = 0$ but $dx/d\theta$ is NOT ZERO. If *both* $dy/d\theta = 0$ and $dx/d\theta = 0$, then dy/dx is *undefined*. If you examine $dy/d\theta$ and $dx/d\theta$ individually for $\theta = 0$ and $\theta = \pi$, then you will see they are *both* zero at those locations.

Similarly, both $dy/d\theta$ and $dx/d\theta$ are zero for $\theta = \pi$ and $\theta = 3\pi/2$.

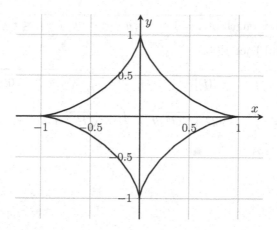

Fig. B.6   The parametric curve $x = \cos^3 \theta, y = \sin^3 \theta$.

All in all, the fact that dy/dx simplified to $-\tan \theta$ was not only irrelevant, it was actually some mathematical misdirection. If you examine the full expressions above for $dy/d\theta$ and $dx/d\theta$, you will see there are no locations where one is zero while the other is not. Therefore, we have our answer to (c): there are cusps at the four points where dy/dx will be undefined: $(\pm a, 0)$ and $(0, \pm a)$. This curve is shown in Fig. B.6, plotted for $a = 1$. ∎

---

(2) Find the area inside the curve $x = a\cos^3 \theta; y = a\sin^3 \theta$.

☐ A complete circuit of this curve happens on $0 \le t \le 2\pi$. Therefore the area inside is, by Useful Fact 12.2:

$$A = \int_a^b y(t)x'(t)\,dt = \int_0^{2\pi} (a\sin^3 \theta)(-3a\cos^2 \theta \sin \theta)\,d\theta$$

$$= -3a^2 \int_0^{2\pi} \sin^4 \theta \cos^2 \theta\,d\theta = -\frac{3\pi a^2}{8}$$

The area comes back as negative because this curve is traversed counterclockwise from $t = 0$ to $t = 2\pi$, whereas the integral needs a clockwise orientation. So the integral really should have been

$$\int_{2\pi}^0 (a\sin^3 \theta)(-3a\cos^2 \theta \sin \theta)\,d\theta$$

The area inside the curve is $3\pi a^2/8$. ∎

(3) Find the arc length of $x = 1 + 3t^2; y = 4 + 2t^3$ for $0 \le t \le 1$.

□ by Useful Fact 12.3:

$$L = \int_a^b \sqrt{[x'(t)]^2 + [y'(t)]^2}\, dt = \int_0^1 \sqrt{(6t)^2 + (6t^2)^2}\, dt$$

$$= \int_0^1 \sqrt{36t^2 + 36t^4}\, dt = \int_0^1 6t\sqrt{t^2 + 1}\, dt$$

$$= 4\sqrt{2} - 2 \quad \blacksquare$$

B.6.3 *Intro to Polar Coords. — Challenge — Solved*

---

(1) Describe the region given by the polar bounds $-1 \leq r \leq 1$, $\pi/4 < \theta < 3\pi/4$.

☐ This is a bowtie-shaped region inside a circle of radius 1. We start at an angle of $\pi/4$ above the positive x-axis and sweep counterclockwise to an angle of $\pi/4$ past the positive y-axis. This covers $0 < r \leq 1$. As we sweep through this angular region again and apply negative r values, we get the mirror image of this region across the origin. The entire region is shaped like a bowtie. (See Fig. B.7.) ∎

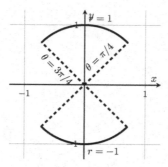

Fig. B.7 Polar bow-tie $-1 \leq r \leq 1$, $\dfrac{\pi}{4} < \theta < \dfrac{3\pi}{4}$.

---

(2) Convert $r = \tan\theta\sec\theta$ to Cartesian form and identify it.

☐ The conversion equations in Useful Fact 12.4 say that $\cos\theta = x/r$, so that $\sec\theta = r/x$. We also know $\tan\theta = y/x$. So, we get

$$r = \tan\theta\sec\theta = \frac{y}{x}\cdot\frac{r}{x}$$

The r cancels from each side, and we're left with $1 = y/x^2$, or $y = x^2$. So our curve is the parabola $y = x^2$. ∎

---

(3) Convert $x + y = 9$ to polar form.

☐ Using the basic conversion equations, $x + y = 9$ becomes $r\cos\theta + r\sin\theta = 9$. We can't do much else with this, except perhaps write it as:

$$r = \frac{9}{\cos\theta + \sin\theta}$$

🔲 FFT: The line $x+y = 9$ is a continuous function. When we put the equation in polar form and solve that form for r, as shown in the final line, are there any values of θ that result in $\cos\theta = -\sin\theta$? Because if so, those are points where r is not defined, and the curve will not be continuous. So either there are no such points, or maybe we should not always be so eager to solve an equation for one of the variables. What's your opinion? 🔲   ■

B.6.4 *Calc. of Polar Coords. — Challenge — Solved*

(1) Find the slope of the line tangent to $r = \ln\theta$ at $\theta = e$.

By Useful Fact 12.5,

$$\frac{dy}{dx} = \frac{\frac{dr}{d\theta}\sin\theta + r\cos\theta}{\frac{dr}{d\theta}\cos\theta - r\sin\theta} = \frac{\frac{1}{\theta}\sin\theta + \ln\theta\cos\theta}{\frac{1}{\theta}\cos\theta - \ln\theta\sin\theta} = \frac{\sin\theta + \theta\ln\theta\cos\theta}{\cos\theta - \theta\ln\theta\sin\theta}$$

So when $\theta = e$,

$$\frac{dy}{dx} = \frac{\sin e + e\ln e\cos e}{\cos e - e\ln e\sin e} = \frac{\sin e + e\cos e}{\cos e - e\sin e} \approx 1.02$$

The polar curve $r = \ln\theta$ is shown in Fig. B.8, along with the tangent line (dashed) at $\theta = e$. FFT: Note the large swoop off to the left and out of the figure, going towards $-\infty$. This is NOT a plotting error. Do you know why it is there? Hint: Consider what happens to r as θ approaches 0 from the positive side. ■

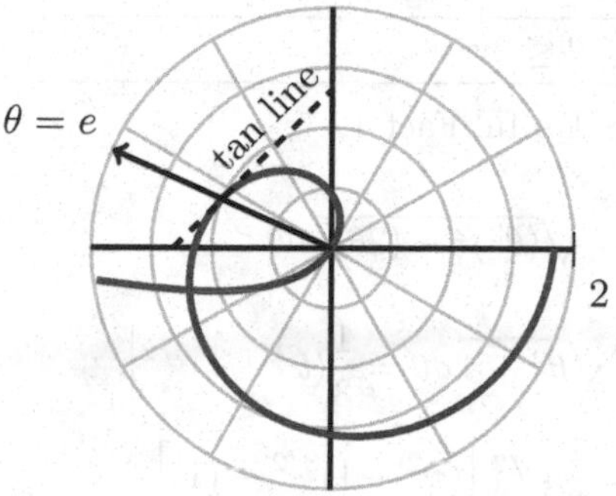

Fig. B.8 The polar curve $r = \ln\theta$ with tangent line at $\theta = e$.

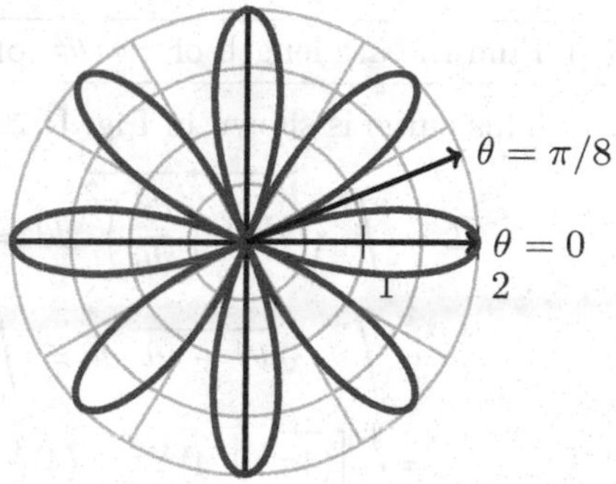

Fig. B.9 $r = 2\cos 4\theta$ and bounding θ values for one loop.

(2) Find the area inside one loop of $r = 2\cos 4\theta$.

□ This polar curve is shown in Fig. B.9. One loop is delineated by two consecutive θ values which produce $r = 0$. Here, we need $4\theta = -\pi/2$ and $4\theta = \pi/2$, i.e. $\theta = -\pi/8, \pi/8$. So,

$$A = \int_a^b \frac{1}{2}r^2\, d\theta = \int_{-\pi/8}^{\pi/8} \frac{1}{2}(2\cos 4\theta)^2\, d\theta = \frac{\pi}{4}$$

Alternately, by symmetry you can also set

$$A = 2\int_0^{\pi/8} \frac{1}{2}(2\cos 4\theta)^2\, d\theta = \frac{\pi}{4} \quad ■$$

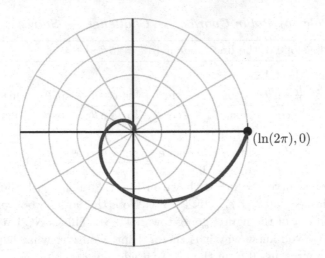

$(\ln(2\pi), 0)$

Fig. B.10   The curve $r = \ln\theta$ for $\theta = 0$ to $\theta = 2\pi$.

---

(3) Find the arc length of $r = \theta^2$ for $0 \le \theta \le 2\pi$.

This curve is shown in Fig. B.10. By Useful Fact 12.7,

$$L = \int_a^b \sqrt{r^2 + \left(\frac{dr}{d\theta}\right)^2}\, d\theta = \int_0^{2\pi} \sqrt{(\theta^2)^2 + (2\theta)^2}\, d\theta$$

$$= \int_0^{2\pi} \sqrt{\theta^4 + 4\theta^2}\, d\theta = \int_0^{2\pi} \theta\sqrt{\theta^2 + 4}\, d\theta = \frac{1}{3}(\theta^2 + 4)^{3/2}\Big|_0^{2\pi}$$

$$= \frac{1}{3}\left[(4\pi^2 + 4)^{3/2} - (4)^{3/2}\right] = \frac{1}{3} 4^{3/2}\left[(\pi^2 + 1)^{3/2} - (1)\right]$$

$$= \frac{8}{3}\left[(\pi^2 + 1)^{3/2} - 1\right] \quad \blacksquare$$

Index

Printed in the United States
by Baker & Taylor Publisher Services

Printed in the United States
by Baker & Taylor Publisher Services